T0183075

Lecture Notes in Computer Science 9130

Commenced Publication in 1973
Founding and Former Series Editors:
Gerhard Goos, Juris Hartmanis, and Jan van Leeuwen

More information about this series at http://www.springer.com/series/7407

Jianxin Wang · Chee Yap (Eds.)

Frontiers in Algorithmics

9th International Workshop, FAW 2015
Guilin, China, July 3–5, 2015
Proceedings

 Springer

Editors
Jianxin Wang
Central South University
Changsha
China

Chee Yap
Courant Institute
New York University
New York, NY
USA

ISSN 0302-9743   ISSN 1611-3349   (electronic)
Lecture Notes in Computer Science
ISBN 978-3-319-19646-6   ISBN 978-3-319-19647-3   (eBook)
DOI 10.1007/978-3-319-19647-3

Library of Congress Control Number: 2015940411

LNCS Sublibrary: SL1 – Theoretical Computer Science and General Issues

Springer Cham Heidelberg New York Dordrecht London

Printed on acid-free paper

Springer International Publishing AG Switzerland is part of Springer Science+Business Media
(www.springer.com)

Preface

The 9th International Frontiers of Algorithmics Workshop (FAW 2015) was held during July 3–5, 2015, at Guilin, China. The workshop brings together researchers working on all aspects of computer science for the exchange of ideas and results.

FAW 2015 was the ninth conference in the series. The previous eight meetings were held during August 1–3, 2007, in Lanzhou, June 19–21, 2008, in Changsha, June 20–23, 2009, in Hefei, August 11–13, 2010, in Wuhan, May 28–31, 2011, in Jinhua, May 14–16, 2012, in Beijing, June 26–28, 2013, in Dalian, June 28–30, 2014, in Zhangjiajie.

In all, 65 submissions were received from more than 12 countries and regions. The FAW 2015 Program Committee selected 28 papers for presentation at the conference. In addition, we had two plenary speakers: Fedor Fomin (University of Bergen, Norway) and Chee Yap (New York University, USA). Many thanks for their contributions to the conference and proceedings.

We would like to thank the Program Committee members and external reviewers for their hard work in reviewing and selecting papers. We are also very grateful to all the authors who submitted their work to FAW 2015. Finally, we would like to thank the editors at Springer and the local organization chairs for their hard work in the preparation of this conference.

July 2015

Jianxin Wang
Chee Yap

Organization

Program Committee

Hee-Kap Ahn	Pohang University of Science and Technology, Korea
Hans L. Bodlaender	Utrecht University, The Netherlands
Yixin Cao	Hong Kong Polytechnic University, China
Marek Chrobak	University of California, Riverside, USA
Barry Cooper	University of Leeds, UK
Qilong Feng	Central South University, China
Henning Fernau	University of Trier, Germany
Mordecai J. Golin	Hong Kong University of Science and Technology, China
Jiong Guo	Shandong University, China
Gregory Z. Gutin	University of London, UK
Kun He	Huazhong University of Science and Technology, China
Tomio Hirata	Nagoya University, Japan
Hiro Ito	University of Electro-Communications, Japan
Klaus Jansen	University of Kiel, Germany
Iyad Kanj	DePaul University, USA
Ming-Yang Kao	Northwestern University, USA
Naoki Katoh	Kyoto University, Japan
Michael A. Langston	University of Tennessee, USA
Guohui Lin	University of Alberta, Canada
Tian Liu	Peking University, China
Daniel Lokshtanov	University of California San Diego, USA
Dániel Marx	Hungarian Academy of Sciences, Hungary
Venkatesh Raman	Institute of Mathematical Sciences, India
Ulrike Stege	University of Victoria, Canada
Xiaoming Sun	Institute of Computing Technology, CAS, China
Jianxin Wang (Co-chair)	Central South University, China
Gerhard J. Woeginger	Eindhoven University of Technology, The Netherlands
Ge Xia	Lafayette College, USA
Ke Xu	Beihang University, China
Boting Yang	University of Regina, Canada
Chee Yap (Co-chair)	New York University, USA
Binhai Zhu	Montana State University, USA
Xiaofeng Zhu	Guangxi Normal University, China

General Conference Chairs

John Hopcroft   Cornell University, USA
Shichao Zhang   University of Technology, Sydney, Australia

Program Chairs

Jianxin Wang   Central South University, China
Chee Yap   New York University, USA

Publication Chairs

Yixin Cao   Hong Kong Polytechnic University, China
Xiaofeng Zhu   Guangxi Normal University, China

Organizing Chairs

Qilong Feng   Central South University, China
Zhenjun Tang   Guangxi Normal University, China

Additional Reviewers

Andro-Vasko, James	Kawahara, Jun	Panolan, Fahad
Bae, Sang Won	Kawamura, Akitoshi	Park, Dongwoo
Barbero, Florian	Kehua, Guo	Pferschy, Ulrich
Belmonte, Rémy	Kellerer, Hans	Phillips, Charles
Brodal, Gerth Stølting	Kim, Min-Gyu	Proietti, Guido
Buchanan, Austin	Kim, Sang-Sub	Reinhardt, Klaus
Bulteau, Laurent	Kullmann, Oliver	Seto, Kazuhisa
Cai, Leizhen	Li, Minming	Sheng, Bin
Chen, Xujin	Li, Wenjun	Solis-Oba, Roberto
Cho, Dae-Hyeong	Liu, Xingwu	Stee, Rob Van
Dutot, Pierre-Francois	Lu, Allan Yuping	Strusevich, Vitaly
Epstein, Leah	Lu, Songjian	Teruyama, Junichi
Fata, Elaheh	M.S., Ramanujan	Thorup, Mikkel
Fujiwara, Hiroshi	Majumdar, Diptapriyo	Trevisan, Vilmar
Gagarin, Andrei	Mastrolilli, Monaldo	Wakatsuki, Mitsuo
Hagan, Ron	Mertzios, George	Wang, Kai
Higashikawa, Yuya	Misra, Neeldhara	Yoon, Sang-Duk
Jacobs, David	Miyazawa, Flavio K.	You, Jie
Jiang, Minghui	Nielsen, Jesper Sindahl	Zhang, Jialin
Jones, Mark	Oh, Eunjin	
Kaluza, Maren	Okamoto, Yoshio	

Contents

Invited Talks

Graph Modification Problems: A Modern Perspective

Fedor V. Fomin[1]([✉]), Saket Saurabh[1], and Neeldhara Misra[2]

[1] Department of Informatics, University of Bergen, Bergen, Norway
{fomin,saket}@ii.uib.no
[2] Department of Computer Science and Automation, Indian Institute of Science,
Bangalore, India

Abstract. We give an overview of recent results and techniques in parameterized algorithms for graph modification problems.

In network (or graph) modifications problem we have to modify (repair, improve, or adjust) a network to satisfy specific required properties while keeping the cost of the modification to the minimum. The commonly adapted mathematical model in the study of network problems is the *graph modification* problem. This is a fundamental unifying problem with a tremendous number of applications in various disciplines like machine learning, networking, sociology, data mining, computational biology, computer vision, and numerical analysis, and many others. We start with three generic examples of graph modification.

Our first example is the *connectivity augmentation* problem. Here the graph models an existing network (say, a telecommunication network) and the goal is to enhance the network to ensure resilience against link failures. In other words, by adding a few links between nodes we wish to obtain a network with better connectivity. This is a special case of the graph modification problem where we want to improve the connectivity of a graph.

The second example is *graph clustering*. This is the fundamental problem of identification of closely related objects that have many interactions within themselves and few with the rest of the system. A group of objects of this type is known as a *cluster* (or *community*). In Fig. 1 one can find a clustering of the scientific-collaboration network of the members of the Algorithms groups at the University of Bergen. One of the most popular common to clustering is to model a system as a weighted graph. Then the task is to identify a set of low-cost edges (insignificant interactions) which removal partition the graph into clusters and this is again a special case of the graph modification problem.

The third example is the fundamental problem arising in sparse matrix computations. During Gaussian eliminations of large sparse matrices new non-zero elements called fill can replace original zeros thus increasing storage requirements and running time needed to solve the system. The problem of finding the right

Supported by the European Research Council (ERC) via grant Rigorous Theory of Preprocessing, reference 267959.

J. Wang and C. Yap (Eds.): FAW 2015, LNCS 9130, pp. 3–6, 2015.
DOI: 10.1007/978-3-319-19647-3_1

elimination ordering minimizing the number of fill elements can be expressed as the problem of adding the minimum number of edges transforming the graph into a *chordal* graph, i.e. the graph without chordless cycles of length more than four. Graph modification problems resulting in a graph with a nice combinatorial characterization like being an interval, perfect, or planar graph, is a common theme in various applications.

Fig. 1. Example of clustering

Unfortunately, all mentioned examples of graph modifications are NP-hard problems. NP-hardness is one of the most fundamental concepts in computational complexity, capturing the notion of computational intractability [5]. Although NP-hard problems frequently arise in practice (e.g., class scheduling, protein folding, or designing faster microprocessors) and have been attacked from many algorithmic angles [2–4,6,9–12], no polynomial time (efficient) algorithm exists to date for any NP-hard problem. It is widely believed that NP-hard problems cannot be solved in polynomial time and thus are intractable. A proof of this conjecture would solve the famous P *vs* NP question, one of the deepest and most difficult problems in mathematics and science.

The systematic algorithmic study of the graph editing problems can be traced to the classical work of Lewis and Yannakakis [8] from the 1980s. They investigated the complexity of the vertex deletion problems, where the aim is to obtain a graph that satisfies a given hereditary non-trivial property.

In this talk we give an overview of several recent results dealing with the intractability of modification problems from the angle of parameterized complexity and algorithms. The framework of parameterized complexity was introduced by Downey and Fellows [2]. Parameterized complexity is basically a two-dimensional generalization of "P *vs* NP" where in addition to the overall input size n, one studies how a relevant secondary measurement affects the computational complexity of problem instances. The philosophy of parameterized complexity is that besides overall input size, in many scenarios there are other secondary measurements that fundamentally affect the computational complexity of problems. These secondary measurements capture additional relevant information about problems, like structural restrictions on the input distribution considered, such as a bound on the treewidth of an input graph, and provide opportunities for designing efficient algorithms. In parameterized complexity each problem instance comes with a parameter k. A parameterized problem that can be solved in $\mathcal{O}(f(k)n^c)$ time is said to be fixed-parameter tractable (FPT). Above FPT, there exists a hierarchy of complexity classes, known as the W-hierarchy. Just as NP-hardness is used as an evidence that a problem is probably not polynomial time solvable, showing that a parameterized problem is hard for one of these classes gives evidence to the belief that the problem is unlikely to be fixed-parameter tractable. The principal analogue of the classical intractability class NP is W1. In particular, this means that an FPT algorithm for any W1-hard problem would yield an $\mathcal{O}(f(k)n^c)$ time algorithm for every problem in the class W1.

Graph modification problems were studied from parameterized complexity perspective too. For example, Cai [1] proved that for any property defined by a finite set of forbidden induced subgraphs, the corresponding modification problem is FPT. Further results for hereditary properties were obtained by Khot and Raman [7]. While most of the results in the area concern hereditary properties, for non-hereditary properties like for different types of connectivity demands, density, or specific vertex degree requirements, until very recently, not much was known. The situation changed drastically within the last 3–4 yr, when a number of novel results paved several directions for further expansions of the area.

We give an overview of the most recent trends in parameterized algorithms for graph modification problems including kernelization and subexponential algorithms.

Acknowledgement. We thank Pål Grønås Drange for the figure.

References

1. Cai, L.: Fixed-parameter tractability of graph modification problems for hereditary properties. Inf. Process. Lett. **58**(4), 171–176 (1996)
2. Downey, R.G., Fellows, M.R.: Fundamentals of Parameterized Complexity. Springer, Texts in Computer Science (2013)
3. Flum, J., Grohe, M.: Parameterized Complexity Theory. Springer, Texts in Theoretical Computer Science. An EATCS Series (2006)
4. Fomin, F.V., Kratsch, D.: Exact Exponential Algorithms. Texts in Theoretical Computer Science. An EATCS Series. Springer, Heidelberg (2010)
5. Garey, M.R., Johnson, D.S.: Computers and Intractability: A Guide to the Theory of NP-Completeness. W.H.Freeman, New York (1979)
6. Golumbic, M.C.: Algorithmic Graph Theory and Perfect Graphs. Annals of Discrete Mathematics, vol. 57, 2nd edn. Elsevier, Amsterdam (2004)
7. Khot, S., Raman, V.: Parameterized complexity of finding subgraphs with hereditary properties. Theor. Comput. Sci. **289**(2), 997–1008 (2002)
8. Lewis, J.M., Yannakakis, M.: The node-deletion problem for hereditary properties is np-complete. J. Comput. Syst. Sci. **20**(2), 219–230 (1980)

9. Michalewicz, Z., Fogel, D.B.: How to Solve It: Modern Heuristics. Springer, Heidelberg (2004)
10. Motwani, R., Raghavan, P.: Randomized Algorithms. Cambridge University Press, New York (1995)
11. Niedermeier, R.: Invitation to Fixed-Parameter Algorithms. Oxford Lecture Series in Mathematics and its Applications, vol. 31. Oxford University Press, Oxford (2006)
12. Vazirani, V.V.: Approximation Algorithms. Springer, Heidelberg (2001)

Soft Subdivision Search in Motion Planning, II: Axiomatics

Chee K. Yap[(⊠)]

Department of Computer Science, Courant Institute of Mathematical Sciences,
New York University, New York, NY 10012, USA
yap@cs.nyu.edu

Abstract. We propose to design motion planning algorithms with a strong form of resolution completeness, called **resolution-exactness**. Such planners can be implemented using **soft predicates** within the subdivision paradigm. The advantage of softness is that we avoid the Zero problem and other issues of exact computation. **Soft Subdivision Search** (SSS) is an algorithmic framework for such planners. There are many parallels between our framework and the well-known Probabilistic Road Map (PRM) framework. Both frameworks lead to algorithms that are practical, flexible, extensible, with adaptive and local complexity. Our several recent papers have demonstrated these favorable properties on various non-trivial motion planning problems. In this paper, we provide a general axiomatic theory underlying these results. We also address the issue of subdivision in non-Euclidean configuration spaces, and how exact algorithms can be recovered using soft methods.

1   Introduction

Motion planning has been studied for over 30 years, and remains a central problem in robotics. Path planning is the most basic form of motion planning in which we only consider kinematics, ignoring issues of timing, dynamics, non-holonomic constraints, sensing and mapping. In the algorithmic study of path planning, the problem is reduced to connectivity or reachability in some configuration space. There are three main approaches here: Exact, Sampling and Subdivision. Divergent paths have been taken: theoreticians favor the Exact Approach [2], but practical roboticists prefer the Sampling and Subdivision Approaches [9,11]. For two decades, the Sampling Approach has dominated the field. According to Choset et al. [9, p. 201], *"PRM, EST, RRT, SRT, and their variants have changed the way path planning is performed for high-dimensional robots. They have also paved the way for the development of planners for problems beyond basic path planning."* The premise of this paper is that subdivision has many merits over sampling, and this power has not been fully exploited. But to open

---

C.K. Yap—Plenary Talk at the 9th Int'l. Frontiers of Algorithmics Workshop (FAW 2015) in Guilin, China, July 3–5. This work is supported by NSF Grants CCF-0917093 and CCF-1423228.

© Springer International Publishing Switzerland 2015
J. Wang and C. Yap (Eds.): FAW 2015, LNCS 9130, pp. 7–22, 2015.
DOI: 10.1007/978-3-319-19647-3_2

up this exploitation, we need to give it a sound foundation. This paper will provide one such foundation. We formulate the **Soft Subdivision Search** or SSS to unify and generalize our several recent papers [12,13,20,21] in which we designed and implemented subdivision planners for several classes of robots. These SSS planners are relatively easy to design and implement. In our experiments, they outperform random sampling methods.

To introduce our approach, we compare the notion of correctness according to the three approaches. In the path planning problem, the robot R_0 is fixed, and each input instance is (Ω, α, β) where $\Omega \subseteq \mathbb{R}^k$ ($k = 2, 3$) is a description of the obstacles, and $\alpha, \beta \in \mathcal{C}space(R_0)$ are the start and goal configurations. In exact algorithms, the planner must return a path if one exists, and must return NO-PATH otherwise. In sampling, the input has an extra parameter N that bounds the maximum number of samples; the planner is said to be "sampling complete" if the planner returns a path with high probability when one exists and N is sufficiently large. In subdivision, the input has an extra resolution parameter $\varepsilon > 0$, and the planner is "resolution complete" if the planner returns a path when the ε is small enough. Thus sampling and (current) subdivision planners are similar in that their behaviors are only prescribed when there is a path. If there is no path, nothing is prescribed. In computability, such one-sided prescription of algorithmic behavior is well-known and is called "partial completeness". To make the completeness "total", we [20] introduce the concept of **resolution-exact** planners. Such a planner has an **accuracy constant** $K > 1$ (independent of input) such that:

(P) If there is a path of clearance $K\varepsilon$, it returns a path.
(N) If there is no path of clearance ε/K, it returns NO-PATH.

Thus the NO-PATH output guarantees that there is no path of clearance $K\varepsilon$. But the true innovation is the gap between the clearance bounds $K\varepsilon$ and ε/K: our planner could either return a path or NO-PATH when the optimal clearance lies in this gap. This "indeterminacy", unavoidable in some sense [20], has a big payoff — resolution-exact planners can be implemented with purely numerical approximations. As all the standard fundamental constants[1] of Physics are known to no more than 8 digits of accuracy, and no robot dimension, actuator control, sensors or environment is known to nearly such accuracy, we should not see this indeterminacy as a limitation.

Our paper [25] is a companion to the present paper, providing background and other motivations. It presents SSS alongside PRM [10] as two general algorithmic "frameworks" based on a small number of subroutines and data structures. We get specific algorithms by instantiating these subroutines and data structures. As framework, "PRM" can cover many of its known variants. These two frameworks share many favorable properties, all lacking in exact algorithms. But we claim one advantage of SSS over PRM: *PRM has a halting problem which SSS does not have.* We clarify this remark: under the usual idea that NO-PATH

---

[1] Except speed of light which is exactly known, by definition.

means "non-existence of paths", PRM cannot halt when there is no path. But suppose PRM adopts our viewpoint that NO-PATH means "no path of sufficient clearance". Now, PRM *could* halt[2] but this amounts to exhaustive (exponential) search. In effect, exponential search amounts to non-halting. But our subdivision approach need not suffer from exponential behavior because we are able to eliminate large regions of the configuration space with a single test. Conceivably, there are adaptive search strategies that guarantee polynomial size search trees. For example, such results are known in our subdivision work on root isolation [6,18,19]: here, the worst-case subdivision tree sizes is provably linear (resp., quadratic) in terms of tree depth for real (resp., complex) root isolation.

1. Overview: In Sect. 2, we describe the SSS Framework. In Sect. 3, we provide the abstract elements of SSS: configuration spaces are replaced by metric spaces and Non-Euclidean spaces are subdivided via charts and atlases. Section 4 proves properties of SSS planners that satisfy some general axioms. Section 5 shows that exact algorithms can be recovered with SSS planners. We conclude in Sect. 6. For reasons of space, some proofs are deferred to the full paper. Figures 1 and 2 are in color.

2   The SSS Framework

What sets Subdivision Search apart from sampling or grid methods is that its predicates are not point-based but region-based. Suppose each $\gamma \in Cspace$ has a classification as FREE, STUCK, or MIXED. Write $C(\gamma)$ for the classification of γ. We extend the classification to a set (or region) $B \subseteq Cspace$ as follows: define $C(B) = $ FREE (resp., $=$ STUCK) iff each $\gamma \in B$ is FREE (resp., STUCK); otherwise $C(B) = $ MIXED. A classification function $\widetilde{C}$ is a **soft predicate** (relative to C) if it is conservative (i.e., $\widetilde{C}(B) \neq $ MIXED implies $C(B) = \widetilde{C}(B)$) and convergent (i.e., if $\lim_{i \to \infty} B_i \to \gamma \in Cspace$ then $\widetilde{C}(B_i) = C(\gamma)$ for i large enough). Here we write $\lim_{i \to \infty} B_i \to \gamma$ to mean that $\{B_i : i \geq 0\}$ is a monotone decreasing sequence of sets that converge to γ.

Let us now use soft predicates for path planning. Fixed robot R_0. The motion planning input is $(\Omega, \alpha, \beta, \varepsilon)$ as above. It is standard (and without much loss) to also specify an initial box $B_0 \subseteq Cspace$ to confine our sought-for path. Our main data structure is a subdivision tree, T. It is useful to initially imagine $Cspace \subseteq \mathbb{R}^d$, and T as the standard multidimensional version of quadtrees, rooted at B_0. But bear in mind our goal of extending $Cspace$ to non-Euclidean spaces, and B to non-box geometries. The SSS planner amounts to a loop that "grows" T in each iteration by expanding some leaf until we find a path or conclude NO-PATH. There are two supporting data structures and three key routines:

- (Priority Queue) Q is a priority queue comprising those MIXED-leaves with length $\ell(B)$ (defined below) is at least ε.

---

[2] To do this, it would have to detect (probabilistically) that the sampling is dense enough, a non-trivial extension of the current PRM formulations.

- (Union-Find) D is a union-find data structure to maintain the connected components of the FREE boxes. As soon as we find a new FREE box, we form its union with the other adjacent FREE boxes. Boxes B, B' are **adjacent** if $B \cap B'$ is a $d - 1$ dimensional set.
- (Classifier) The routine $\widetilde{C}$ is a soft predicate that classifies each node in $\mathcal{T}$ as FREE/STUCK/MIXED.
- (Search Strategy) This is represented by the queue's Q.getNext() that returns a box in Q of highest priority.
- (Expander) The subroutine Expand(B) subdivides B into two or more subboxes. These subboxes become the children of B in $\mathcal{T}$. In general, Expand(B) represents a splitting strategy because it may have to choose from one or more alternative expansions.
- For $\gamma \in \mathcal{C}space$, let $Box(\gamma)$ denote any leaf in $\mathcal{T}$ that contains γ. Also, $Find(\gamma)$ denote the box returned by the find operation of D when it is given $Box(\gamma)$. Thus, a path is found as soon as we discover $Find(\alpha) = Find(\beta)$.

Putting them together, we get our SSS framework:

SSS FRAMEWORK
1.   ▷ *Initialization.*
       While ($\widetilde{C}(Box(\alpha)) \neq$ FREE)
            If $Box(\alpha)$ has length $< \varepsilon$, Return (NO-PATH)
            Else Expand($Box(\alpha)$)
       While ($\widetilde{C}(Box(\beta)) \neq$ FREE)
            ... do the same for β ...
2.   ▷ *Main Loop:*
       While ($Find(\alpha) \neq Find(\beta)$)
            If Q is empty, Return(NO-PATH)
            $B \leftarrow Q$.getNext()
            Expand(B)
3.   Compute a FREE channel P from $Box(\alpha)$ to $Box(\beta)$
       Generate and return the "canonical path" $\overline{P}$ inside P.

This framework has been used successfully to implement our disc and triangle planners [20], and our 2-link planner [12] including an interesting variant where self-crossing is not allowed [13]. Illustrating the power of subdivision and softness, we can easily generalize all these examples by fattening the robots and/or the polygonal obstacles. Notice that such extensions would be difficult for exact methods (to our knowledge, exact algorithms are unknown for such extensions). Of course many variants of this framework has appeared in the subdivision literature; conversely, some of these algorithms can be recaptured within SSS. E.g., the hierarchical search of Zhu and Latombe [28], Barbehenn and Hutchinson [1], or Zhang, Kim and Manocha (2008) [27]. One major difference is that these papers expand along a "mixed channels" (i.e., path comprising FREE or MIXED boxes). We could modify our getNext to achieve similar behavior; one advantage of this approach is that NO-PATH could be detected before emptying the queue. This abstract description hides an important feature of our technique: our computation of $\widetilde{C}$ is deeply intertwined with the expansion of $\mathcal{T}$ (see [8]). Steve LaValle (insightfully) described this as "opening up the blackbox" of collision testing.

3   Generalized Setting for SSS

Once the SSS framework has been instantiated with specific routines, we have an SSS planner. How do we know that the planner is resolution-exact? Our goal is to prove this under general "axiomatic" conditions. Designing a short list of such axioms is very useful: first, it gives us a uniform way to check that any proposed SSS algorithm is resolution-exact, just by checking the axioms. We could for instance apply this to our previous planners [12,13,20]. Second, because planning is a complex task, and we expect that SSS will suffer many variants, we must know the boundaries of the variations. The axioms serve as boundary markers.

The starting point is to replace $\mathcal{C}space$ by a metric space X, and replace $\mathcal{C}free$ by an open set $Y \subseteq X$. Points in the boundary ∂Y of Y are said to be **semi-free**. Let $C_Y : X \to \{+1, 0, -1\}$ denote the (exact) **classifier** for Y: $C_Y(\gamma) := +1/0/-1$ iff γ belongs to $Y/\partial(Y)/X \setminus \overline{Y}$ where $\overline{Y}$ is the closure of Y. Note that we have performed a simple (non-essential) translation in our classification values: FREE $\to +1$, MIXED $\to 0$, and STUCK $\to -1$.

We extend the classification of points to classification of sets. There are two general ways to extend any function to a function on sets: let $f : S \to T$ be a function. The **set extension** of f (still denoted f) is the function $f : 2^S \to 2^T$ such that for $B \subseteq S$, $f(B) = \{f(b) : b \in B\}$. Here 2^S denotes the power set of S. Another general method applies to any geometric[3] predicate $g : S \to \{+1, 0, -1\}$. The **set extension** of g (still denoted g) is the geometric predicate $g : 2^S \to \{+1, 0, -1\}$ such that for any definite value $v \in \{+1, -1\}$, $g(B) = v$ iff $g(b) = v$ for all $v \in B$; otherwise $g(B) = 0$.

Although the set extension of the classifier $C_Y : X \to \{+1, 0, -1\}$ is applicable to any subset $B \subseteq X$, in practice, we need B is be "nice" in order to carry out our algorithm: B must be able to support subdivision, $C_Y(B)$ must be (softly) computable, and we should be able to discuss the limits of such sets, $\lim_{i \to \infty} B_i$. We next capture these properties using "test cells".

2. Test Cells and Subdivision Trees: Consider an Euclidean set $B \subseteq \mathbb{R}^d$. It is called a **test cell** if it is a full-dimensional, compact and convex polytope. For $d = 1$ ($d = 2$), test cells are intervals (convex polygons). Our subdivision of the metric space X will be carried out using such test cells.

Let the **width** $w(B)$ (resp., **length** $\ell(B)$) refer to the minimum (resp., maximum) length of an edge of B. The unique smallest ball containing B is called the **circumball** of B; its center and radius are denoted $c(B)$ and $r(B)$. Note that $c(B)$ need not lie in the interior of B. The **inner radius** $r_0(B)$ of B is the

---

[3] A **geometric predicate** is a 3-valued function, with a distinguished value 0 called the **indefinite value**. The others are called **definite values**. This is in contrast to a **logical predicate** which is 2-valued.

largest radius of a ball contained in B. Let $ic(B)$ comprises the centers of balls of radius $r_0(B)$ that are contained in B. E.g., if B is a rectangle, then $ic(B)$ is a line segment. Clearly, $ic(B)$ is convex. Then $c(ic(B))$ is called the **inner center** of B, denoted $c_0(B)$. Unlike $c(B)$, we now have $c_0(B)$ in the interior of B. We use $c_0(B)$ as follows: for any $\alpha > 0$, αB will mean scaling B by a factor α relative to the center $c_0(B)$. If $\alpha > 1$ (< 1) this amounts to growing (shrinking) B. The inverse operation is denoted B/α. Thus $(\alpha B)/\alpha = B$. The **aspect ratio** of B is $\rho(B):=r(B)/r_0(B) > 1$.

By a **subdivision** of a test cell B, we mean any finite set of test cells $\{B_1, \ldots, B_m\}$ such that $B = \bigcup_{i=1}^{m} B_i$ and $\dim(B_i \cap B_j) < d$ for all $i \neq j$. We denote the subdivision relationship as $B = B_1 \uplus B_2 \uplus \cdots \uplus B_m$.

Let $\Box\mathbb{R}^d$ denote some set of test cells. For instance, $\Box\mathbb{R}^d$ may the set of all boxes, or the set of all simplices. Let the function $\texttt{Expand} : \Box\mathbb{R}^d \to 2^{\Box\mathbb{R}^d}$ return a subdivision $\texttt{Expand}(B)$ of B. In general, $\texttt{Expand}$ is a non-deterministic function[4] and we may call it an "expansion scheme". Using an expansion scheme, we can grow subdivision trees rooted in any $B \in \Box\mathbb{R}^d$, by repeated expansion at any chosen leaf. We note some concrete schemes:

- **Longest Edge Bisection:** let $\Box\mathbb{R}^d$ be simplices and $\texttt{Expand}(B)$ returns a subdivision of B into two simplices by bisecting the longest edge in B (see [17]).
- **Box Subdivision Scheme:** let $\Box\mathbb{R}^d$ be the set of all (axes-parallel) boxes and $\texttt{Expand}(B)$ return a set of 2^i congruent boxes (for some $i = 1, \ldots, d$). This set is defined by introducing i axes-parallel hyperplanes through the center of B. There are $\binom{d}{i}$ ways to choose these hyperplanes. So there are $2^d - 1$ possible expansions.
- **Dyadic Schemes:** We call a scheme is **dyadic** if, for any test cell B, each vertex of a subcell $B' \in \texttt{Expand}(B)$ is either a vertex of B or the midpoint of an edge of B. The previous two examples are dyadic schemes. The significance of such schemes is that they can be exactly and efficiently computed: recall that a **dyadic number** (or BigFloat) is a rational number of the form $m2^n$ ($m, n \in \mathbb{Z}$). The operations $+, -, \times$ on dyadic numbers are very efficient and division by 2 is exact. Vertices of test cells in a dyadic subdivision tree have the form $\sum_{i=1}^{k} c_i v_i$ where c_i are dyadic numbers and $v_1, \ldots, v_k$ are the vertices of the root. The bit size of the c_i's grows linearly with the depth, not exponentially.

3. Subdivision Atlases for Non – euclidean Spaces: Note that if we have a point or ball robot in Euclidean space, then the resolution-exactness of SSS algorithms is indeed trivial. But configuration spaces are rarely Euclidean. Subdivision in non-Euclidean spaces is a nontrivial problem. Likewise, sampling in such spaces is also a research issue (Yershova et al. [26]). Our approach is to borrow the language of charts and atlases from differential geometry. Suppose

---

[4] We use the notation in, e.g., [3]. This means there is a set, denoted $set-\texttt{Expand}(B)$, of subdivisions of B, and $\texttt{Expand}(B)$ denotes (non-deterministically) any element of this set. We assume $set-\texttt{Expand}(B)$ is non-empty so that $\texttt{Expand}(B)$ is a total function.

the metric space X has the property $X = X_1 \cup X_2 \cup \cdots \cup X_m$ such that for each X_t, we have an onto homeomorphism $\mu_t : B_t \to X_t$ where B_t is a test cell, and $\dim(\mu_t^{-1}(X_t \cap X_s)) < d$ for all $t \neq s$. We call each μ_t a **chart** and the set $\{\mu_t : t = 1, \ldots, m\}$ is called an **subdivision atlas** for X.

The subdivision of X is thus reduced to subdivision in each X_t, carried out vicariously, via the chart μ_t. More precisely, let $\texttt{Expand}_t : \square B_t \to 2^{\square B_t}$ be an expansion scheme where $\square B_t \subseteq 2^{B_t}$ is a set of test cells. Call $\mu_t(B) := \{\mu_t(\gamma) : \gamma \in B\}$ $(B \in \square B_t)$ a test cell **induced** by μ_t. Let $\square X$ denote the set of induced test cells. Finally, let $\square X$ denote the disjoint union of the $\square X_t$'s (for all $t = 1, \ldots, m$) and let $\texttt{Expand}_X : \square X \to 2^{\square X}$ denote the induced expansion defined by $\texttt{Expand}_X(\mu_t(B)) = \mu_t(\texttt{Expand}_t(B))$. We have thus achieved subdivision in X. In the following, we might say "B/α" (scaling), "$c(B)$" (center), etc. But it should be understood that we mean $\mu(B'/\alpha)$, $\mu(c(B'))$, etc., where $B = \mu(B')$ for some test cell B'.

Call the intersection $X_t \cap X_s$ $(s \neq t = 1, \ldots, m)$ an **atlas transition** if $\dim(X_t \cap X_s) = d - 1$. For motion planning, recall that two cells are adjacent if they share a face of codimension 1. Thus atlas transitions yield adjacencies between cells in $\square X_s$ and in $\square X_t$. Thus we have two kinds of adjacencies: those that arise from the subdivision of test cells, and from atlas transitions.

4. Subdivision Atlases for S^2 and $SO(3)$:

We now give consider two non-Euclidean metric spaces, S^2 and $SO(3)$. We will identify $SO(3)$ with the unit quaternions, $q = (a, b, c, d) = a + ib + jc + kd$ with $a^2 + b^2 + c^2 + d^2 = 1$. Then $SO(3)$ is a metric space with a metric $d(\cdot, \cdot)$ given by the angle $d(q, q') := \cos^{-1}(|q \cdot q'|)$ between two unit quaternions q, q' (see [26]). Likewise, we can treat S^2 as a metric space with the great circle distance.

We are interested in the 2-sphere S^2 because the configuration spaces of several simple rigid robots living in $\mathbb{R}^3$ is given by $\mathbb{R}^3 \times S^2$: a rod (1D), a cone or bullet (3D), a disc (2D) and a ring (1D). See Fig. 1(a). The ring is interesting because it is the simplest rigid robot that is not simply-connected. Despite the simplicity of their configuration spaces (being 5-DOF), it seems that no complete exact planners have been designed for them. The reason seems to be related to the difficulties of exact algorithms for the "Voronoi Quest" [22]. We are currently designing and implementing a resolution-exact planner for a rod [21]. It would test the practicality of our theory. We can make the rod, ring and disc into **thick robots** by taking their Minkowski sum with a 3D-ball. But we expect that any SSS planner for thin robots will extend relatively easily to thick analogues (similar to the situation in the plane [12]).

Note that S^2 is not a subgroup, but a quotient group of $SO(3)$ (this is clear from the Hopf fibration of $SO(3)$ [26]). To create a subdivision atlas for S^2, let $I^3 = I \times I \times I$ be the 3-cube where $I = [-1, 1]$. Its boundary ∂I^3 can be subdivided into 6 squares denoted $S_{\pm \delta}$ where $\delta \in \{x, y, z\}$. See Fig. 1(b). For instance, $S_{+z} = \{(x, y, 1) : x, y \in I\}$ and $S_{-z} = \{(x, y, -1) : x, y \in I\}$. We obtain a subdivision chart of S^2 by using 6 charts: $\mu_{\pm \delta} : S_{\pm \delta} \to S^2$ where $\mu_{\pm \delta}(q) = q/\|q\|$ where $\|q\|$ is the Euclidean norm. Note that $\mu_{\pm \delta}$ does not depend

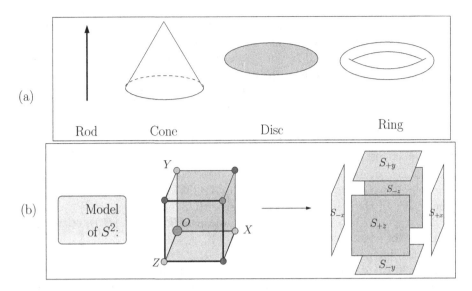

Fig. 1. 3D rigid robots with 5-DOF (Color figure online)

on $\pm\delta$ and so there is really only one function $\mu(q)$ for all the charts. The inverse map $\mu^{-1} : S^2 \to \partial I^3$ is also easy: $\mu^{-1}(\gamma) = \gamma/\|\gamma\|_\infty$ where $\|q\|_\infty$ is the infinity norm.

Call this construction the **cubic atlas** for S^2. We now construct a similar cubic atlas for $SO(3)$ (it was mentioned in Nowakiewicz [14]).

Begin with the 4-cube I^4: it has eight 3-dimensional cubes as faces. After identifying the opposite faces, we have four faces denoted $C_w^3, C_x^3, C_y^3, C_z^3$ (see Fig. 2). We define the chart: $\mu_t : C_t^3 \to SO(3)$ given by $\mu_t(q) = q/\|q\|$ (where $t = w, x, y, z$). As noted above, we must keep track of the adjacencies that arise from our atlas. In our case, this arise from the identification of antipodal points, $q \sim -q$ in S^3. In our cubic model, this information is transferred to identification of 2-dimensional faces among of C_t^3.

A chart $\mu : B_t \to X_t$ is **good** if there exists a **chart constant** $C_0 > 0$ such that for all $q, q' \in B_t$, $1/C_0 \le \frac{d_X(\mu(q), \mu(q'))}{\|q - q'\|} \le C_0$ where $d_X(\cdot, \cdot)$ is the metric in X_t. The subdivision atlas is **good** if there is an **atlas constant** C_0 that is common to its charts. Note that good atlases can be used to produce nice sampling sequences: since our test cells are Euclidean sets, we can exploit sampling of Euclidean sets. Alternatively, we can produce a "uniform" subdivision into sufficiently test cells, and pick the center of each test cell as sample point. The following is immediate:

Lemma 1. *The cubic subdivision atlases for S^2 and $SO(3)$ are good.*

5. Soft Predicates: We define soft predicates in the space X. Let $Y \subseteq X$. We call $\widetilde{C} : \Box X \to \{+1, 0, -1\}$ a **soft classifier** of Y if it satisfies two properties:

– (conservative) for all $B \in \Box X$, $\widetilde{C}(B) \ne 0$ implies $\widetilde{C}(B) = C_Y(\mu(B))$.

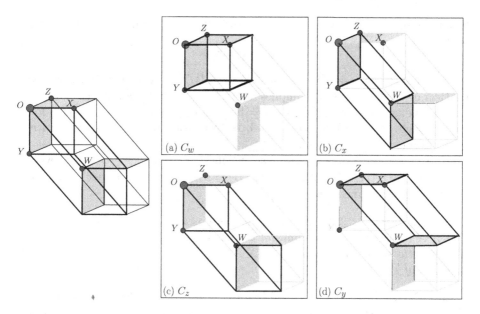

Fig. 2. The Cubic Atlas for $SO(3)$ (Color figure online)

– (convergent) if $q = \lim_{i \to \infty} B_i$ then $\widetilde{C}(B_i) = C_Y(\mu(q))$ for i large enough.

For resolution-exactness, we need another property: a soft classifier $\widetilde{C}$ is **effective** if there is an **effectivity factor** $\sigma > 1$ such that if $\widetilde{C}(B) = +1$ then $\widetilde{C}(B/\sigma) = +1$. For instance, we see that effectivity of $\widetilde{C}$ implies it is convergent. Note we do not require $C_Y(B) = -1$ to imply $C_Y(B/\sigma) = -1$.

Given $\alpha, \beta \in X$ and $Y \subseteq X$, the exact planning problem is finding a path from α to β in Y if they belong to the same connected component of Y, and NO-PATH otherwise. The resolution-exact version will require a connection between the metric in configuration space X and the metric in physical space $\mathbb{R}^k$. For this purpose, recall the concepts of footprint and separation of Euclidean sets (see [20,25]): Our robot R_0 lives in physical space $\mathbb{R}^k$ ($k = 2$ or $k = 3$) amidst an obstacle set $\Omega \subseteq \mathbb{R}^k$. The **footprint map** is $Fprint : Cspace \to 2^{\mathbb{R}^k}$ where $Cspace = Cspace(R_0)$ is the configuration space. Intuitively, $Fprint(\gamma) \subseteq \mathbb{R}^k$ is the physical space occupied by robot R_0 in configuration γ. The **clearance function**, $Cl : Cspace \to \mathbb{R}_{\geq 0}$ is given by $Cl(\gamma) := Sep(Fprint(\gamma), \Omega)$, where $Sep(A, B) := \inf \{\|a - b\| : a \in A, b \in B\}$ is the **separation** between two Euclidean sets in $\mathbb{R}^k$. We say γ is **free** if $Cl(\gamma) > 0$. A **motion** is a continuous function $\pi : [0, 1] \to Cspace$; its **clearance** is $\inf \{Cl(\pi(t)) : t \in [0, 1]\}$. Call π a **path** if it has positive clearance.

In our abstract formulation, we postulate the existence of a continuous function $Cl : X \to \mathbb{R}$ without reference to the underlying footprint or Ω. Moreover, this is called a **generalized clearance function** because we now allow negative clearance, interpreted as "penetration depth" (e.g., [8,27]). Call Cl a **clearance**

function for Y if $Y = \{\gamma \in X : C\ell(\gamma) > 0)\}$. We then consider interval functions of the form

$$\square C\ell : \square X \to \square \mathbb{R}.$$

(Recall that $\square \mathbb{R}$ is a set of intervals.) We call $\square C\ell$ a **conservative approximation** of $C\ell$ if $\square C\ell(B) \neq 0$ implies $\square C\ell(B) = C\ell(B)$ for all $B \in \square X$. We say $\square C\ell$ **converges** to $C\ell$ if whenever $\gamma = \lim_{i \to \infty} B_i$, then $\square C\ell(B_i) = C\ell(\gamma)$ for i large enough. Finally, $\square C\ell$ is called a **box function** for $C\ell$ if it is conservative and convergent relative to $C\ell$.

Note that $\square C\ell$ defines a classification function $\widetilde{C} : \square X \to \{+1, 0, -1\}$ where $\widetilde{C}(B) = 0$ iff $0 \in \square C\ell(B)$; otherwise, $\widetilde{C}(B) = \mathtt{sign}(\square C\ell(B))$ (either $+1$ or -1). The following is immediate:

Lemma 2. *Let $C\ell : X \to \mathbb{R}$ be a clearance function for $Y \subseteq X$, and suppose $\square C\ell : \square X \to \square \mathbb{R}$ is a box function for $C\ell$.*
Then the classification function $\widetilde{C} : \square X \to \{+1, 0, -1\}$ defined by $\square C\ell$ is a soft classifier of Y.

6. Soft Predicates for Complex Robots: An example of a robot with complex geometry is the gear robot of Zhang et al. [27]. Such robots pose difficulties for exact algorithms. We show that soft predicates for complex robots can be decomposed. Let $G_0 \subseteq \mathbb{R}^2$ be the gear robot. We write it as a union $G_0 = \cup_{j=1}^{m} T_j$ of triangles T_j. The free space of G_0 can be written as the intersections of the freespaces of T_j, provided the T_j's are expressed in a common coordinate system. This proviso requires a slight generalization of the soft predicate for triangles in [8]. The next theorem shows how to obtain a soft predicate for G_0 from those of the T_j's:

Theorem 1 (Decomposability of Soft Predicates). *Suppose $Y = Y_1 \cap \cdots \cap Y_m$. If $\widetilde{C}_i : \square X \to \{+1, 0, -1\}$ is a soft classifier for Y_i, then the following is a soft classifier for Y:*

$$\widetilde{C}(B) := \begin{cases} +1 & \text{if } (\forall j)[\widetilde{C}_j(B) = +1] \\ -1 & \text{if } (\exists i)[\widetilde{C}_i(B) = -1], \\ 0 & \text{else.} \end{cases}$$

If each $\widetilde{C}_j$'s has effectivity factor σ, then $\widetilde{C}(B)$ has effectivity factor σ.

Proof. We easily check that $\widetilde{C}(B)$ is safe. To show convergence, suppose that $B_i \to p$. If $p \in Y$, then $p \in Y_j$ for each j. That means $\widetilde{C}_j(B_i) = 1$ for i large enough. I.e., $\widetilde{C}(B_i) = 1$ for i large enough. This proves $\lim_{i \geq 0} \widetilde{C}(B_i) = +1 = C(p)$. If $p \in X \setminus \overline{Y}$, then $p \in X \setminus \overline{Y_j}$ for some Y_j. This means $\widetilde{C}_j(B_i) = -1$ for i large enough, and therefore $\widetilde{C}(B_i) = -1$ for i large enough. Again, $\lim_{i \geq 0} \widetilde{C}(B_i) = -1 = C(p)$. Suppose $p \in \partial Y$. Then $p \in \partial Y_j$ for some j and for all $k \neq j$, $p \in \overline{Y_k}$. That implies that $\widetilde{C}_j(B_i) \in \{+1, 0\}$ and Again, $\lim_{i \geq 0} \widetilde{C}(B_i) = 0 = C(p)$. This proves the softness of the predicate $\widetilde{C}(B)$.

Assume each $\widetilde{C}_j$ has an effectivity factor $\sigma > 1$. Let $C_j(B)$ be the exact box predicate for Y_j. Suppose $C(B)$ is free. This means each $C_j(B)$ is free. By definition of effectivity, each $\widetilde{C}_j(B/\sigma)$ is free. Hence $\widetilde{C}(B/\sigma)$ is free.   **Q.E.D.**

4   Axiomatic Properties of SSS

We prove general properties of SSS planners using basic assumptions which we call **axioms**. The proofs are instructive because they reveal how these axioms and properties of SSS are used. We introduce 5 axioms, beginning with these four:

- **(A0: Softness)**
  $\widetilde{C}$ is a soft classifier for $Cfree = \{\gamma \in X : C\ell(\gamma) > 0\}$.
- **(A1: Bounded dyadic expansion)**
  The expansion scheme is dyadic, and there is a constant $D_0 > 2$ such that $\texttt{Expand}(B)$ splits B into at most D_0 subcells, each with at most D_0 vertices, with the ratio $\ell(B)/w(B)$ at most D_0.
- **(A2: Clearance is Lipschitz)**
  There is a constant $L_0 > 0$ such that for all $\gamma, \gamma' \in Cfree$, $|C\ell(\gamma) - C\ell(\gamma')| < L_0 \cdot d_X(\gamma, \gamma')$ where $d_X(\cdot, \cdot)$ is the metric on X.
- **(A3: Good Atlas)**
  The subdivision atlas has a atlas constant $C_0 \geq 1$.

Note that these axioms concern about the clearance $C\ell : X \to \mathbb{R}$, classification $\widetilde{C} : \square X \to \square \mathbb{R}$ and the $\texttt{Expand}$ scheme. We have no axioms about $\texttt{getNext}$ because the needed properties are embedded in the SSS framework, namely $\texttt{getNext}$ returns a MIXED-leaf with length $\ell(B) \geq \varepsilon$ if any exist. Recall that in general, $B \in \square X$ is induced via our charts μ_t, and so the metrics such as $\ell(B)$ and $w(B)$ are induced from the Euclidean sets B' where $\mu_t(B') = B$, i.e., $\ell(B)$ refers to $\ell(\mu_t^{-1}(B)) = \ell(B')$, etc. Note that (A1) does not bound the aspect ratio $r(B)/r_0(B)$ and these may be unbounded (slivers can arise). (A2) relates clearance to the metric on X: this is a non-trivial axiom in non-Euclidean spaces. (A3) is necessary since subdivision of X is done via charts $\{\mu_t : t = 1, \ldots, m\}$.

Theorem 2 (Halting). *Every SSS planner halts. When a path is output, it is valid.*

Proof. In any infinite path $\{B_i : i \geq 0\}$, (A1) implies $\lim_i \ell(B_i) \to 0$. Since we do not subdivide a box if "$\ell(B) < \varepsilon$", halting is assured. At termination, we either report a path or output NO-PATH. If we report a path, it meant we found a "free channel" from $B(\alpha)$ to $B(\beta)$. We check that SSS ensures that the channel is truly free: the dyadic scheme (A1) ensures that test cells are computed exactly, and thus adjacencies are computed without error. Each cell in the channel is classified as FREE, and this truly free because (A0) ensures a conservative classifier $\widetilde{C}$. Finally, output paths are valid as they are contained in free channels.   **Q.E.D.**

This theorem depends only on (A0) and (A1). Although our goal in (A0) is soft classifiers, it is a useful preliminary to consider the case where $\widetilde{C}$ is the exact classifier. In this case, we say our planner is **exact**. This preliminary step is captured in the next result:

Theorem 3 (Exact SSS). *Assuming an exact SSS planner:*
(a) If there is no path, the planner outputs NO-PATH.
(b) If there is a path with clearance $\geq 2C_0 D_0 \varepsilon L_0$, *the planner outputs a path.*

Proof. Part(a) is essentially the contrapositive of the above Halting theorem. For part(b), let $\mathcal{T}$ be the subdivision tree at termination. The nodes of $\mathcal{T}$ are induced cells of $\square X$. Each $B \in \square X$ comes from an Euclidean test cell $\mu^{-1}(B) \in \square \mathbb{R}^d$ for some chart μ. Euclidean distance $\| \cdot \|_2$ in $\mu^{-1}(B)$ and the metric $d_X(\cdot, \cdot)$ of X are related via the chart constant C_0. Let $\pi : [0, 1] \to X$ be a path from α to β with clearance $2C_0 D_0 \varepsilon L_0$. By way of contradiction, suppose SSS outputs NO-PATH. This implies that every mixed leaf satisfies $\ell(B) < \varepsilon$. Consider the set $\mathcal{A}$ of leaves of $\mathcal{T}$ that intersect $\pi[0, 1]$ (the range of π). If $B \in \mathcal{A}$, there exists $t \in [0, 1]$ such that $\pi(t) \in B$. This implies B is either free or mixed. We claim that B is free. If B is mixed, then $\ell(\mu^{-1}(B)) < \varepsilon$ and there is a point $p' \in B$ that is semi-free. Thus $\|p - q\|_2 < D_0 \varepsilon$ for any two Euclidean points p, q in $\mu^{-1}(B)$. Using the chart μ, we conclude that $d_X(\mu(p), \mu(q)) < C_0 D_0 \varepsilon$. Therefore $d_X(\pi(t), p') \leq d_X(\pi(t), c(B)) + d_X(c(B), p') < 2C_0 D_0 \varepsilon$. By (A2), $|C\ell(\pi(t)) - C\ell(p')| < 2C_0 D_0 \varepsilon L_0$. Thus $C\ell(p') > C\ell(\pi(t)) - 2C_0 L_0 \varepsilon L_0 \geq 0$, i.e., p' is free. This contradicts the assumption that p' is semi-free; so B must be free. Thus we obtain a channel of free cells from α to β using cells in $\mathcal{A}$. The existence of such a channel implies that our union-find data structure in SSS would surely detect a path.   **Q.E.D.**

Our goal is not to produce the sharpest constants but to reveal their roles in our framework. Notice that this theorem has a gap: if the optimal clearance lies in $(0, 2C_0 D_0 \varepsilon L_0)$, the exact Planner may output either a path or NO-PATH.

7. Three Desiderata: The literature invariably assumes exactness in its analysis, such as in Theorem 3. But there are three desiderata beyond such a result. The first is to remove the exactness assumption. Second, we would like to strength the hypothesis of Theorem 3(a) to "*if there is no path with clearance ε/K*" for some input-independent $K > 1$. In other words, NO-PATH ought to mean no path of "sufficient clearance", a reasonable idea in view of the inherent uncertainty of physical devices. Third, we may want to strengthen the conclusion of Theorem 3(b) so that the output path has clearance $\geq \varepsilon/K$.

The first desiderata calls for soft predicates. We say that the SSS planner is **effective** if the soft predicate $\widetilde{C}$ has an effectivity constant $\sigma > 1$. In applications, it is useful to assume that $\widetilde{C}$ is **isotone**[5] i.e., $\widetilde{C}(B) \neq 0$ and $B' \subseteq B$ implies $\widetilde{C}(B') \neq 0$. The proof of part(b) in the previous theorem can be extended to show:

---

[5] This term is from the interval literature. Though not strictly necessary, but it simplifies some arguments.

Theorem 4 (Effective SSS). *Assume an SSS planner with effectivity $\sigma > 1$.*
(a) If there is no path, the planner outputs NO-PATH.
(b) If there is a path with clearance $\geq C_0 D_0 \varepsilon (1 + \sigma) L_0$, the planner outputs a path.

The indeterminacy gap is slightly widened to $(0, C_0 D_0 \varepsilon (1 + \sigma) L_0)$ by the soft predicate.

The second desiderata amounts to asking for a resolution-exact planner. As defined in the Introduction, such planners has an accuracy constant $K > 1$. So we seek to narrow indeterminacy gap by raising the gap lower bound from 0 to ε/K. The fundamental issue is to infer a lower bound on the clearance of a path inside a free channel. This requires a new axiom:

- **(A4: Translational Cells)**
  There is a constant $K_0 > 0$ such that if $B \in \square X$ is free, then its inner center $c_0 = c_0(B)$ has clearance $C\ell(c_0) \geq K_0 \cdot r_0(B)$. Such cells are said to be **translational**.

Like (A2), axiom (A4) relates the clearance to the metric space (via the chart μ). The "translational" terminology is based on the analogy that if X is purely translational, then (A4) is true. But in fact, it will be true in all the common motion planning scenarios.

Theorem 5 (Resolution-Exact SSS). *Assuming (A0–A4), SSS planners are resolution-exact.*

This proof is more involved and will appear in the full paper. The third desiderata requires that we strengthen condition (P) in the definition of resolution-exactness as follows:

(P') If there is a path of clearance $K\varepsilon$, then return a path of clearance ε/K.

See [20, 25] where (P') is used. The combination of (P) and (N) implies that whenever a path is output, we are assured that *there exists a path of clearance* ε/K. So (P') attempts to turn this existential guarantee into a constructive guarantee. Unfortunately, this requires additional effort as in [20, 25]. We will not attempt an axiomatic treatment to achieve (P') here.

5   What About Exactness?

Can the SSS framework produce[6] exact algorithms? The answer is yes, but as always, only in the algebraic case. First, here is a non-solution: *using an exact SSS planner with the resolution parameter $\varepsilon = 0$.* Using an exact SSS reintroduces the need for algebraic computation. By setting $\varepsilon = 0$, indeterminacy

---

[6] We are indebted to Steve LaValle for asking this question at the IROS 2011 Workshop in San Francisco.

is removed, but at a high price: if there is no path, then SSS will not halt. Even if there is a path, SSS may not halt; but this could be fixed by imposing a "generalized BFS" property on `getNext`. For these reasons, our normal formulation of SSS requires $\varepsilon > 0$ and $K > 1$. We now present a solution within the SSS framework using an effective soft predicate. The idea is to exploit the theory of constructive zero bounds [24].

Proposition 3. *If R_0, Ω are semi-algebraic sets, and the parameters α, β are algebraic, then there is an effectively computable number $\delta = \delta(R_0, \Omega, \alpha, \beta) > 0$ such that: if there is a path from α to β, then there is one with clearance δ.*

One way to derive such a δ is to use the general retraction theory in [15,16,23]: there is a "retract" $V \subseteq \mathcal{C}free = \mathcal{C}free(R_0, \Omega)$ and a retraction map $Im :$ $\mathcal{C}free \rightarrow V$ with this property: for all $\alpha, \beta \in \mathcal{C}free$, we have that α, β are path-connected in $\mathcal{C}free$ iff $Im(\alpha), Im(\beta)$ are path-connected in V. Here V is a Voronoi diagram and we can subdivide V into semi-algebraic Voronoi cells. The minimum clearance on V serves as δ, and this can be lower bounded using the degree and height of the semi-algebraic sets [5]. The upshot is this:

Theorem 6. *Suppose we have a resolution-exact planner with accuracy constant $K > 1$. If we choose ε to be $\delta(R_0, \Omega, \alpha, \beta)/K$, then our SSS planner is exact: it outputs* `NO-PATH` *iff there is no path.*

6   Conclusion

In this paper, we described the SSS framework for designing resolution-exact algorithms. We argued [25] that it shares many of the attractive properties of the successful PRM framework. Subdivision algorithms are as old as the history of path planning [4]. But to our knowledge, the simple properties of soft classifiers have never been isolated, nor have concepts of resolution-limited computation been carefully scrutinized. We believe focus on these "simple ideas" will open up new classes of algorithms that are practical *and* theoretically sound. This has implications beyond motion planning. Our work in SSS is not just abstract, as we have validated these ideas in several non-trivial planners [12,13,20].

There are many open questions concerning SSS framework. Perhaps the biggest challenge for SSS is the conventional wisdom that PRM can provide practical solutions for problems with high degrees-of-freedom, while resolution methods can only reach medium DOF, generally regarded as 5–8 DOF (Choset et al. [9, p. 202]). Likewise, in Nowakiewicz [14], "*[subdivision methods] are not suitable for 6-DOF rigid body motion planning due to the large expected number of cells ... We believe that in high-dimensional spaces it has little practical value.*"

The other major challenge is a theoretical one: how to do complexity analysis of adaptive subdivision in Motion Planning (cf. [18]). Here are some other topics:

– The current SSS framework detects `NO-PATH` by exhaustion. It is a challenge to design efficient techniques (related to maintaining homology) to allow fast detection of `NO-PATH`. A promising new work by Kerber and Cabello [7] shows how to do this when $\mathcal{C}space = \mathbb{R}^2$.

- Beyond kinematic spaces, subdivision in state spaces for kinodynamic planning seems quite challenging.
- Design and analysis of good adaptive search strategies, including the Voronoi heuristic [23], or randomized or hybrid ones. E.g., efficient updates for dynamic A-star search [1] seems open.

Acknowledgments. I am indebted to Yi-Jen Chiang, Danny Halperin, Steve LaValle, and Vikram Sharma for many helpful discussions.

References

1. Barbehenn, M., Hutchinson, S.: Toward an exact incremental geometric robot motion planner. In: Proceedings of Intelligent Robots and Systems 1995, vol. 3, pp. 39–44 (1995). 1995 IEEE, RSJ International Conference, Pittsburgh. PA, USA, pp. 5–9, August 1995
2. Basu, S., Pollack, R., Roy, M.-F.: Algorithms in Real Algebraic Geometry. Algorithms and Computation in Mathematics, vol. 10, 2nd edn. Springer, Heidelberg (2006)
3. Beyersdorff, O., Köbler, J., Messner, J.: Nondeterministic functions and the existence of optimal proof systems. Theor. Comput. Sci. **410**(38–40), 3839–3855 (2009)
4. Brooks, R.A., Lozano-Perez, T.: A subdivision algorithm in configuration space for findpath with rotation. In: Proceedings of the 8th IJCAI, San Francisco, CA, USA, vol. 2, pp. 799–806. Morgan Kaufmann Publishers Inc. (1983)
5. Brownawell, W.D., Yap, C.K.: Lower bounds for zero-dimensional projections. In: 2009 34th International Symposium on Symbolic and Algebraic Computation (ISSAC 2009), pp. 79–86. KIAS, Seoul, Korea, 28–31 July 2009
6. Burr, M., Krahmer, F.: SqFreeEVAL: an (almost) optimal real-root isolation algorithm. J. Symb. Comput. **47**(2), 153–166 (2012)
7. Cabello, S., Kerber, M.: Semi-dynamic connectivity in the plane. In: Algorithms and Data Structure Symposium (WADS 2015) (to appear, 2015). arXiv:1502.03690
8. Chiang, Y.-J., Yap, C.: Numerical subdivision methods in motion planning. 2011 Poster, IROS Workshop on Progress and Open Problems in Motion Planning, San Francisco, 30 September 2011
9. Choset, H., Lynch, K.M., Hutchinson, S., Kantor, G., Burgard, W., Kavraki, L.E., Thrun, S.: Principles of Robot Motion: Theory, Algorithms, and Implementations. MIT Press, Boston (2005)
10. Kavraki, L., Švestka, P., Latombe, C., Overmars, M.: Probabilistic roadmaps for path planning in high-dimensional configuration spaces. IEEE Trans. Robot. Autom. **12**(4), 566–580 (1996)
11. LaValle, S.M.: Planning Algorithms. Cambridge University Press, Cambridge (2006)
12. Luo, Z., Chiang, Y.-J., Lien, J.-M., Yap, C.: Resolution exact algorithms for link robots. In: 2014 Proceedings of the 11th WAFR, Boğaziçi University, Istanbul, Turkey, 3–5 August 2014. (to appear in a Springer Tracts in Advanced Robotics (STAR))
13. Luo, Z., Yap, C.: Resolution exact planner for non-crossing 2-link robot (2015, Submitted)

14. Nowakiewicz, M.: MST-based method for 6DOF rigid body motion planning in narrow passages. In: 2010 Proceedings of IEEE/RSJ International Conference on Intelligent Robots and Systems, Taipei, Taiwan, pp. 5380–5385, 18–22 October 2010

15. Ó'Dúnlaing, C., Sharir, M., Yap, C.K.: Retraction: a new approach to motion-planning. ACM Symp. Theor. Comput. **15**, 207–220 (1983)

16. Ó'Dúnlaing, C., Yap, C.K.: A "retraction" method for planning the motion of a disc. J. Algorithms **6**, 104–111 (1985)

17. Rivara, M.-C.: Lepp-bisection algorithms, applications and mathematical properties. Appl. Numer. Math. **59**(9), 2218–2235 (2009)

18. Sagraloff, M., Yap, C.K.: A simple but exact and efficient algorithm for complex root isolation. In: Emiris, I.Z. (ed.) 36th International Symposium on Symbolic and Algebraic Computation, San Jose, California, pp. 353–360, 8–11 June 2011

19. Sharma, V., Yap, C.: Near optimal tree size bounds on a simple real root isolation algorithm. In: 2012 37th International Symposium on Symbolic and Algebraic Computation (ISSAC 2012), Grenoble, France, pp. 319–326, 22–25 July 2012

20. Wang, C., Chiang, Y.-J., Yap, C.: On soft predicates in subdivision motion planning. In: 2014 Computational Geometry: Theory and Applications, Special Issue for SoCG, Rio de Janeiro, Brazil, 17–20 June 2013

21. Wei, Z., Yap, C.: Soft subdivision planner for a rod (2015. in preparation)

22. Yap, C., Sharma, V., Lien, J.-M.: Towards exact numerical voronoi diagrams. In: 2012 9th Proceedings of the International Symposium of Voronoi Diagrams in Science and Engineering (ISVD), Rutgers University, NJ, pp. 2–16. IEEE, 27–29 June 2012. Invited Talk

23. Yap, C.K.: Algorithmic motion planning. In: Schwartz, J., Yap, C. (eds.) Advances in Robotics. Algorithmic and Geometric Issues, vol. 1, pp. 95–143. Lawrence Erlbaum Associates, Hillsdale (1987)

24. Yap, C.K.: Robust geometric computation. In: Goodman, J.E., O'Rourke, J. (eds.) Handbook of Discrete and Computational Geometry, 2nd edn, pp. 927–952. Chapman & Hall/CRC, Boca Raton (2004)

25. Yap, C.K.: Soft subdivision search in motion planning. In: Aladren, A., et al. (eds.) In: Proceedings of 1st Workshop on Robotics Challenge and Vision (RCV 2013), A Computing Community Consortium (CCC) Best Paper Award, Robotics Science and Systems Conference (RSS 2013), Berlin (2013). arXiv:1402.3213

26. Yershova, A., Jain, S., LaValle, S.M., Mitchell, J.C.: Generating uniform incremental grids on SO(3) using the Hopf fibration. IJRR **29**(7), 801–812 (2010)

27. Zhang, L., Kim, Y.J., Manocha, D.: Efficient cell labeling and path non-existence computation using C-obstacle query. Int. J. Robot. Res. **27**(11–12), 1325–1349 (2008)

28. Zhu, D., Latombe, J.-C.: New heuristic algorithms for efficient hierarchical path planning. IEEE Trans. Robot. Autom. **7**, 9–20 (1991)

Contributed Papers

On r-Gatherings on the Line

Toshihiro Akagi and Shin-ichi Nakano[✉]

Gunma University, Kiryu 376-8515, Japan
nakano@cs.gunma-u.ac.jp

Abstract. In this paper we study a recently proposed variant of the facility location problem, called the r-gathering problem. Given an integer r, a set C of customers, a set F of facilities, and a connecting cost $co(c, f)$ for each pair of $c \in C$ and $f \in F$, an r-gathering of customers C to facilities F is an assignment A of C to *open facilities* $F' \subset F$ such that r or more customers are assigned to each open facility. We give an algorithm to find an r-gathering with the minimum cost, where the cost is $\max_{c_i \in C}\{co(c_i, A(c_i))\}$, when all C and F are on the real line.

Keywords: Algorithm · Facility location · Gathering

1 Introduction

The facility location problem and many of its variants are studied [5,6]. In the basic facility location problem we are given (1) a set C of customers, (2) a set F of facilities, (3) an opening cost $op(f)$ for each $f \in F$, and (4) a connecting cost $co(c, f)$ for each pair of $c \in C$ and $f \in F$, then we open a subset $F' \subset F$ of facilities and find an assignment A from C to F' so that a designated cost is minimized.

In this paper we study a recently proposed variant of the problem, called the r-gathering problem [4]. An r-gathering of customers C to facilities F is an assignment A of C to *open facilities* $F' \subset F$ such that r or more customers are assigned to each open facility. This means each open facility has enough number of customers. We assume $|C| \geq r$ holds. Then we define the cost of (the *max* version of) a gathering as $\max_{c_i \in C}\{co(c_i, A(c_i))\}$. (We assume $op(f_j) = 0$ for each $f_j \in F$ in this paper.) The min-max version of the r-*gathering problem* finds an r-gathering having the minimum cost. For the min-sum version see the brief survey in [4].

Assume that F is a set of locations for emergency shelters, and $co(c, f)$ is the time needed for a person $c \in C$ to reach a shelter $f \in F$. Then an r-gathering corresponds to an evacuation assignment such that each opened shelter serves r or more people, and the r-gathering problem finds an evacuation plan minimizing the evacuation time span.

Armon [4] gave a simple 3-approximation algorithm for the r-gathering problem and proves that with assumption $P \neq NP$ the problem cannot be approximated within a factor of less than 3 for any $r \geq 3$. In this paper we give an

© Springer International Publishing Switzerland 2015
J. Wang and C. Yap (Eds.): FAW 2015, LNCS 9130, pp. 25–32, 2015.
DOI: 10.1007/978-3-319-19647-3_3

$O((|C| + |F|) \log(|C| + |F|))$ time algorithm to solve the r-gathering problem when all C and F are on the real line.

The remainder of this paper is organized as follows. Section 2 gives an algorithm to solve a decision version of the r-gathering problem. Section 3 contains our main algorithm for the r-gathering problem. Sections 4 and 5 present two more algorithms to solve two similar problems. Finally Sect. 6 is a conclusion.

2   (k,r)-Gathering on the Line

In this section we give a linear time algorithm to solve a decision version of the r-gathering problem [3].

Given customers $C = \{c_1, c_2, \cdots, c_{|C|}\}$ and facilities $F = \{f_1, f_2, \cdots, f_{|F|}\}$ on the real line (we assume they are distinct points and appear in those order from left to right respectively) and two numbers k and r, then problem $P(C, F, j, i)$ finds an assignment A of customers $C_i = \{c_1, c_2, \cdots, c_i\}$ to open facilities $F'_j \subset F_j = \{f_1, f_2, \cdots, f_j\}$ such that (1) r or more customers are assigned to each open facility, (2) $co(c_i, A(c_i)) \leq k$ for each $c_i \in C_i$ and (3) $f_j \in F'_j$. (2) means each customer is assigned to a near facility, and (3) means the rightmost facility is forced to open. We assume that $co(c, f)$ is the distance between $c \in C$ and $f \in F$, and for each $f_j \in F$ interval $[f_j - k, f_j + k]$ contains r or more customers, otherwise we can remove such f_j from F since such f_j never open.

An assignment A of C_i to F_j is called *monotone* if, for any pair $c_{i'}, c_i$ of customers with $i' < i$, $A(c_{i'}) \leq A(c_i)$ holds. In a monotone assignment the interval induced by the assigned customers to a facility never intersects other interval induced by the assigned customers to another facility. We can observe that if $P(C, F, j, i)$ has a solution then $P(C, F, j, i)$ also has a monotone solution. Also we can observe that if $P(C, F, j, i)$ has a solution and $co(c_{i+1}, f_j) \leq k$ then $P(C, F, j, i + 1)$ also has a solution.

If $P(C, F, j, i)$ has a solution for some i then let $s(f_j)$ be the minimum i such that $P(C, F, j, i)$ has a solution. Note that (3) $f_j \in F'_j$ means $c_{s(f_j)}$ is located in interval $[f_j - k, f_j + k]$. We define $P(C, F, j)$ to be the problem to find such $s(f_j)$ and a corresponding assignment. If $P(C, F, j, i)$ has no solution for every i then we say $P(C, F, j)$ has no solution, otherwise we say $P(C, F, j)$ has a solution.

Lemma 1. *For any pair $f_{j'}$ and f_j in F with $j' < j$, $s(f_{j'}) \leq s(f_j)$ holds.*

Proof. Assume otherwise. Then $s(f_{j'}) > s(f_j)$ holds. Modify the assignment corresponding to $s(f_j)$ as follows. Reassign the customers assigned to f_j to $f_{j'}$ then close f_j. The resulting assignment is an r-gathering of $C_{s(f_j)}$ to $F_{j'}$ and now $s(f_{j'}) = s(f_j)$. A contradiction.   □

Assume that $P(C, F, j)$ has a solution and $c_1 < f_j - k$. Then the corresponding solution has one or more open facilities except for f_j. Choose the solution of $P(C, F, j)$ having the minimum second rightmost open facility, say $f_{j'}$. We say $f_{j'}$ is the *mate* of f_j and write $mate(f_j) = f_{j'}$. We have the following three cases based on the condition of the mate $f_{j'}$ of f_j.

Case 1: $P(C, F, j')$ has a solution, $f_{j'} + k < f_j - k$, the interval $(f_{j'} + k, f_j - k)$ has no customer and the interval $[f_j - k, f_j + k]$ has r or more customers.

Case 2: $P(C, F, j')$ has a solution, $c_{s(f_{j'})} \geq f_j - k$ and interval $(c_{s(f_{j'})}, f_j + k]$ has r or more customers.

Case 3: $P(C, F, j')$ has a solution, $c_{s(f_{j'})} < f_j - k$ and interval $[f_j - k, f_j + k]$ has r or more customers.

For each f_j by checking the three conditions above for every possible mate $f_{j'}$ one can design $O(|F|^2 + |C|)$ time algorithm based on a dynamic programming approach. However we can omit the most part of the checks by the following lemma.

Lemma 2. *(a) Assume $P(C, F, j)$ has a solution. If $P(C, F, j + 1)$ also has a solution then $mate(f_j) \leq mate(f_{j+1})$ holds.*
(b) For $f_j \in F$, let f_{min} be the minimum $f_{j'}$ such that (i)$P(C, F, j')$ has a solution and (ii)$f_{j'} + k \geq f_j - k$, if such f_{min} exists. If $P(C, F, j)$ has no solution with the second rightmost open facility f_{min}, then (b1) any $f_{j''}$ satisfying $f_{min} < f_{j''} < f_j$ is not the mate of f_j, and $P(C, F, j)$ has no solution, and (b2) $f_{min} \leq mate(f_{j+1})$ holds if $mate(f_{j+1})$ exists.

Proof. (a) Assume otherwise. If $mate(f_{j+1}) + k < f_j - k$ holds then $mate(f_{j+1})$ is also the mate of f_j, a contradiction. If $mate(f_{j+1}) + k \geq f_j - k$ holds then by Lemma 1 $mate(f_{j+1})$ is also the mate of f_j, a contradiction. (b1) Immediate from Lemma 1. (b2) Assume otherwise. If $mate(f_{j+1}) + k < f_j - k$ holds then $mate(f_{j+1})$ is also the mate of f_j, a contradiction. If $mate(f_{j+1}) + k \geq f_j - k$ holds then f_{min} is $mate(f_{j+1})$ not $mate(f_j)$, a contradiction. □

Lemma 2 means after searching for the mate of f_j upto some $f_{j'}$ the next search for the mate of f_{j+1} can start at the $f_{j'}$. Based on the lemma above we can design algorithm **find(k, r)-gathering**.

In the preprocessing we compute, for each $f_j \in F$, (1) the index of the first customer in interval $(f_j + k, c_{|C|})$, (2) the index of the first customer in interval $[f_j - k, c_{|C|})$ and (3) the index of the r-th customer in interval $[f_j - k, c_{|C|})$. Also we store the index $s(f_j)$ for each $f_j \in F$. Those needs $O(|C| + |F|)$ time. After the preprocessing the algorithm runs in $O(|F|)$ time since $j' \leq j$ always holds the most inner part to compute $s(f_j)$ executes at most $2|F|$ times.

We have the following lemma.

Lemma 3. *One can solve the (k, r)-gathering problem in $O(|C| + |F|)$ time.*

3   r-Gathering on the Line

In this section we give an $O((|C| + |F|) \log(|C| + |F|))$ time algorithm to solve the r-gathering problem when all C and F are on the real line.

Algorithm 1. find(k, r)-gathering (C, F, k)

$j = 1$
// One open facility Case //
while interval $[f_j - k, f_j + k]$ has both c_1 and c_r **do**
    set $s(f_j)$ to be the r-th customer c_r
    $j = j + 1$
end while
// Two or more open facilities Case//
$j' = 1$
while $j \leq |F|$ **do**
    $flag = $ off
    while $flag =$off and $s(f_j)$ is not defined yet and $j' < j$ **do**
        if $P(C, F, f_{j'})$ has a solution and $f_{j'} + k < f_j - k$, interval $(f_{j'} + k, f_j - k)$ has
        no customer **then**
            set $s(f_j)$ to be the r-th customer in the interval $[f_j - k, f_j + k]$
        else if $P(C, F, f_{j'})$ has a solution and $f_{j'} + k \geq f_j - k$ **then**
            $flag = $ on
            if $s(f_{j'}) \geq f_j - k$ and interval $(s(f_{j'}), f_j + k]$ has r or more customers **then**
                set $s(f_j)$ to be the r-th customer in the interval $(s(f_{j'}), f_j + k]$
            else if $P(C, F, f_{j'})$ has a solution, $s(f_{j'}) < f_j - k$ and interval $[f_j - k, f_j + k]$
            has r or more customers **then**
                set $s(f_j)$ to be the r-th customer in the interval $[f_j - k, f_j + k]$
            end if
        end if
        $j' = j' + 1$
    end while
    $j = j + 1$
end while
if some f_j with defined $s(f_j)$ has $c_{|C|}$ within distance k **then**
    output YES
else
    output NO
end if

Our strategy is as follows. First we can observe that the minimum cost k^* of a solution of an r-gathering problem is some $co(c, f)$ with $c \in C$ and $f \in F$. Since the number of distinct $co(c, f)$ is at most $|C||F|$, sorting them needs $O(|C||F| \log(|C||F|))$ time. Then find the smallest k such that the (k, r)-gathering problem has a solution by binary search, using the linear-time algorithm in the preceding section $\log(|C||F|)$ times. Those part needs $O((|C| + |F|) \log |C||F|)$ time. Thus the total running time is $O(|C||F| \log(|C||F|))$.

However by using the sorted matrix searching method [7] (See the good survey in [2, Section 3.3]) we can improve the running time to $O((|C|+|F|) \log(|C|+ |F|))$. Similar technique is also used in [8,9] for a fitting problem. Now we explain the detail in our simplified version.

First let M_C be the matrix in which each element is $m_{i,j} = c_i - f_j$. Then $m_{i,j} \geq m_{i,j+1}$ and $m_{i,j} \leq m_{i+1,j}$ always holds, so the elements in the rows and columns are sorted respectively. Similarly let M_F be the matrix in which each element is $m'_{i,j} = f_j - c_i$. The minimum cost k^* of an optimal solution of an r-gathering problem is some positive element in those two matrices. We can find the smallest k in M_C for which the (k, r)-gathering problem has a solution, as follows.

Let n be the smallest integer which is (1) a power of 2 and (2) larger than or equal to $\max\{|C|, |F|\}$. Then we append the largest element $m_{|C|,1}$ to M_C as the elements in the lowest rows and the leftmost columns so that the resulting matrix has exactly n rows and n columns. Note that the elements in the rows and columns are still sorted respectively. Let M_C be the resulting matrix. Our algorithm consists of stages $s = 1, 2, \cdots, \log n$, and maintains a set L_s of submatrices of M_C possibly containing k^*. Hypothetically first we set $L_0 = \{M_C\}$. Assume we are now starting stage s.

For each submatrix M in L_{s-1} we partite M into the four submatrices with $n/2^s$ rows and $n/2^s$ columns and put them into L_s.

Let k_{min} be the median of the upper right corner elements of the submatrices in L_s. Then for the $k = k_{min}$ we solve the (k, r)-gathering problem. We have two cases.

If the (k, r)-gathering problem has a solution then we remove from L_s each submatrix with the upper right corner element (the smallest element) greater than k_{min}. Since $k_{min} \geq k^*$ holds each removed submatrix has no chance to contain k^*. Also if L_s has several submatrices with the upper right corner element equal to k_{min} then we remove them except one from L_s. Thus we can remove $|L_s|/2$ submatrices from L_s.

Otherwise if the (k, r)-gathering problem has no solution then we remove from L_s each submatrix with the lower left corner element (the largest element) smaller than k_{min}. Since $k_{min} < k^*$ holds each removed submatrix has no chance to contain k^*. Now we can observe that, for each "chain" of submatrices, which is the sequence of submatrices in L_s with lower-left to upper-right diagonal on the same line, the number of submatrices (1) having the upper right corner element smaller than k_{min} (2) but remaining in L_i is at most one (since the elements on "the common diagonal line" are sorted). Thus, if $|L_s|/2 > D_s$, where $D_s = 2^{s+1}$ is the number of chains plus one, then we can remove at least $|L_s|/2 - D_s$ submatrices from L_s.

Similarly let k_{max} be the median of the lower left corner elements of the submatrices in L_s, and for the $k = k_{max}$ we solve the (k, r)-gathering problem and similarly remove some submatrices from L_s. This ends stage s.

Now after stage $\log n$ each matrix in $L_{\log n}$ has just one element, then we can find the k^* using a binary search with the linear-time decision algorithm.

We can prove that at the end of stage s the number of submatrices in L_s is at most $2D_s$, as follows.

First L_0 has 1 submatrix and $1 \leq 2D_0 = 2 \cdot 2^{0+1}$ submatrix. By induction assume L_{s-1} has $2D_{s-1} = 2 \cdot 2^s$ submatrices.

At stage s we first partite each submatrix in L_{s-1} into four submatrices then put them into L_s. Now the number of submatrices in L_s is $4 \cdot 2D_{s-1} = 4D_s$. We have four cases.

If the (k, r)-gathering problem has a solution for $k = k_{min}$ then we can remove at least a half of the submatrices from L_s, and so the number of the remaining submatrices in L_s is at most $2D_s$, as desired.

If the (k, r)-gathering problem has no solution for $k = k_{max}$ then we can remove at least a half of the submatices from L_s, and so the number of the remaining submatices in L_s is at most $2D_s$, as desired.

Otherwise if $|L_s|/2 \leq D_s$ then the number of the submatices in L_s (even before the removal) is at most $2D_s$, as desired.

Otherwise (1) after the check for $k = k_{min}$ we can remove at least $|L_s|/2 - D_s$ submatices (consisting of too small elements) from L_s, and (2) after the check for $k = k_{max}$ we can remove at least $|L_s|/2 - D_s$ submatices (consisting of too large elements) from L_s, so the number of the remaining submatices in L_s is at most $|L_s| - 2(|L_s|/2 - D_s) = 2D_s$, as desired.

Thus at the end of stage s the number of submatrices in L_s is always at most $2D_s$.

Now we consider the running time. We implicitly treat each submatrix as the index of the upper right element in M_C and the number of lows. Except for the calls of the linear-time decision algorithm for the (k, r)-gathering problem, we need $O(|L_{s-1}|) = O(D_{s-1})$ time for each stage $s = 1, 2, \cdots, \log n$, and $D_0 + D_1 + \cdots + D_{\log n - 1} = 2 + 4 + \cdots + 2^{\log n} < 2 \cdot 2^{\log n} = 2n$ holds, so this part needs $O(n)$ time in total. (Here we use the linear time algorithm to find the median.)

Since each stage calls the linear-time decision algorithm twice this part needs $O(n \log n)$ time in total.

After stage $s = \log n$ each matrix has just one element, then we can find the k^* among the $|L_{\log n}| \leq 2D_{\log n} = 4n$ elements using a binary search with the linear-time decision algorithm at most $\log 4n$ times. This part needs $O(n \log n)$ time.

Then we similarly find the smallest k in M_F for which the (k, r)-gathering problem has a solution. Finally we output the smaller one among the two.

Thus the total running time is $O((|C| + |F|) \log(|C| + |F|))$.

Theorem 1. *One can solve the r-gathering problem in $O((|C| + |F|) \log(|C| + |F|))$ time when all C and F are on the real line.*

4 r-Gather Clustering

In this section we give an algorithm to solve a similar problem by modifying the algorithm in Sect. 3.

Given a set C of n points on the plane an *r-gather-clustering* is a partition of the points into clusters such that each cluster has at least r points. The r-gather-clustering problem [1] finds an r-gather-clustering minimizing the maximum radius among the clusters, where the radius of a cluster is the minimum radius of the disk which can cover the points in the cluster. A polynomial time 2-approximation algorithm for the problem is known [1].

When all C are on the real line, in any solution of any r-gather-clustering problem, we can assume that the center of each disk is at the midpoint of some pair of points, and the radius of an optimal r-gather-clustering is the half of the distance between some pair of points in C.

Given C and two numbers k and r the decision version of the r-gather-clustering problem find an r-gather-clustering with the maximum radius k. We can assume that in any solution of the problem the center of each disk is at $c-k$ for some $c \in C$. Thus, by introducing the set of all such points as F, we can solve the decision version of the r-gather-clustering problem as the (k,r)-gathering problem. Using the algorithm in Sect. 2 we can solve the problem in $O(|C|)$ time.

Now we explain our algorithm to solve the r-gather-clustering problem. First sort C in $O(|C| \log |C|)$ time. Let $c_1, c_2, \cdots, c_{|C|}$ be the resulting non decreasing sequences and let M be the matrix in which each element is $m_{i,j} = (c_i - c_j)/2$. Note that the optimal radius is in M and this time M has $|C|$ rows and columns. Now $m_{i,j} \geq m_{i,j+1}$ and $m_{i,j} \geq m_{i+1,j}$ holds, so the elements in the rows and columns are sorted respectively. Then as in Sect. 3 we can find the optimal radius by the sorted matrix searching method. The algorithm calls the decision algorithm $O(\log |C|)$ times and the decision algorithm runs in $O(|C|)$ time, and in the stages the algorithm needs $O(|C|)$ time in total except for the calls. Finally we needs $O(|C| \log |C|)$ time for the last binary search. Thus the total running time is $O(|C| \log |C|)$.

Theorem 2. *One can solve the r-gather-clustering problem in $O(|C| \log |C|)$ time when all points in C are on the real line.*

5   Outlier

In this section we consider a generalization of the r-gathering problem where at most h customers are allowed to be not assigned.

An *r-gathering with h-outliers* of customers C to facilities F is an assignment A of $C \backslash C'$ to *open facilities* $F' \subset F$ such that r or more customers are assigned to each open facility and $|C'| \leq h$. The *r-gathering with h-outliers problem* finds an r-gathering with h-outliers having the minimum cost.

Given customers $C = \{c_1, c_2, \cdots, c_{|C|}\}$ and facilities $F = \{f_1, f_2, \cdots, f_{|F|}\}$ on the real line and three numbers k and r and h, problem $P(C, F, j, i, h)$ finds an *r-gathering with h-outliers* of $C_i = \{c_1, c_2, \cdots, c_i\} \backslash C_i'$ to $F_j' \subset F_j = \{f_1, f_2, \cdots, f_j\}$ such that (1) r or more customers are assigned to each open facility, (2) $co(c_i, A(c_i)) \leq k$ for each $c_i \in C_i \backslash C_i'$, (3) $f_j \in F_j'$ and (4) $|C_i'| \leq h$. For designated j and h' if $P(C, F, j, i, h')$ has a solution for some i then let $s(f_{j,h'})$ be the minimum i such that $P(C, F, j, i, h')$ has a solution. We define $P(C, F, j, h')$ to be the problem to find such $s(f_{j,h'})$ and a corresponding assignment.

By a dynamic programming approach one can compute $P(C, F, j, h')$ for each $j = 1, 2, \cdots, |F|$ and $h' = 1, 2, \cdots, h$ in $O(|C| + h^2 |F|)$ time in total. Then one can decide whether an *r-gathering with h-outliers problem* has a solution with cost k.

Lemma 4. *One can decide whether an r-gathering with h-outliers problem has a solution with cost k in $O(|C| + h^2|F|)$ time.*

The minimum cost k^* of a solution of an r-gathering with h-outliers problem is again some $co(c, f)$ with $c \in C$ and $f \in F$. By the sorted matrix searching method using the $O(|C| + h^2|F|)$ time decision algorithm above one can solve the problem with outliers in $O((|C| + h^2|F|)\log(|C| + |F|))$ time.

Theorem 3. *One can solve the r-gathering with h-outliers problem in $O((|C| + h^2|F|)\log(|C| + |F|))$ time when all C and F are on the real line.*

6   Conclusion

In this paper we have presented an algorithm to solve the r-gathering problem when all C and F are on the real line. The running time of the algorithm is $O((|C| + |F|)\log(|C| + |F|))$. We also presented two more algorithm to solve two similar problems.

Can we design a linear time algorithm for the r-gathering problem when all C and F are on the real line?

References

1. Aggarwal, G., Feder, T., Kenthapadi, K., Khuller, S., Panigrahy, R., Thomas, D., Zhu, A.: Achieving anonymity via clustering, Tranaction on Algorithms, vol. 6, Article No.49, pp. 49:1-49:19 (2010)
2. Agarwal, P., Sharir, M.: Efficient algorithms for geometric optimization. Comput. Surv. **30**, 412–458 (1998)
3. Akagi, T., Nakano, S.: On (k, r)-gatherings on a road. In: Proceedings of Forum on Information Technology, FIT 2013, RA-001 (2013)
4. Armon, A.: On min-max r-gatherings. Theoret. Comput. Sci. **412**, 573–582 (2011)
5. Drezner, Z.: Facility Location: A Survey of Applications and Methods. Springer, New York (1995)
6. Drezner, Z., Hamacher, H.W.: Facility Location: Applications and Theory. Springer, Heidelberg (2004)
7. Frederickson, G., Johnson, D.: Generalized selection and ranking: sorted matrices. SIAM J. Comput. **13**, 14–30 (1984)
8. Fournier, H., Vigneron, A.: Fitting a step function to a point set. In: Halperin, D., Mehlhorn, K. (eds.) ESA 2008. LNCS, vol. 5193, pp. 442–453. Springer, Heidelberg (2008)
9. Liu, J.-Y.: A randomized algorithm for weighted approximation of points by a step function. In: Wu, W., Daescu, O. (eds.) COCOA 2010, Part I. LNCS, vol. 6508, pp. 300–308. Springer, Heidelberg (2010)

A New Algorithm for Intermediate Dataset Storage in a Cloud-Based Dataflow

Jie Cheng[1], Daming Zhu[2(✉)], and Binhai Zhu[3]

[1] School of Mechanical, Electrical and Information Engineering,
Shandong University, Weihai, China
[2] School of Computer Science and Technology, Shandong University, Jinan, China
{chjie,dmzhu}@sdu.edu.cn
[3] Department of Computer Science, Montana State University, Bozeman,
MT 59717-3880, USA
bhz@cs.montana.edu

Abstract. Running a dataflow in a cloud environment usually generates many useful intermediate datasets. A strategy for running a dataflow is to decide which datasets should be stored, while the rest of them are regenerated. The intermediate dataset storage (IDS) problem asks to find a strategy for running a dataflow, such that the total cost is minimized. The current best algorithm for linear-structure IDS takes $O(n^4)$ time, where "linear-structure" means that the structure of the datasets in the dataflow is a pipeline. In this paper, we present a new algorithm for this problem, and improve the time complexity to $O(n^3)$, where n is the number of datasets in the pipeline.

1 Introduction

A cloud-based dataflow is a data-driven workflow deployed in a cloud computing environment. In a cloud-based dataflow, there are usually a large number of datasets, including initial dataset, output dataset and a large volume of intermediate datasets generated during the execution. The intermediate datasets often contain valuable intermediate results, thus would be frequently traced back for re-analyzing or re-using [1]. Since the dataflow systems are executed in a cloud computing environment, all the resources used need to be paid for. As indicated in [2], storing all of the intermediate datasets may induce a high storage cost, while if all the intermediate datasets are deleted and regenerated when needed, the computation cost of the system may also be very high. Hence, an optimal strategy is needed to store some datasets and regenerate the rest of them when needed so as to minimize the total cost of the whole workflow system [3,4], which is called the intermediate dataset storage (IDS) problem.

In a cloud dataflow system, when a deleted dataset needs to be regenerated, the computation cost will involve not only itself but its direct predecessors, if these predecessors are also deleted. Hence, the computation cost of a sequence of deleted datasets needs to be accumulated, which leads to the IDS problem. In [2], Yuan *et al.* presented the background of the IDS problem in scientific

© Springer International Publishing Switzerland 2015
J. Wang and C. Yap (Eds.): FAW 2015, LNCS 9130, pp. 33–44, 2015.
DOI: 10.1007/978-3-319-19647-3_4

workflows and proposed an intermediate data dependency graph (IDG). Based on IDG, they presented two algorithms as the minimum cost benchmark of the IDS problem, a linear CTT-SP algorithm for linear workflow which takes $O(n^4)$ time, and a general CTT-SP algorithm for parallel structure workflow which takes $O(n^9)$ time. Besides [2], there have been some related research. Zohrevandi and Bazzi [5] presented a branch-and-bound algorithm for the common intermediate dataset storage between two scientific workflows, which is related to the IDS problem. Adams *et al.* [3] proposed a model balancing the computation cost and the storage cost. The approach proposed by Han *et al.* [6] is to support automatic intermediate data reusing for large-scale cloud dataflow based on Perti-nets. As far as we know, the current best exact algorithm for the IDS problem is the one proposed by Yuan *et al.* in [2].

This paper focuses on the IDS problem for linear-structure cloud dataflow systems. We present a binary tree model that is called *S-C* tree for the IDS problem. In an *S-C* tree, a vertex represents a choice of a dataset, which could be storage or computation, and the price of a vertex represents the generation cost for the choice. Based on the *S-C* tree model, the optimal solution to the IDS problem can be converted to searching for an optimal full path in the *S-C* tree with the minimum path cost. To reduce the searching space, we propose a group of pruning strategies, by which, more than $\frac{k-1}{2k}$ of the branches will be pruned off at each level k. Therefore, with the increasing of the searching level, the searching space grows linearly. Using these pruning strategies, we present an exact algorithm for the linear-structure IDS problem and prove that the algorithm takes $O(n^3)$ time.

The rest of the paper is organized as follows. Section 2 introduces the IDS problem and some related concepts are defined there. The *S-C* tree model of the IDS problem, including the proof of some theorems, are presented in Sect. 3. Section 4 describes the algorithm based on the em S-C tree and the corresponding analysis. Section 5 concludes the paper.

2   The IDS Problem

In this section, we first introduce some related concepts, and then give the definition of the IDS problem.

Definition 1. *A linear-structure cloud dataflow F can be expressed as $F = (DS, TS)$, where,*

- *$DS = \{d_0, d_1, \cdots, d_n\}$ is a set of datasets, where n is the number of intermediate datasets. d_0 denotes the initial dataset, and d_n is the output dataset of F. For each $d_i, 0 < i < n$, d_{i-1} is the direct predecessor of d_i, and d_{i+1} is the direct successor of d_i;*
- *$DT = \{t_1, t_2, \cdots, t_n\}$ is a set of tasks, where $t_i, 0 \leq i \leq n$, is a logical computation unit executed using d_{i-1} as the input and outputs the dataset d_i. Given a dataset $d_i, 0 \leq i \leq n$, t_i is called the execution task of d_i.*

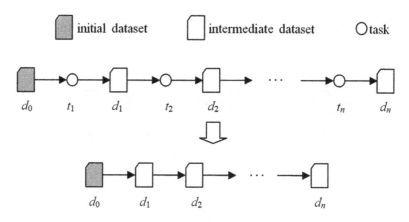

Fig. 1. The exemplar graph of a linear dataflow.

For simplicity, the linear-structure cloud dataflow is simply called *dataflow* throughout the rest of the paper. Since this paper focuses on datasets, a dataflow can also be simplified as a sequence of datasets, denoted as $F = \{d_0, d_1, \cdots, d_n\}$, as shown in Fig. 1.

As mentioned in [1], there are two basic types of resources in cloud computing: storage and computation. Normally, the service price of a cloud platform is proportional to the size of storage resource and also to the instance-hour for computation resource.

Given a dataflow F, we say that a dataset d is a storage dataset if d is selected to be stored; otherwise, it is a computation dataset. Thus F can be separated into two subsets, denoted as $F = S \cup C$, where S is the set of storage datasets and C is the set of computation datasets. We assume that the initial dataset d_0 is a storage dataset.

In a dataflow $F = \{d_0, d_1, \cdots, d_n\}$, there are two ways to generate an intermediate dataset $d_i, (0 < i \leq n)$, storage and computation. That is, if $d_i \in S$, it is available when needed; otherwise, it is deleted and has to be regenerated by computation. Therefore, there have two kinds of costs related to d_i, which are storage cost $x(d_i)$ if $d_i \in S$ and computation cost $y(d_i)$ if $d_i \in C$. In general, $x(d_i)$ is proportional to the size of d_i, and $y(d_i)$ is proportional to the running time of the execution task t_i.

Definition 2. *Given a dataset $d_j(j > 0)$, we say that the dataset d_i is the* **storage-prior** *of $d_j, 0 \leq i < j$, denoted as $d_i \mapsto d_j$, if $d_i \in S$ and for any $i < k < j, d_k \in C$. That is, $d_i \mapsto d_j$ means that d_i is the nearest predecessor storage dataset of d_j, as shown in Fig. 2.*

As indicated in [2], when we want to regenerate a computation dataset d_j, we have to find its direct predecessor d_{j+1} which may also be deleted, so we have to further trace the nearest stored predecessor, the storage-prior dataset of d_j.

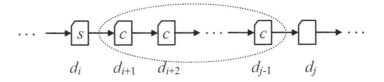

$\boxed{s}$ storage dataset $\boxed{c}$ computation dataset

Fig. 2. An exemplar graph of $d_i \mapsto d_j$

Hence, for any intermediate dataset d_j, its generation cost is defined as:

$$G\_cost(d_j) = \begin{cases} x(d_j), & \text{if } (d_j \in S) \\ \sum_{k=i+1}^{j} y(d_k), & \text{if } (d_j \in C) \wedge (d_i \mapsto d_j) \end{cases} \qquad (2\text{-}1)$$

Based on the above concepts, the IDS problem is defined as follows.

Input: given a dataflow $F = \{d_0, d_1, \cdots, d_n\}$, for each intermediate dataset d_i, its storage cost $x(d_i)$ and its computation cost $y(d_i)$.
Output: the set of storage dataset S and the set of computation dataset C.
Objective: the total cost of F, $\sum_{k=1}^{n} G\_cost(d_k)$, is minimized.

3   Binary Tree Model for the IDS Problem

The objective of the IDS problem is to find an optimal mapping between the intermediate datasets and the set C or S. Since a dataset has only two choices, we apply a binary tree as the problem model.

3.1   S-C Tree Model

Definition 3. *An* **S-C tree** *of a given dataflow* $F = \{d_0, d_1, \cdots, d_n\}$, *denoted as* $Tree^F$, *is full binary tree with* $n + 1$ *levels, in which:*

(1) The root represents the initial dataset d_0.
(2) The set of nodes at the ith *level in* $Tree^F, 0 \leq i \leq n$, *denoted as* $N|_{Tree^F}^{i}$, *is mapped to the dataset* d_i.
(3) Any node $\tau \in N|_{Tree^F}^{i}, 0 \leq i < n$, *its left child* $left(\tau)$ *and right child* $right(\tau)$ *represent that the dataset* d_{i+1} *is selected to be stored and deleted respectively.*

Figure 3 shows a 5-level *S-C* tree. According to Definition 3, the set of nodes in $Tree^F$ can be separated into $S = \{s_0, s_1, \cdots, s_{2^n-1}\}$ and $C = \{c_0, c_1, \cdots, c_{2^n-1}\}$, which are mapped respectively to the set of storage datasets and set of computation datasets in F. We can see that set S is composed of the root s_0 and all the left-child nodes, and set C consists of all the right-child nodes. The nodes of set S and C are also simply called *S*-nodes and *C*-nodes respectively.

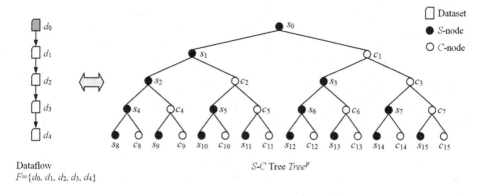

Dataflow
$F=\{d_0, d_1, d_2, d_3, d_4\}$

S-C Tree $Tree^F$

Fig. 3. An exemplar graph of 5-level S-C tree.

Definition 4. *Given an S-C tree $Tree^F$, the **S-prior** of a C node τ, denoted as $\widehat{\tau}$, is the nearest S-ancestor of τ. We use $Path_{\tau}^{\hookrightarrow}$ to denote the ordered set of nodes of the path that is from $\widehat{\tau}$ to τ.*

For example, in Fig. 3, $\widehat{c_4} = s_2, \widehat{c_5} = s_1$. $Path_{c_5}^{\hookrightarrow} = \{s_1, c_2, c_5\}, Path_{c_7}^{\hookrightarrow} = \{s_0, c_1, c_3, c_7\}$. As we can see, in $Path_{\tau}^{\hookrightarrow}$, only the first node $\widehat{\tau}$ is an S-node, others are C-nodes.

Definition 5. *Given a dataflow $F = \{d_0, d_1, \cdots, d_n\}$ and its S-C tree $Tree^F$, assume that $\tau \in N \mid_{Tree^F}^{i}, 0 \leq i < n$, the **weight** and **cost** of τ are defined as follows.*

$$w(\tau) = \begin{cases} 0, & \text{if } \tau \in S \\ y(d_i), & \text{if } \tau \in C \end{cases} \tag{3-1}$$

$$cost(\tau) = \begin{cases} x(d_i), & \text{if } \tau \in S \\ \sum_{\alpha \in Path_{\tau}^{\hookrightarrow}} w(\alpha), & \text{if } \tau \in C \end{cases} \tag{3-2}$$

We assume that $cost(s_0) = 0$.

Definition 6. *Given an S-C tree T with $n + 1$ levels, a **path** $P = \{p_a, p_{a+1}, p_{a+2}, \cdots, p_b\}, 0 \leq a \leq b \leq n$, is an ordered set of nodes from p_a to p_b, in which $p_i \in N \mid_T^i$, $a \leq i < b$, and p_i is the parent node of p_{i+1}. The cost of path P is defined as: $Cost(P) = \sum_{i=a}^{b} cost(p_i)$.*

Definition 7. *Given an S-C tree T with $n+1$ levels, a k- **path** $P^k = \{p_0, p_1, p_2, \cdots, p_k\}, 0 \leq k \leq n$, is a path in which the first node is the root of T. An n-path is also called a **full path**. A full path Λ is called an **optimal full path** if and only if for any full path Λ', $Cost(\Lambda) \leq Cost(\Lambda')$.*

For example, in Fig. 3, $\{s_0, s_1, c_2, c_5\}$ is a 3-path, and $\{s_0, s_1, c_2, c_5, c_{11}\}$ is a full path.

Given a k-path $P^k = \{p_0, p_1, p_2, \cdots, p_k\}$ and a path $P = \{p_k, p_{k+1}, p_{k+2}, \ldots, p_{k+i}\}$, we use $link(P^k, P)$ to denote the $(k + i)$-path: $\{p_0, p_1, p_2, \cdots, p_k, p_{k+1}, p_{k+2}, \cdots, p_{k+i}\}$. In addition, if τ is a child node of p_k, we also use $link(P^k, \tau)$ to denoted the $(k + 1)$-path: $\{p_0, p_1, p_2, \ldots, p_k, \tau\}$.

Definition 8. *Given an S-C tree T with $n + 1$ levels, let $\tau \in N \mid_T^k, 0 \leq k \leq n$, the sub-tree which is rooted at τ is called a k-**subtree**. A k-subtree δ is called an **optimal k-subtree** if and only if δ contains an optimal full path from the kth level to the nth level.*

For example, an *S-C* tree itself is an optimal 0-subtree. Assume that $\Lambda = \{p_0, p_1, p_2, \cdots, p_n\}$ is an optimal full path, then the sub-tree which is rooted at $p_i (0 \leq i \leq n)$ is an optimal i-subtree.

Based on the *S-C* tree model, the IDS problem can be converted as: given a dataflow F and its *S-C* tree $Tree^F$, find an optimal full path of $Tree^F$.

3.2 Proofs of the Theorems

For convenience, given an *S-C* tree T, assume that a node $\tau \in N \mid_T^k$, we use $subtree(\tau)$ to denote a k-subtree which is rooted at τ.

Definition 9. *Given an S-C tree, let τ and τ' be the nodes at the same level. We say that τ is **equivalent to** τ', denoted as $\tau \equiv \tau'$, if and only if $((\tau, \tau' \in S) \vee (\tau, \tau' \in C)) \wedge (cost(\tau) = cost(\tau'))$. If $((\tau, \tau' \in S) \vee (\tau, \tau' \in C)) \wedge (cost(\tau) < cost(\tau'))$, we say that τ is **superior to** τ', denoted as $\tau \prec \tau'$.*

The equivalent and superior relations between nodes are both transitive.

Lemma 1. *Given a dataflow $F = \{d_0, d_1, \ldots, d_n\}$ and its S-C tree $Tree^F$, if τ and τ' are both S-nodes at the same level, then $\tau \equiv \tau'$.*

Proof. Assume that $\tau, \tau' \in N \mid_{Tree^F}^i, 0 < i \leq n$. Since τ and τ' are both S-nodes, according to Definition 5, $cost(\tau) = cost(\tau') = x(d_i)$, thus $\tau \equiv \tau'$. □

Definition 10. *Assume that δ and δ' are k-subtrees, τ and τ' are nodes belonging to δ and δ' respectively. We say that τ' is the **corresponding node** of τ about δ and δ', denoted as $\tau \leftrightarrow \tau' \mid (\delta, \delta')$, if one of the following conditions is satisfied:*

(1) $(\delta = subtree(\tau)) \wedge (\delta' = subtree(\tau'))$;
(2) $\exists (v \leftrightarrow v' \mid (\delta, \delta')) \wedge (\tau = left(v)) \wedge (\tau' = left(v'))$;
(3) $\exists (v \leftrightarrow v' \mid (\delta, \delta')) \wedge (\tau = right(v)) \wedge (\tau' = right(v'))$.

Definition 11. *Assume that δ and δ' are both k-subtrees, Λ and Λ' are full path of δ and δ' respectively. Let $\tau_i \in (N \mid_\delta^i \cap \Lambda)$ and $\tau_i' \in (N \mid_{\delta'}^i \cap \Lambda')$. We say that Λ' is the **corresponding path** of Λ about δ and δ', denoted as $\Lambda \leftrightarrow \Lambda' \mid (\delta, \delta')$, if and only if $\tau_i \leftrightarrow \tau_i' \mid (\delta, \delta')$ for any $0 \leq i \leq k$.*

In Fig. 3, $\{s_1, c_2, s_5, c_{10}\} \leftrightarrow \{c_1, c_3, s_7, c_{14}\} | (subtree(s_1), (subtree(c_1))$.

Definition 12. *Assume that δ and δ' are both k-subtrees, we say that δ is* **equivalent to** δ', *denoted as $\delta \equiv \delta'$, if and only if for each node τ in δ and its corresponding node τ' in $\delta', \tau \equiv \tau'$. We say that δ is* **superior to** δ', *denoted as $\delta \prec \delta'$, if and only if the set of nodes of δ can be separated into two subsets, A and B, which satisfy the following conditions:*

(1) $A = \{\tau | (\tau \leftrightarrow \tau' | (\delta, \delta')) \wedge (\tau \equiv \tau')\}$;
(2) $B = \{\tau | (\tau \leftrightarrow \tau' | (\delta, \delta')) \wedge (\tau \prec \tau')\}$;
(3) B is nonempty.

That is, each node in A is equivalent to its corresponding node in δ', and each node in B is superior to its corresponding node in δ'. The equivalent and superior relations between k-subtrees are both transitive.

Definition 13. *Given an S-C tree T, let $P_1^k = \{p_{1,0}, p_{1,1}, \ldots, p_{1,k}\}$ and $P_2^k = \{p_{2,0}, p_{2,1}, \ldots, p_{2,k}\}$ be k-paths of $T, 0 < k \leq n$. We say that P_1^k is* **equivalent to** P_2^k, *denoted as $P_1^k \equiv P_2^k$, if and only if $p_{1,i} \equiv p_{2,i}$ for any $0 \leq i \leq n$. We say that P_1^k is* **superior to** P_2^k, *denoted as $P_1^k \prec P_2^k$, if and only if P_1^k can be separated into two nonempty subsets, $P_1^k = A \cup B$, such that $A = \{p_{1,i} | p_{1,i} \equiv p_{2,i}, 0 \leq i < n\}$ and $B = \{p_{1,i} | p_{1,i} \prec p_{2,i}, 0 \leq i < n\}$.*

The equivalent and superior relations between k-paths are also transitive.

Lemma 2. *Given a workflow $F = \{d_0, d_1, \ldots, d_n\}$ and its S-C tree $Tree^F$, let $\tau, \tau' \in N|_{Tree^F}^i, 0 < i \leq n$, if $\tau \equiv \tau'$, then $subtree(\tau) \equiv subtree(\tau')$.*

Proof. Since $\tau, \tau' \in N|_{Tree^F}^i$, and $left(\tau)$ and $left(\tau')$ are both S-nodes, based on Lemma 1, we have: $left(\tau) \equiv left(\tau')$. $\hspace{2cm}$ (a-1)

As $right(\tau)$ and $right(\tau')$ are both C-nodes, $cost(right(\tau)) = \sum_{\alpha \in Path_{right(\tau)}^{\hookrightarrow}} \omega(\alpha) = \sum_{\alpha \in Path_\tau^{\hookrightarrow}} \omega(\alpha) + \omega(right(\tau)) = cost(\tau) + y(d_{i+1})$, and $cost(right(\tau')) = \sum_{\alpha \in Path_{right(\tau')}^{\hookrightarrow}} \omega(\alpha) = \sum_{\alpha \in Path_{\tau'}^{\hookrightarrow}} \omega(\alpha) + \omega(right(\tau')) = cost(\tau') + y(d_{i+1})$. Due to $\tau \equiv \tau'$, we have $cost(\tau) = cost(\tau')$, so $cost(right(\tau)) = cost(right(\tau'))$, then: $right(\tau) \equiv right(\tau')$. $\hspace{2cm}$ (a-2)

Summarizing (a-1) and (a-2), both $left(\tau)$ and $right(\tau)$ are respectively equivalent to $left(\tau')$ and $right(\tau')$. By this analogy, the rest nodes of $subtree(\tau)$ and $subtree(\tau')$ can be dealt with in the same manner. Hence, any node of $subtree(\tau)$ is equivalent to its corresponding node of $subtree(\tau')$. That is: $subtree(\tau) \equiv subtree(\tau')$. $\hspace{1cm}\square$

Corollary 1. *Let τ and τ' be two nodes at the same level in an S-C tree T, then $subtree(left(\tau)) \equiv subtree(left(\tau'))$.*

Proof. Assume that $\tau, \tau' \in N|_T^i$, then $left(\tau) \in N|_T^{i+1}$ and $left(\tau') \in N|_T^{i+1}$. According to Lemma 1, $left(\tau) \equiv left(\tau')$. Based on Lemma 2, we can obtain: $subtree(left(\tau)) \equiv subtree(left(\tau'))$. $\hspace{1cm}\square$

Lemma 3. *Let τ and τ' be two nodes at the same level in an S-C tree T, if $\tau, \tau' \in C$ and $\tau \prec \tau'$, then $subtree(\tau) \prec subtree(\tau')$.*

Proof. Assume that $\tau, \tau' \in N|_T^i$. Based on Lemma 1, we have:

$$left(\tau) \equiv left(\tau'). \tag{b-1}$$

Similar to the proof of Lemma 2, we have: $cost(right(\tau)) = \sum_{\alpha \in Path_{right(\tau)}^{\rightarrow}}$
$\omega(\alpha) = \sum_{\alpha \in Path_\tau^{\rightarrow}} \omega(\alpha) + \omega(right(\tau)) = cost(\tau) + y(d_{i+1})$, and $cost(right(\tau')) = \sum_{\alpha \in Path_{right(\tau')}^{\rightarrow}} \omega(\alpha) = \sum_{\alpha \in Path_{\tau'}^{\rightarrow}} \omega(\alpha) + \omega(right(\tau')) = cost(\tau') + y(d_{i+1})$. Due to $\tau \prec \tau'$, we have $cost(\tau) < cost(\tau')$, thus $cost(right(\tau)) < cost(right(\tau'))$, then: $right(\tau) \prec right(\tau')$. (b-2)

Summarizing (b-1) and (b-2), we can separate the set of nodes of $subtree(\tau)$ into two subsets, A and B. We add the nodes of $subtree(left(\tau))$ into subset A, and $right(\tau)$ into subset B. By this analogy, $right(\tau)$ can be dealt with in the same manner like τ until all the nodes are contained in A or B. Each node in A is equivalent to its corresponding node of $subtree(\tau')$, and each node in B is superior to its corresponding node of $subtree(\tau')$. Therefore, $subtree(\tau) \prec subtree(\tau')$. □

Theorem 1. *Given an S-C tree T, let $P_i^k = \{p_{i,0}, p_{i,1}, p_{i,2}, \ldots, p_{i,k}\}$ and $P_j^k = \{p_{j,0}, p_{j,1}, p_{j,2}, \ldots, p_{j,k}\}$ be any two k-paths of T, $i \neq j$. If $Cost(P_j^k) < Cost(P_i^k)$, then $subtree(left(p_{i,k}))$ is not an optimal $(k+1)$-subtree.*

Proof. Following Corollary 1, we have $subtree(left(p_{i,k})) \equiv subtree(left(p_{j,k}))$. Assume that P is the optimal full path of $subtree(left(p_{i,k}))$, as shown in Fig. 4, there must exist a corresponding path P' in $subtree(left(p_{j,k}))$ which satisfies $P' \equiv P$. Since $Cost(P_j^k) < Cost(P_i^k)$, then the full path $link(P_j^k, P') \prec link(P_i^k, P)$. Thus $link(P_i^k, P)$ must not be the optimal full path of T. That is, $subtree(left(p_{i,k}))$ is not the sub-tree through which the optimal full path passes. Hence, $subtree(left(\tau))$ is not an optimal $(k+1)$-subtree. □

Theorem 2. *Given an S-C tree of a dataflow $F = \{d_0, d_1, \ldots, d_n\}$, let $\Omega = \{P_1^k, P_2^k, \ldots, P_m^k\}$ be a set of k-paths, where $P_i^k = \{p_{i,0}, p_{i,1}, p_{i,2}, \ldots, p_{i,k}\}, 1 \leq$*

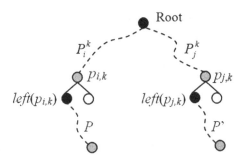

Fig. 4. An exemplar graph of Theorem 1.

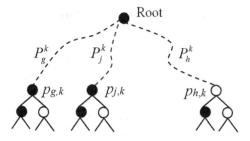

Fig. 5. An exemplar graph of Theorem 2.

$i \leq m$. *Assume that Ω contains an optimal full path from the root to the kth level, then if $Cost(P_j^k) = \min_{p^k \in \Omega} Cost(P^k)$ and $p_{j,k} \in S, subtree(p_{j,k})$ must be an optimal k-subtree.*

Proof. As shown in Fig. 5, let P_g^k and P_h^k $(g \neq h \neq i)$ be any two k-paths taken from Ω such that $p_{g,k} \in S$ and $p_{h,k} \in C$. According to the precondition, we have $Cost(P_j^k) < Cost(P_g^k)$ and $Cost(P_j^k) < Cost(P_h^k)$.

(1) Since $p_{g,k}$, $p_{j,k} \in S$, based on Lemmas 1 and 2, we have $p_{g,k} \equiv p_{j,k}$, thus $subtree(p_{j,k}) \equiv subtree(p_{g,k})$. Let P be the optimal full path of $subtree(p_{g,k})$, then there must exist a corresponding path P' in $subtree(p_{j,k})$, which satisfies $P' \equiv P$. Since $Cost(P_j^k) < Cost(P_g^k)$, we have $link(P_j^k, P') \prec link(P_g^k, P)$. Thus $link(P_g^k, P)$ must not be the optimal full path of $Tree^F$. That is, $subtree(p_{g,k})$ is not the sub-tree which the optimal full path passes through, thus $subtree(p_{g,k})$ is not an optimal k-subtree.

(2) Based on Corollary 1, we have:

$$subtree(left(p_{j,k})) \equiv subtree(left(p_{h,k})). \tag{c-1}$$

Since $P_{j,k} \in S$, thus $cost(right(p_{j,k})) = y(d_{i+1})$. While due to $P_{h,k} \in C$, we also have $cost(right(p_{h,k})) = \sum_{\alpha \in Path_{right(P_{h,k})}^{\rightarrow}} \omega(\alpha)$ which is equal to $\sum_{\alpha \in Path_{P_{h,k}}^{\rightarrow}} \omega(\alpha) + \omega(right(p_{h,k})) = cost(p_{h,k}) + y(d_{k+1})$. That is, $subtree(right(p_{j,k})) \prec subtree(right(p_{h,k}))$. As $right(p_{j,k})$ and $right(p_{h,k})$ are both C-nodes, according to Lemma 3, we have:

$$subtree(right(p_{j,k})) \prec subtree(right(p_{h,k})). \tag{c-2}$$

Due to (c-1), (c-2) and $Cost(P_j^k) < Cost(P_h^k)$, we can obtain that, assuming P is the optimal full path of $subtree(p_{h,k})$, there must exist a corresponding path P' in $subtree(p_{j,k})$, which satisfies $P' \prec P$, so we have $link(P_j^k, P') \prec link(P_h^k, P)$. Hence, $subtree(p_{h,k})$ is not the sub-tree through which the optimal full path of $tree^F$ passes, that is, $subtree(p_{g,k})$ is not an optimal k-subtree.

Summarizing (1) and (2), if Ω contains an optimal full path from the root to the kth level, the optimal full path must pass through $subtree(p_{j,k})$, thus $subtree(p_{j,k})$ must be an optimal k-subtree.

4   Algorithm for the IDS Problem Based on the S-C Tree

Based on the *S-C* tree model, the IDS problem is converted to searching an optimal full path of a given dataflow *S-C* tree. Using Theorems 1 and 2, we can obtain the following pruning strategies. By these strategies, the search space can be greatly reduced.

(1) Search for the optimal full path of the given *S-C* tree from top to bottom by level. At each level k, the search space is set to $\Omega = \{P_1^k, P_2^k, \ldots, P_m^k\}$, in which $P_i^k = \{p_{i,0}, p_{i,1}, p_{i,2}, \ldots, p_{i,k}\}, 1 \leq i \leq m$, is the k-path that has not been pruned off.

(2) At each level k, let P_j^k be the current best k-path which satisfies $Cost(P_j^k)$ $= \min_{p^k \in \Omega} Cost(P^k)$, then:

  (a) If $p_{j,k} \in C$, based on Theorem 1, for any $p_{i,k}, i \neq j, subtree(left(p_{i,k}))$ is not an optimal $(k + 1)$-subtree thus can be pruned off, so all

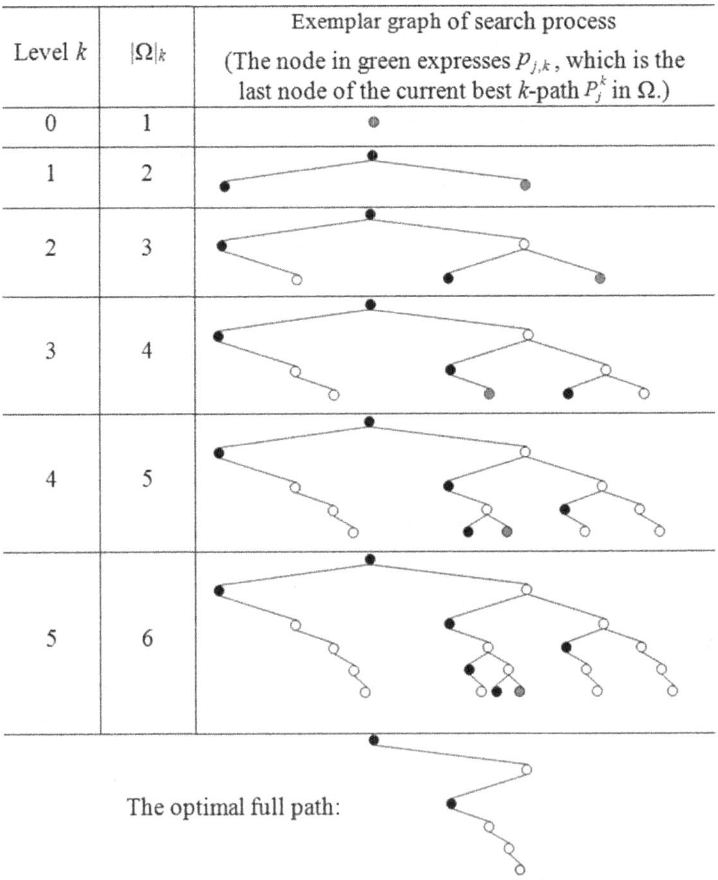

Level k	$|\Omega|_k$	Exemplar graph of search process (The node in green expresses $P_{j,k}$, which is the last node of the current best k-path P_j^k in Ω.)
0	1	
1	2	
2	3	
3	4	
4	5	
5	6	

The optimal full path:

Fig. 6. An example of Algorithm 1.

```
Algorithm 1. IDS algorithm based on S-C tree
Input: An C-S tree of a given dataflow F=(d₀, d₁, d₂,..., dₙ);
        for any dᵢ, 0<i≤n, x(dᵢ) and y(dᵢ) are available.
Output: The optimal full path of tree^F and the optimal cost.
Begin
1      Initial Ω←{d₀};   min_cost←0;
2      For (k =1 to n)
3          min_cost←min_cost+ x(dₖ);
4          For (each P_temp ∈ Ω)
5              If (Cost(P_temp)≤min_cost)
6                  min_cost←Cost(P_temp);
7                  P←P_temp;
8                  τ←tail(P_temp);
9          If (k = n)   Then return P and min_cost;
10         Else
11             If (τ∈ S )
12                     Ω←{link(P, left(τ)), link(P, right(τ))};
13             Else
14                 For (each P_temp ∈ Ω)
15                     P_temp←link(P_temp, right(tail(P_temp)));
16                 Ω←Ω∪ {link(P, left(τ))};
End
```

Fig. 7. The IDS algorithm based on S-C tree.

$link(P_i^k, right(p_{i,k}))$, $i \neq j$, as well as $link(P_j^k, left(p_{j,k}))$ and $link(P_j^k, right(p_{j,k}))$ will be contained in Ω for the next round of search.

(b) If $p_{j,k} \in S$, according to Theorem 2, $subtree(p_{j,k})$ is the optimal k-subtree, so any $subtree(p_{i,k})$, $i \neq j$, can be pruned off, so only $link(P_j^k, left(p_{j,k}))$ and $link(P_j^k, right(p_{j,k}))$ can be contained into Ω for the next round of search.

Figure 6 shows an example of the searching process. We can see that more than $\frac{k-1}{2k}$ of the branches are pruned off at each level of searching. Based on the strategies above, we present the IDS algorithm in Fig. 7.

In line 8 and 15, the function $tail(P_{temp})$ means the last node of P_{temp}.

Theorem 3. *The searching space Ω increases linearly with the level of the S-C tree in Algorithm 1.*

Proof. Let $|\Omega|_k$ denote the size of Ω in the searching of the kth level, $0 \leq k \leq n$. According to Algorithm 1, we have:

(1) When $k = 0$, $\Omega = \{d_0\}$, thus $|\Omega|_0 = 1$.
(2) When $k > 0$, if $\tau \in S$, $|\Omega|_{k+1} = 2$; else $\tau \in C$, then for each $P_{temp} \in \Omega$, P_{temp} will be replaced by $link(P_{temp}, right(tail(P_{temp})))$, and $link(P, left(\tau))$ will

be the only additional new comer of Ω in the next round of searching, hence, $|\Omega|_{k+1} = |\Omega|_k + 1$.

Therefore, in the worst-case, $|\Omega|_{k+1} = |\Omega|_k + 1$, then we have $|\Omega|_k = |\Omega|_{k-1} + 1$, $|\Omega|_{k-1} = |\Omega|_{k-2} + 1, \ldots, |\Omega|_1 = |\Omega|_0 + 1$, so we can obtain that $|\Omega|_k = k + 1$. That is, Ω increases linearly with the level of the S-C tree. □

For a k-path $P^k = \{p_0, p_1, p_2, \ldots, p_k\}, 0 \le k \le n$, the calculation of $Cost(P^k)$ takes $O(k)$ time. Following Theorem 3, Algorithm 1 takes $O(n^3)$ time in the worst case. Furthermore, since Ω is composed of n n-paths, thus the space complexity of Algorithm 1 is $O(n^2)$.

5   Conclusions

In this paper, we solved the IDS problem for linear-structure dataflow by using an S-C tree model. The running time of our algorithm is $O(n^3)$, which improves the previous bound of $O(n^4)$. In the near future, we will study the IDS problems for cloud dataflow with a non-linear structure, such as parallel structure and non-structure dataflows.

Acknowledgements. This paper is supported by national natural science foundation of China: 61472222, and natural science foundation of Shandong province: ZR2012Z002.

References

1. Deelman, E., Chervenak, A.: Data management challenges of data-intensive scientific workflows. In: IEEE International Symposium on Cluster Computing and the Grid (CCGrid 2008), pp. 687–692, Lyon, France (2008)
2. Yuan, D., Yang, Y., Liu, X., Zhang, G., Chen, J.: On-demand minimum cost benchmarking for intermediate data storage in scientific cloud workflow systems. J. Parallel Distrib. Comput. **71**(2), 316–332 (2011)
3. Adams, I., Long, D.D.E., Miller, E.L., Pasupathy, S., Storer, M.W.: Maximizing efficiency by trading storage for computation. In: Workshop on Hot Topics in Cloud Computing (HotCloud 2009), pp. 1–5, San Diego, CA (2009)
4. Yuan, D., Yang, Y., Liu, X., Zhang, G., Chen, J.: A data dependency based strategy for intermediate data storage in scientific cloud workflow systems. Concurr. Comput. Pract. Exp. **24**(9), 956–976 (2010)
5. Zohrevandi, M., Bazzi, R.A.: The bounded data reuse problem in scientific workflows. In: 2013 IEEE 27th International Symposium on Parallel & Distributed Processing, pp. 1051–1062 (2013)
6. Han, L.X., Xie, Z., Baldock, R.: Automatic data reuse for accelerating data intensive applications in the Cloud. In: The 8th International Conference for Internet Technology and Secured Transactions (ICITST-2013), pp. 596–600 (2013)

Efficient Computation of the Characteristic Polynomial of a Threshold Graph

Martin Fürer[(✉)]

Department of Computer Science and Engineering, Pennsylvania State University,
University Park, State College, PA 16802, USA
furer@cse.psu.edu
http://www.cse.psu.edu/~furer

Abstract. An efficient algorithm is presented to compute the characteristic polynomial of a threshold graph. Threshold graphs were introduced by Chvátal and Hammer, as well as by Henderson and Zalcstein in 1977. A threshold graph is obtained from a one vertex graph by repeatedly adding either an isolated vertex or a dominating vertex, which is a vertex adjacent to all the other vertices. Threshold graphs are special kinds of cographs, which themselves are special kinds of graphs of clique-width 2. We obtain a running time of $O(n \log^2 n)$ for computing the characteristic polynomial, while the previously fastest algorithm ran in quadratic time.

Keywords: Efficient algorithms · Threshold graphs · Characteristic polynomial

1 Introduction

The characteristic polynomial of a graph $G = (V, E)$ is defined as the characteristic polynomial of its adjacency matrix A, i.e. $\chi(G, \lambda) = \det(\lambda I - A)$. The characteristic polynomial is a graph invariant, i.e., it does not depend on the enumeration of the vertices of G. The complexity of computing the characteristic polynomial of a matrix is the same as that of matrix multiplication [10,13] (see [2, Chap.16]), currently $O(n^{2.376})$ [4]. For special classes of graphs, we expect to find faster algorithms for the characteristic polynomial. Indeed, for trees, a chain of improvements [12,16] resulted in an $O(n \log^2 n)$ time algorithm [7]. The determinant and rank of the adjacency matrix of a tree can even be computed in linear time [5]. For threshold graphs (defined below), Jacobs et al. [9] have designed an $O(n^2)$ time algorithm to compute the characteristic polynomial. Here, we improve the running time to $O(n \log^2 n)$. As usual, we use the algebraic complexity measure, where every arithmetic operation counts as one step. Throughout this paper, $n = |V|$ is the number of vertices of G.

Threshold graphs [3,8] are defined as follows. Given n and a sequence $b = (b_1, \ldots, b_{n-1}) \in \{0, 1\}^{n-1}$, the threshold graph $G_b = (V, E)$ is defined by

M. Fürer—Research supported in part by NSF Grant CCF-1320814.

J. Wang and C. Yap (Eds.): FAW 2015, LNCS 9130, pp. 45–51, 2015.
DOI: 10.1007/978-3-319-19647-3_5

$V = [n] = \{1, \ldots, n\}$, and for all $i < j$, $\{i, j\} \in E$ iff $b_i = 1$. Thus G_b is constructed by an iterative process starting with the initially isolated vertex n. In step $j > 1$, vertex $n - j + 1$ is added. At this time, vertex j is isolated if b_j is 0, and vertex j is adjacent to all other (already constructed) vertices $\{j+1, \ldots, n\}$ if $b_j = 1$. It follows immediately that G_b is isomorphic to $G_{b'}$ iff $b = b'$. G_b is connected if $b_1 = 1$, otherwise vertex 1 is isolated. Usually, the order of the vertices being added is $1, 2, \ldots, n$ instead of $n, n-1, \ldots, 1$. We choose this unconventional order to simplify our main algorithm.

Threshold graphs have been widely studied and have several applications from combinatorics to computer science and psychology [11].

In the next section, we study determinants of weighted threshold graph matrices, a class of matrices containing adjacency matrices of threshold graphs. In Sect. 3, we design our efficient algorithm to compute the characteristic polynomial of threshold graphs. We also look at its bit complexity in Sect. 4, and finish with open problems.

2   The Determinant of a Weighted Threshold Graph Matrix

We are concerned with adjacency matrices of threshold graphs, but we consider a slightly more general class of matrices. We call them weighted threshold graph matrices. Let $M_{b_1 b_2 \ldots b_{n-1}}^{d_1 d_2 \ldots d_n}$ be the matrix with the following entries.

$$\left(M_{b_1 b_2 \ldots b_{n-1}}^{d_1 d_2 \ldots d_n}\right)_{ij} = \begin{cases} b_i & \text{if } i < j \\ b_j & \text{if } j < i \\ d_i & \text{if } i = j \end{cases}$$

Thus, the weighted threshold matrix for $(b_1 b_2 \ldots b_{n-1}; d_1 d_2 \ldots d_n)$ looks like this.

$$M_{b_1 b_2 \ldots b_{n-1}}^{d_1 d_2 \ldots d_n} = \begin{pmatrix} d_1 & b_1 & b_1 & \ldots & b_1 & b_1 \\ b_1 & d_2 & b_2 & \ldots & b_2 & b_2 \\ b_1 & b_2 & d_3 & \ldots & b_3 & b_3 \\ \vdots & \vdots & \vdots & \ddots & \vdots & \vdots \\ b_1 & b_2 & b_3 & \ldots & d_{n-1} & b_{n-1} \\ b_1 & b_2 & b_3 & \ldots & b_{n-1} & d_n \end{pmatrix}$$

In order to compute the determinant of $M_{b_1 b_2 \ldots b_{n-1}}^{d_1 d_2 \ldots d_n}$, we subtract the penultimate row from the last row and the penultimate column from the last column. In other words, we do a similarity transformation with the following regular matrix

$$P = \begin{pmatrix} 1 & 0 & 0 & \ldots & 0 & 0 \\ 0 & 1 & 0 & \ldots & 0 & 0 \\ 0 & 0 & 1 & \ldots & 0 & 0 \\ \vdots & \vdots & \vdots & \ddots & \vdots & \vdots \\ 0 & 0 & 0 & \ldots & 1 & -1 \\ 0 & 0 & 0 & \ldots & 0 & 1 \end{pmatrix},$$

i.e.,

$$P_{ij} = \begin{cases} 1 & \text{if } i = j \\ -1 & \text{if } i = n \text{ and } j = n - 1 \\ 0 & \text{otherwise.} \end{cases}$$

The row and column operations applied to $M_{b_1 b_2 \ldots b_{n-1}}^{d_1 d_2 \ldots d_n}$ produce the similar matrix

$$P^T M_{b_1 b_2 \ldots b_{n-1}}^{d_1 d_2 \ldots d_n} P = \begin{pmatrix} d_1 \, b_1 \, b_1 \, \ldots & b_1 & 0 \\ b_1 \, d_2 \, b_2 \, \ldots & b_2 & 0 \\ b_1 \, b_2 \, d_3 \, \ldots & b_3 & 0 \\ \vdots \, \vdots \, \vdots \, \ddots & \vdots & \vdots \\ b_1 \, b_2 \, b_3 \, \ldots & d_{n-1} & b_{n-1} - d_{n-1} \\ 0 \, 0 \, 0 \, \ldots \, b_{n-1} - d_{n-1} & d_n + d_{n-1} - 2b_{n-1} \end{pmatrix}$$

Naturally, the determinant of P is 1, implying

$$\det\left(P^T M_{b_1 b_2 \ldots b_{n-1}}^{d_1 d_2 \ldots d_n} P \right) = \det\left(M_{b_1 b_2 \ldots b_{n-1}}^{d_1 d_2 \ldots d_n} \right).$$

Furthermore, we observe that $P^T M_{b_1 b_2 \ldots b_{n-1}}^{d_1 d_2 \ldots d_n} P$ has a very nice pattern.

$$P^T M_{b_1 b_2 \ldots b_{n-1}}^{d_1 d_2 \ldots d_n} P = \begin{pmatrix} & & & 0 \\ & & & 0 \\ & M_{b_1 b_2 \ldots b_{n-2}}^{d_1 d_2 \ldots d_{n-1}} & & 0 \\ & & & \vdots \\ & & & 0 \\ & & & b_{n-1} - d_{n-1} \\ 0 \, 0 \, 0 \, \ldots \, 0 \, b_{n-1} - d_{n-1} & d_n + d_{n-1} - 2b_{n-1} \end{pmatrix}$$

To further compute the determinant of $P^T M_{b_1 b_2 \ldots b_{n-1}}^{d_1 d_2 \ldots d_n} P$, we use Laplacian expansion by minors applied to the last row.

$$\det\left(M_{b_1 b_2 \ldots b_{n-1}}^{d_1 d_2 \ldots d_n} \right) = \det\left(P^T M_{b_1 b_2 \ldots b_{n-1}}^{d_1 d_2 \ldots d_n} P \right)$$

$$= (d_n + d_{n-1} - 2b_{n-1}) \det\left(M_{b_1 b_2 \ldots b_{n-2}}^{d_1 d_2 \ldots d_{n-1}} \right) - (b_{n-1} - d_{n-1})^2 \det\left(M_{b_1 b_2 \ldots b_{n-3}}^{d_1 d_2 \ldots d_{n-2}} \right)$$

By defining the determinant of the 0×0 matrix $M_{b_1 b_2 \ldots b_{n-1}}^{d_1 d_2 \ldots d_n}$ with $n = 0$ to be 1, and checking the determinants for $n = 1$ and $n = 2$ directly, we obtain the following result.

Theorem 1. $D_n = \det\left(M_{b_1 b_2 \ldots b_{n-1}}^{d_1 d_2 \ldots d_n} \right)$ *is determined by the recurrence equation*

$$D_n = \begin{cases} 1 & \text{if } n = 0 \\ d_1 & \text{if } n = 1 \\ (d_n + d_{n-1} - 2b_{n-1})D_{n-1} - (b_{n-1} - d_{n-1})^2 D_{n-2} & \text{if } n \geq 2 \end{cases}$$

$\square$

This has an immediate implication, as we assume every arithmetic operation takes only 1 step.

Corollary 1. *The determinant of an $n \times n$ weighted threshold graph matrix can be computed in time $O(n)$.*

Proof. Every step of the recurrence takes a constant number of arithmetic operations. $\qquad\qquad\square$

For arbitrary matrices, the tasks of computing matrix products, matrix inverses, and determinants are all equivalent [2, Chap.16], currently $O(n^{2.376})$ [4]. For weighted threshold graph matrices, they all seem to be different. We have just seen that the determinant can be computed in linear time, which is optimal, as this time is already needed to read the input. The same lower bound holds for computing the characteristic polynomial, and we will show an $O(n \log^2 n)$ algorithm. It is not hard to see that the multiplication of weighted threshold graph matrices can be done in quadratic time. This is again optimal, because the product is no longer a threshold graph matrix, and its output requires quadratic time.

3   Computation of the Characteristic Polynomial of a Threshold Graph

The adjacency matrix A of the n-vertex threshold graph G defined by the sequence $(b_1, \ldots, b_{n-1})$ is the matrix $M^{0\,0\ldots0}_{b_1 b_2 \ldots b_{n-1}}$, and the characteristic polynomial of this threshold graph is

$$\chi(G, \lambda) = \det(\lambda I - A) = \det\left(M^{\lambda\,\lambda\ldots\lambda}_{-b_1 -b_2 \cdots -b_{n-1}}\right).$$

This immediately implies that any value of the characteristic polynomial can be computed in linear time.

The characteristic polynomial itself can be computed by the recurrence equation of Theorem 1. Here all $d_i = \lambda$, and D_n, as the characteristic polynomial of an n-vertex graph, obviously is a polynomial of degree n in λ. Now, the computation of D_n from D_{n-1} and D_{n-2} according to the recurrence equation is a multiplication of polynomials. It takes time $O(n)$, as one factor is always of constant degree. The resulting total time is quadratic. The same quadratic time is achieved, when we compute the characteristic polynomial $\chi(G, \lambda)$ for n different values of λ and interpolate to obtain the polynomial $\chi(G, \lambda)$.

We want to do better. Therefore, we write the recurrence equation of Theorem 1 in matrix form.

$$\begin{pmatrix} D_n \\ D_{n-1} \end{pmatrix} = \begin{pmatrix} d_n + d_{n-1} - 2b_{n-1} & -(b_{n-1} - d_{n-1})^2 \\ 1 & 0 \end{pmatrix} \begin{pmatrix} D_{n-1} \\ D_{n-2} \end{pmatrix}$$

Noticing that $D_0 = 1$ and $D_1 = \lambda$, and all $d_i = \lambda$, we obtain the following matrix recurrence immediately.

Theorem 2. *For*

$$B_i = \begin{pmatrix} 2(\lambda - b_i) & -(b_i - \lambda)^2 \\ 1 & 0 \end{pmatrix} \text{ for } i = 1, \ldots, n-1,$$

we have

$$\begin{pmatrix} D_n \\ D_{n-1} \end{pmatrix} = B_{n-1} B_{n-2} \cdots B_1 \begin{pmatrix} \lambda \\ 1 \end{pmatrix}$$

□

This results in a much faster way to compute the characteristic polynomial $\chi(G, \lambda)$.

Corollary 2. *The characteristic polynomial $\chi(G, \lambda)$ of a threshold graph G with n vertices can be computed in time $O(n \log^2 n)$.*

Proof. For every i, all the entries in the 2×2 matrix B_i are polynomials in λ of degree at most 2. Therefore, products of any k such factors have entries which are polynomials of degree at most $2k$. To be more precise, actually the degree bound is k, because by induction on k, one can easily see that the degree of the i, j-entry of any such k-fold product matrix is at most

$$k \text{ for } i = 1 \text{ and } j = 1,$$
$$k + 1 \text{ for } i = 1 \text{ and } j = 2,$$
$$k - 1 \text{ for } i = 2 \text{ and } j = 1,$$
$$k \text{ for } i = 2 \text{ and } j = 2,$$

But the bound of $2k$ would actually be sufficient for our purposes. W.l.o.g., we may assume that $n - 1$ (the number of factors) is a power of 2. Otherwise, we could fill up with unit matrices. Now the product $B_{n-1} B_{n-2} \cdots B_1$ is computed in $\log(n - 1)$ rounds of pairwise multiplication to reduce the number of factors by half each round. We use the FFT (Fast Fourier Transform) to compute the product of two polynomials of degree n in time $O(n \log n)$ [1].

In the rth round $(r = 1, \ldots, \log(n - 1))$, we have $(n - 1)2^{-r}$ pairs of matrices with entries of degree at most $2^{r-1} + 1$, requiring $O(n2^{-r})$ multiplications of polynomials of degree at most $2^{r-1} + 1$. With FFT this can be done in time $O(n2^{-r})(2^{r-1} + 1) \log(2^{r-1} + 1) = O(nr)$. Summing over all rounds $r = 1, \ldots, \log(n - 1)$, results in a running time of $O(n \log^2 n)$.   □

Omitting the simplification of $d_i = \lambda$ in Theorem 2, we see immediately, that also the characteristic polynomial of a weighted threshold graph matrix can be computed in the same asymptotic time of $O(n \log^2 n)$.

4   Complexity in the Bit Model

By definition, the characteristic polynomial of an n-vertex graph can be viewed as a sum of $n!$ monomials with coefficients from $\{-1, 0, 1\}$. Thus all coefficients of the characteristic polynomial have absolute value at most $n!$, and can therefore be represented by binary numbers of length $O(n \log n)$.

The coefficients of the characteristic polynomials of some graphs can be of this order of magnitude. For an example, one can start with an $n \times n$ Hadamard matrix with only 1's in the first row. Its determinant has an absolute value of $n^{n/2}$. Adding the first row to all other rows and dividing all other rows by 2 results in a 0-1-matrix whose determinant has an absolute value of $2^{-n+1}n^{n/2}$. The bipartite graph G with this bipartite adjacency matrix has a determinant with absolute value $\left(2^{-n/2+1}(n/2)^{n/4}\right)^2 = 4(n/8)^{n/2}$. Thus the constant coefficient of the characteristic polynomial of G has length $\Omega(n \log n)$.

With such long coefficients, the usual assumption of arithmetic operations in constant time is actually unrealistic for large n. Therefore, the bit model might be more useful. We can use the Turing machine time, because our algorithm is sufficiently uniform. No Boolean circuit is known to compute such things with asymptotically fewer operations than the number of steps of a Turing machine.

We use the fast $m \log m 2^{O(\log^* m)}$ integer multiplication algorithm [6] (where m is the length of the factors) to compute the FFT for the polynomials. A direct implementation, just using fast integer multiplication everywhere during a fast polynomial multiplication, results in time

$$(nr)(2^r r^2 2^{O(\log^* r)}) = n2^r r^3 2^{O(\log^* r)}$$

for the rth round, where $O(n2^{-r})$ pairs of polynomials of degree $O(2^r)$ are multiplied. The coefficients of these polynomials have length $O(2^r r)$. As the coefficients and the degrees of the polynomials increase at least geometrically, only the last round with $r = \log n$ counts asymptotically. The resulting time bound is $n^2 \log^3 n \, 2^{O(\log^* n)}$. Using Schönhage's [14] idea of encoding numerical polynomials into integers in order to do polynomial multiplication, a speed-up is possible. Again only the last round matters. Here a constant number of polynomials of degree $O(n)$ with coefficients of length $O(n \log n)$ are multiplied. For this purpose, each polynomial is encoded into a number of length $O(n^2 \log n)$, resulting in a computation time of

$$n^2 \log^2 n \, 2^{O(\log^* n)}.$$

Actually, because the lengths of coefficients are not smaller than the degree of the polynomials, no encoding of polynomials into numbers is required for this speed-up. In this case, one can do the polynomial multiplication in a polynomial ring over Fermat numbers as in Schönhage and Strassen [15]. Then, during the Fourier transforms all multiplications are just shifts. Fast integer multiplication is only used for the multiplication of values. This is not of theoretical importance, as it results in the same asymptotic $n^2 \log^2 n \, 2^{O(\log^* n)}$ computation time, just with a better constant factor.

5 Open Problems

We have improved the time to compute the characteristic polynomial of a threshold graph from quadratic to almost linear (in the algebraic model). The question remains whether another factor of $\log n$ can be removed. More interesting is the question whether similarly efficient algorithms are possible for richer classes of graphs. Of particular interest are larger classes of graphs containing the threshold graphs, like cographs, graphs of clique-width 2, graphs of bounded clique-width, or even perfect graphs.

References

1. Aho, A., Hopcroft, J., Ullman, J.D.: The Design and Analysis of Computer Algorithms. Addison-Wesley, Reading, MA (1974)
2. Bürgisser, P., Clausen, M., Shokrollahi, M.A.: Algebraic Complexity Theory, Grundlehren der Mathematischen Wissenschaften [Fundamental Principles of Mathematical Sciences], vol. 315. Springer, Berlin (1997)
3. Chvátal, V., Hammer, P.L.: Aggregation of inequalities in integer programming. In: Studies in Integer Programming (Proceedings Workshop Bonn, 1975). Annals of Discrete Mathematics, vol. 1, pp. 145–162. North-Holland, Amsterdam (1977)
4. Coppersmith, D., Winograd, S.: Matrix multiplication via arithmetic progressions. J. Symb. Comput. 9(3), 251–280 (1990)
5. Fricke, G.H., Hedetniemi, S., Jacobs, D.P., Trevisan, V.: Reducing the adjacency matrix of a tree. Electron. J. Linear Algebr. 1, 34–43 (1996)
6. Fürer, M.: Faster integer multiplication. SIAM J. Comput. 39(3), 979–1005 (2009)
7. Fürer, M.: Efficient computation of the characteristic polynomial of a tree and related tasks. Algorithmica 68(3), 626–642 (2014). http://dx.doi.org/10.1007/s00453-012-9688-5
8. Henderson, P.B., Zalcstein, Y.: A graph-theoretic characterization of the pv_chunk class of synchronizing primitives. SIAM J. Comput. 6(1), 88–108 (1977). http://dx.doi.org/10.1137/0206008
9. Jacobs, D.P., Trevisan, V., Tura, F.: Computing the characteristic polynomial of threshold graphs. J. Graph Algorithms Appl. 18(5), 709–719 (2014)
10. Keller-Gehrig, W.: Fast algorithms for the characteristic polynomial. Theor. Comput. Sci. 36(2,3), 309–317 (1985)
11. Mahadev, N.V.R., Peled, U.N.: Threshold Graphs and Related Topics. Annals of Discrete Mathematics. Elsevier Science Publishers B.V. (North Holland), Amsterdam-Lausanne-New York-Oxford-Shannon-Tokyo (1995)
12. Mohar, B.: Computing the characteristic polynomial of a tree. J. Math. Chem. 3(4), 403–406 (1989)
13. Pernet, C., Storjohann, A.: Faster algorithms for the characteristic polynomial. In: Brown, C.W. (ed.) Proceedings of the 2007 International Symposium on Symbolic and Algebraic Computation, University of Waterloo, Waterloo, Ontario, Canada, 29 July–1 August 2007, pp. 307–314. ACM Press, pub-ACM:adr (2007)
14. Schönhage, A.: Asymptotically fast algorithms for the numerical multiplication and division of polynomials with complex coeficients. In: Calmet, J. (ed.) EUROCAM 1982. LNCS, vol. 144, pp. 3–15. Springer, Heidelberg (1982)
15. Schönhage, A., Strassen, V.: Schnelle Multiplikation grosser Zahlen. Computing 7, 281–292 (1971)
16. Tinhofer, G., Schreck, H.: Computing the characteristic polynomial of a tree. Computing 35(2), 113–125 (1985)

A Fast and Practical Method to Estimate Volumes of Convex Polytopes

Cunjing Ge[1,2(✉)] and Feifei Ma[2]

[1] Institute of Software, Chinese Academy of Sciences, Beijing, China
[2] University of Chinese Academy of Sciences, Beijing, China
{gecj,maff}@ios.ac.cn

Abstract. The volume is an important attribute of a convex body. In general, it is quite difficult to calculate the exact volume. But in many cases, it suffices to have an approximate value. Volume estimation methods for convex bodies have been extensively studied in theory, however, there is still a lack of practical implementations of such methods. In this paper, we present an efficient method which is based on the Multiphase Monte-Carlo algorithm to estimate volumes of convex polytopes. It uses the coordinate directions hit-and-run method, and employs a technique of reutilizing sample points. We also introduce a new result checking method for performance evaluation. The experiments show that our method can efficiently handle instances with dozens of dimensions with high accuracy.

1 Introduction

Volume computation is a classical problem in mathematics, arising in many applications such as economics, computational complexity analysis, linear systems modeling, and statistics. It is also extremely difficult to solve. Dyer et al. [1] and Khachiyan [2,3] proved respectively that exact volume computation is #P-hard, even for explicitly described polytopes. Büeler et al. [4] listed five volume computation algorithms for convex polytopes. However, only the instances around 10 dimensions can be solved in reasonable time with existing volume computation algorithms, which is quite insufficient in many circumstances. Therefore we turn attention to volume estimation methods.

There are many results about volume estimation algorithms of convex bodies since the end of 1980s. A breakthrough was made by Dyer, Frieze and Kannan [5]. They designed a polynomial time randomized approximation algorithm (Multiphase Monte-Carlo Algorithm), which was then adopted as the framework of volume estimation algorithms by successive works. At first, the theoretical complexity of this algorithm is $O^*(n^{23})$[1], but it was soon reduced to $O^*(n^4)$ by Lovász, Simonovits et al. [6–9]. Despite the polynomial time results and reduced complexity, there is still a lack of practical implementation. In fact, there are

---

[1] "soft-O" notation O^* indicates that we suppress factors of $\log n$ as well as factors depending on other parameters like the error bound.

© Springer International Publishing Switzerland 2015
J. Wang and C. Yap (Eds.): FAW 2015, LNCS 9130, pp. 52–65, 2015.
DOI: 10.1007/978-3-319-19647-3_6

some difficulties in applying the above volume estimation algorithms. First, in theoretical research of randomized volume algorithms, oracles are usually used to describe the convex bodies and the above time complexity results are measured in terms of oracle queries. However, oracles are too complex and oracle queries are time-consuming. Second, there exists a very large hidden constant coefficient in the theoretical complexity [8], which makes the algorithms almost infeasible even in low dimensions. The reason leading to this problem is that the above research works mostly focus on arbitrary dimension and theoretical complexity. To guarantee that Markov Chains mix in high-dimensional circumstance, it is necessary to walk a large constant number of steps before determining the next point.

In this paper, we focus on practical and applicable method. We only consider specific and simple objects, i.e., convex polytopes. On the other hand, the size of problem instances is usually limited in practical circumstances. With such limited scale, we find that it is unnecessary to sample as many points as the algorithm in [8] indicates. We implement a volume estimation algorithm which is based on the Multiphase Monte-Carlo method. The algorithm is augmented with a new technique to reutilize sample points, so that the number of sample points can be significantly reduced. We compare two hit-and-run methods: the hypersphere directions method and the coordinate directions method, and find that the latter method which is employed in our approximation algorithm not only runs faster, but is also more accurate. Besides, in order to better evaluate the performance of our tool, we also introduce a new result checking method. Experiments show that our tool can efficiently handle instances with dozens of dimensions. To the best of our knowledge, it is the first practical volume estimation tool for convex polytopes.

We now outline the remainder of the paper: In Sect. 2, we propose our method in detail. In Sect. 3, we show experimental results and compare our method with the exact volume computation tool VINCI [10]. Finally we conclude this paper in Sect. 4.

2   The Volume Estimation Algorithm

A convex polytope may be defined as the intersection of a finite number of half-spaces, or as the convex hull of a finite set of points. Accordingly there are two descriptions for a convex polytope: half-space representation (H-representation) and vertex representation (V-representation). In this paper, we adopt the H-representation.

An n-dimensional convex polytope P is represented as $P = \{Ax \leq b\}$, where A is an $(m \times n)$ matrix. a_{ij} represents the element at the i-th row and the j-th column of A, and a_i represents the i-th column vector of A. For simplicity, we also assume that P is full-dimensional and not empty. We use $vol(K)$ to represent the volume of a convex body K, and $B(x, R)$ to represent the ball with radius R and center x.

We define ellipsoid $E = E(A, a) = \{x \in \mathbb{R}^n | (x - a)^T A^{-1} (x - a) \leq 1\}$, where A is a symmetric positive definite matrix.

Like the original multiphase Monte-Carlo algorithm, our algorithm consists of three parts: rounding, subdivision and sampling.

2.1 Rounding

The rounding procedure is to find an affine transformation T on polytope Q such that $B(0,1) \subseteq T(Q) \subseteq B(0,r)$ and a constant $\gamma = \frac{vol(Q)}{vol(T(Q))}$. If $r > n$, T can be found by the Shallow-β-Cut Ellipsoid Method [11], where $\beta = \frac{1}{r}$. It is an iterative method that generates a series of ellipsoids $\{E_i(T_i, o_i)\}$ s.t. $Q \subseteq E_i$, until we find an E_k such that $E_k(\beta^2 T_k, o_k) \subseteq Q$. Then we transform the ellipsoid E_k into $B(0,r)$. Note that this method is numerically unstable on even small-sized problems, such as polytopes in 20-dimensions. Therefore, we adopt a modification of Ellipsoid Method which described in [12].

This procedure could take much time when r is close to n, e.g. $r = n+1$. There is a tradeoff between rounding and sampling, since the smaller r is, the more iterations during rounding and the fewer points have to be generated during sampling. We set $r = 2n$ in our implementation. Rounding can handle very "thin" polytopes which cannot be subdivided or sampled directly. We use P to represent the new polytope $T(Q)$ in the sequel.

2.2 Subdivision

To avoid curse of dimensionality(the possibility of sampling inside a certain space in target object decreases very fast while dimension increases), we subdivide P into a sequence of bodies so that the ratio of consecutive bodies is at most a constant, e.g. 2. Place $l = \lceil n \log_2 r \rceil$ concentric balls $\{B_i\}$ between $B(0,1)$ and $B(0,r)$, where

$$B_i = B(0, r_i) = B(0, 2^{i/n}), \quad i = 0, \ldots, l.$$

Set $K_i = B_i \cap P$, then $K_0 = B(0,1)$, $K_l = P$ and

$$vol(P) = vol(B(0,1)) \prod_{i=0}^{l-1} \frac{vol(K_{i+1})}{vol(K_i)} = vol(B(0,1)) \prod_{i=0}^{l-1} \alpha_i. \qquad (1)$$

So we only have to estimate the ratio $\alpha_i = vol(K_{i+1})/vol(K_i)$, $i = 0, \ldots, l-1$. Since $K_i = B_i \cap P \subseteq B_{i+1} \cap P = K_{i+1}$, we get $\alpha_i \geq 1$. On the other hand, $\{K_i\}$ are convex bodies, then

$$K_{i+1} \subseteq \frac{r_{i+1}}{r_i} K_i = 2^{1/n} K_i,$$

we have

$$\alpha_i = \frac{vol(K_{i+1})}{vol(K_i)} \leq 2.$$

Specially, $K_{i+1} = 2^{1/n} K_i$ if and only if $K_{i+1} = B_{i+1}$ i.e. $B_{i+1} \subseteq P$. That is, $1 \leq \alpha_i \leq 2$ and $\alpha_i = 2 \Leftrightarrow B_{i+1} \subseteq P$.

2.3   Hit-and-Run

To approximate α_i, we generate *step_size* random points in K_{i+1} and count the number of points c_i in K_i. Then $\alpha_i \approx$ *step_size*$/c_i$. It is easy to generate uniform distributions on cubes or ellipsoids but not on $\{K_i\}$. So we use a random walk method for sampling. Hit-and-run method is a random walk which has been proposed and studied for a long time [13–15]. The hypersphere directions method and the coordinate directions method are two hit-and-run methods. In the hypersphere directions method, the random direction is generated from a uniform distribution on a hypersphere; in the coordinate directions method, it is chosen with equal probability from the coordinate direction vectors and their negations. Berbee et al. [14] proved the following theorems.

Theorem 1. *The hypersphere directions algorithm generates a sequence of interior points whose limiting distribution is uniform.*

Theorem 2. *The coordinate directions algorithm generates a sequence of interior points whose limiting distribution is uniform.*

Coordinate directions and their negations are special cases of directions generated on a hypersphere, hence the former theoretical research about volume approximation algorithm with hit-and-run methods mainly focus on the hypersphere directions method [8]. In this paper, we apply the coordinate directions method to our volume approximation algorithm. It starts from a point x in K_{k+1}, and generates the next point x' in K_{k+1} by two steps:

Step 1. Select a line L through x uniformly over n coordinate directions, $e_1, \ldots e_n$.
Step 2. Choose a point x' uniformly on the segment in K_{k+1} of line L.

More specifically, we randomly select the dth component x_d of point x and get x_d's bound $[u, v]$ that satisfies

$$x|_{x_d=t} \in K_{k+1}, \ \forall t \in [u, v] \tag{2}$$

$$x|_{x_d=u}, x|_{x_d=v} \in \partial K_{k+1} \tag{3}$$

("∂" denotes the boundary of a set). Then we choose $x'_d \in [u, v]$ with uniform distribution and generate the next point $x' = x|_{x_d=x'_d} \in K_{k+1}$.

Our hit-and-run algorithm is described in Algorithm 1. $R_i = 2^{i/n}$ is the radius of B_i. Note that $K_{k+1} = B_{k+1} \cap P$, so $x' \in B_{k+1}$ and $x' \in P$. We observe that

$$x' \in B_{k+1} \Leftrightarrow |x'| \leq R_{k+1} \Leftrightarrow x'^2_d \leq R^2_{k+1} - \sum_{i \neq d} x^2_i$$

$$x' \in P \Leftrightarrow a_i x' \leq b_i \Leftrightarrow a_{id} x'_d \leq b_i - \sum_{j \neq d} a_{ij} x_j = \mu_i, \ \forall i$$

Algorithm 1. Hit-And-Run Sampling Algorithm

1: **function** WALK(x, k)
2:      $d \leftarrow random(n)$
3:      $c \leftarrow |x|^2 - x_d^2$
4:      $r \leftarrow \sqrt{R_{k+1}^2 - c}$
5:      $max \leftarrow r - x_d$
6:      $min \leftarrow -r - x_d$
7:      **for** $i \leftarrow 1, m$ **do**
8:          $bound_i \leftarrow (b_i - a_i x)/a_{id}$
9:          **if** $a_{id} > 0$ **and** $bound_i < max$ **then**
10:              $max \leftarrow bound_i$
11:          **else if** $a_{id} < 0$ **and** $bound_i > min$ **then**
12:              $min \leftarrow bound_i$
13:          **end if**
14:      **end for**
15:      $x_d \leftarrow x_d + random(min, max)$
16:      **return** x
17: **end function**

Let

$$u = max\{-\sqrt{R_{k+1}^2 - \sum_{i \neq d} x_i^2}, \frac{\mu_i}{a_{id}}\} \ \forall i \ s.t. \ a_{id} < 0$$

$$v = min\{\sqrt{R_{k+1}^2 - \sum_{i \neq d} x_i^2}, \frac{\mu_i}{a_{id}}\} \ \forall i \ s.t. \ a_{id} > 0$$

then interval $[u, v]$ is the range of x_d' that satisfies Formulas (2) and (3), and $u = x_d + min$, $v = x_d + max$ in Algorithm 1.

Usually, $Walk$ function is called millions of times, so it is important to improve its efficiency, such as use iterators in **for** loop and calculation of $|x|$. At the same time, we move the division operation (line 8), which is very slow for double variables, out of $Walk$ function because $(b_i - a_i x)/a_{id} = \frac{b_i}{a_{id}} - \frac{a_i}{a_{id}} x$, i.e., divisions only occur between constants.

2.4   Reutilization of Sample Points

In the original description of the Multiphase Monte Carlo method, it is indicated that the ratios α_i are estimated in natural order, from the first ratio α_0 to the last one α_{l-1}. The method starts sampling from the origin. At the kth phase, it generates a certain number of random independent points in K_{k+1} and counts the number of points c_k in K_k to estimate α_k. However, our algorithm performs in the opposite way: Sample points are generated from the outermost convex body K_l to the innermost convex body K_0, and ratios are estimated accordingly in reverse order.

The advantage of approximation in reverse order is that it is possible to fully exploit the sample points generated in previous phases. Suppose we have already generated a set of points S by random walk with almost uniform distribution in K_{k+1}, and some of them also hit the convex body K_k, denoted by S'. The ratio α_k is thus estimated with $\frac{|S'|}{|S|}$. But these sample points can reveal more information than just the ratio α_k. Since K_k is a sub-region of K_{k+1}, the points in S' are also almost uniformly distributed in K_k. Therefore, S' can serve as part of the sample points in K_k. Furthermore, for any K_i ($0 \le i \le k$) inside K_{k+1}, the points in K_{k+1} that hit K_i can serve as sample points to approximate α_i as well.

Based on this insight, our algorithm samples from outside to inside. Suppose to estimate each ratio within a given relative error, we need as many as $step\_size$ points. At the kth phase which approximates ratio α_{l-k}, the algorithm first calculates the number $count$ of the former points that are also in α_{l-k+1}, then generates the rest ($step\_size - count$) points by random walk.

Unlike sampling in natural order, choosing the starter for each phase in reverse sampling is a bit complex. The whole sampling process in reverse order also starts from the origin point. At each end of the k-th phase, we select a point x in K_{k+1} and employ $x' = 2^{-\frac{1}{n}}x$ as the starting point of the next phase since $2^{-\frac{1}{n}}x \in K_k$.

It's easy to find out that the expected number of reduced sample points with our algorithm is

$$\sum_{i=1}^{l-1} (step\_size \times \frac{1}{\alpha_i}). \tag{4}$$

Since $\alpha_i \le 2$, we only have to generate less than half sample points with this technique. Actually, results of expriments show that we can save over 70 % time consumption on many polytopes.

2.5 Framework of the Algorithm

Now we present the framework of our volume estimation method. Algorithm 2 is the Multiphase Monte-Carlo algorithm with the technique of reutilizing sample points.

The function *Preprocess* represents the rounding procedure and it returns the ratio of γ. In Algorithm 2, the formula $\lceil \frac{n}{2} \log_2 |x| \rceil$ returns index i that $x \in K_i \backslash K_{i-1}$. We use t_i to record the number of sample points that hit $K_i \backslash K_{i-1}$. Furthermore, the sum $count$ of $t_0, \ldots, t_{k+1}$ is the number of reusable sample points that are generated inside K_{k+1}. Then we only have to generate the rest ($step\_size - count$) points inside K_{k+1} in the k-th phase. Then we use $2^{-\frac{1}{n}}x$ as the starting point of the next phase. Finally, according to Eq. (1) and $\gamma = \frac{vol(Q)}{vol(P)}$, we achieve the estimation of $vol(Q)$.

Algorithm 2. The Framework of Volume Estimation Algorithm

1: **function** ESTIMATEVOL
2:     $\gamma \leftarrow Preprocess()$
3:     $x \leftarrow O$
4:     $l \leftarrow \lceil n \log_2 r \rceil$
5:     **for** $k \leftarrow l - 1,\ 0$ **do**
6:         **for** $i \leftarrow count,\ step\_size$ **do**
7:             $x \leftarrow Walk(x, k)$
8:             **if** $x \in B_0$ **then**
9:                 $t_0 \leftarrow t_0 + 1$
10:             **else if** $x \in B_k$ **then**
11:                 $m \leftarrow \lceil \frac{n}{2} \log_2 |x| \rceil$
12:                 $t_m \leftarrow t_m + 1$
13:             **end if**
14:         **end for**
15:         $count \leftarrow \sum_{i=0}^{k} t_i$
16:         $\alpha_k \leftarrow step\_size/count$
17:         $x \leftarrow 2^{-\frac{1}{n}} x$
18:     **end for**
19:     **return** $\gamma \cdot unit\_ball(n) \cdot \prod_{i=0}^{l-1} \alpha_i$
20: **end function**

---

3   Experimental Results

We implement the algorithm in C++ and the tool is named PolyVest (Polytope Volume Estimation). In all experiments, *step_size* is set to $1600\,l$ for the reason discussed in Appendix A and parameter r is set to $2n$. The experiments are performed on a workstation with 3.40 GHz Intel Core i7-2600 CPU and 8 GB memory. Both PolyVest and VINCI use a single core.

3.1   The Performance of PolyVest

Table 1 shows the results of comparison between PolyVest and VINCI. VINCI is a well-known package which implements the state of the art algorithms for exact volume computation of convex polytopes. It can accept either H-representation or V-representation as input. The test cases include: (1) "cube_n": Hypercubes with side length 2, i.e. the volume of "cube_n" is 2^n. (2) "cube_n(S)": Apply 10 times random shear mappings on "cube_n". The random shear mapping can be represented as PQP, with $Q = \begin{pmatrix} I & M \\ 0 & I \end{pmatrix}$, where the elements of matrix M are randomly chosen and P is the products of permutation matrices $\{P_i\}$ that put rows and columns of Q in random orders. This mapping preserves the volume. (3) "rh_n_m": An n-dimensional polytope constructed by randomly choosing m hyperplanes tangent to sphere. (4) "rh_n_m(S)": Apply 10 times random shear mappings on "rh_n_m". (5) "cuboid_n(S)": Scaling "cube_n" by 100 in one direction, and then apply random shear mapping on it once. We use this instance to approximate a "thin stick" which not parallel to any axis. (6) "ran_n_m": An n-dimensional polytope constructed by randomly choosing integer coefficient from -1000 to 1000 of matrix A.

Table 1. Comparison between `PolyVest` and `VINCI`

Instance	n	m	PolyVest Result	Time(s)	VINCI Result	T_{rlass}(s)	T_{hot}(s)	T_{lawnd}(s)
cube_10	10	20	1015.33	0.380	1024	0.004	0.044	0.008
cube_15	15	30	33560.1	1.752	32768	0.300	212.8	0.156
cube_20	20	40	1.08805e+6	4.484	1.04858e+6	—	—	8.085
cube_30	30	60	1.0902e+9	23.197	—	—	—	—
cube_40	40	80	1.02491e+12	72.933	—	—	—	—
cube_10(S)	10	20	1027.1	0.184	1023.86	0.008	0.124	0.024
cube_15(S)	14	28	30898.2	0.784	32766.4	0.428	369.6	0.884
rh_8_25	8	25	793.26	0.132	785.989	0.864	0.160	0.016
rh_10_20	10	20	13710.0	0.240	13882.7	0.284	0.340	0.012
rh_10_25	10	25	5934.99	0.260	5729.52	5.100	1.932	0.072
rh_10_30	10	30	2063.55	0.280	2015.58	660.4[a]	5.772	0.144
rh_8_25(S)	8	25	782.58	0.136	785.984	1.268	0.156	0.032
rh_10_20(S)	10	20	13773.2	0.232	13883.8	0.832	0.284	0.032
rh_10_25(S)	10	25	5667.49	0.252	5729.18	11.949	1.960	0.104
rh_10_30(S)	10	30	2098.89	0.276	2015.87	1251.1[a]	6.356	0.248

[a]Enable the `VINCI` option to restrict memory storage, so as to avoid running out of memory.

In Table 1, T_{rlass}, T_{hot} and T_{lawnd} represent the time consumption of three parameters of methods in `VINCI` respectively. The "rlass" uses Lasserre's method, it needs input of H-representation. The "hot" uses a Cohen&Hikey-like face enumeration scheme, it needs input of V-representation. The "lawnd" uses Lawrence's formula, it is the fatest method in `VINCI` and both descriptions are needed. From "cube_20" to "cube_40", "rlass" and "hot" cannot handle these instances in reasonable time. We did not test instances "cube_30" and "cube_40" by "lawnd", because there are too many vertices in these polytopes.

Observe that the "rlass" and "hot" methods of `VINCI` usually take much more time and space as the scale of the problem grows a bit, e.g. "cube_n($n \geq$ 15)" and "rh_10_30". With H- and V- representations, the "lawnd" method is very fast for instances smaller than 20 dimensions. However, enumerating all vertices of polytopes is non-trivial, as is the dual problem of constructing the convex hull by the vertices. This process is both time-consuming and space-consuming. As a result, "lawnd" method is slower than `PolyVest` for random polytopes around 15 dimensions with only H-representation. The running times

Table 2. Statistical results of `PolyVest`

Instance	Average volume $\overline{v}$	Std Dev σ	95% Confidence interval $\mathcal{I} = [p, q]$	Freq on $\mathcal{I}$	Error $\epsilon = \frac{q-p}{\overline{v}}$
cube_10[a]	1024.91	41 7534	[943.077, 1106.75]	947	15.9695 %
cube_20[a]	1.04551e+6	49092.6	[9.49284e+5, 1.14173e+6]	942	18.4067 %
cube_30	1.06671e+9	5.95310e+7	[9.50024e+8, 1.18339e+9]	96	21.8769 %
cube_40	1.09328e+12	4.85772e+10	[9.98073e+11, 1.18850e+12]	95	17.4175 %
cuboid_10(S)[a]	102258	3162.13	[96060.1, 108456]	953	12.1219 %
cuboid_20(S)[a]	1.04892e+8	388574e+6	[9.72760e+7, 1.12508e+8]	953	14.5217 %
cuboid_30(S)	1.07472e+11	4.42609e+9	[9.87968e+10, 1.16147e+11]	93	16.1440 %
ran_10_30[a]	11.0079	0.413874	[10.1967, 11.8191]	946	14.7383 %
ran_10_50[a]	1.48473	4.81726e-2	[1.39031, 1.57915]	952	12.7186 %
ran_15_30	290.575	12.8392	[265.410, 315.740]	92	17.3208 %
ran_15_50	3.30084	0.145495	[3.01567, 3.58601]	96	17.2787 %
ran_20_50	1.25062	6.60574e-2	[1.12115, 1.38010]	94	20.7053 %
ran_20_100	8.79715e-3	3.144633e-4	[8.18080e-3, 9.41350e-3]	96	14.0125 %
ran_30_60	195.295	10.37041	[174.969, 215.621]	97	20.8157 %
ran_30_100	2.21532e-5	1.13182e-6	[1.99348e-5, 2.43715e-5]	98	20.0276 %
ran_40_100	3.02636e-5	1.76093e-6	[2.68121e-5, 3.3715e-5]	96	22.8091 %

[a] Estimated 1000 times with `POLYVEST`.

of `PolyVest` appear to be more 'stable'. In addition, `PolyVest` only has to store some constant matrices and variable vectors for sampling.

Since `PolyVest` is a volume estimation method instead of an exact volume computation one like `VINCI`, we did more tests on `PolyVest` to see how accurate it is. We estimated 100 times with `PolyVest` for each instance in Table 2 and listed the statistical results. From Table 2, we observe that the frequency on $\mathcal{I}$ is approximately 950 which means $Pr(p \leq \overline{vol(P)} \leq q) \approx 0.95$. Additionally, values of ϵ (ratio of confidence interval's range to average volume $\overline{v}$) are smaller than or around 20 %.

3.2 Result Checking

For arbitrary convex polytopes with more than 10 dimensions, there is no easy way to evaluate the accuracy of `PolyVest` since the exact volumes cannot be computed with tools like `VINCI` in reasonable time. However, we find that a simple property of geometric body is very helpful for verifying the results.

Given an arbitrary geometric body P, an obvious relation is that if P is divided into two parts P_1 and P_2, then we have $vol(P) = vol(P_1) + vol(P_2)$.

For a random convex polytope, we randomly generate a hyperplane to cut the polytope, and test if the results of `PolyVest` satisfy this relation.

Table 3 shows the results of such tests on random polytopes in different dimensions. Each polytope is tested 100 times. Values in column "Freq." are the times that $(vol(P_1) + vol(P_2))$ falls in 95 % confidence interval of $vol(P)$, and these values are all greater than 95. The error $\frac{|Sum - vol(P)|}{vol(P)}$ is quite small. Therefore, the outputs of `PolyVest` satisfy the relation $vol(P) = vol(P_1) + vol(P_2)$. The test results further confirm the reliability of `PolyVest`.

Table 3. Result checking

n	$\overline{vol(P)}$	95 % Confidence interval	$\overline{vol(P_1)}$	$\overline{vol(P_2)}$	Sum	Error	Freq
10	916.257	[847.229, 985.285]	498.394	414.676	913.069	0.348 %	98
20	107.976	[97.4049, 118.548]	50.4808	57.3418	107.823	0.142 %	99
30	261424	[228471, 294376]	40332.7	218637	258969	0.939 %	96
40	5.08e+11	[4.58e+11, 5.57e+11]	9.44e+10	4.15e+11	5.09e+11	0.234 %	98

3.3   The Performance of Two Hit-and-Run Method

In Table 4, t_1 and t_2 represent the time consumption of the coordinate directions and the hypersphere directions method when each method is executed 10 million times. Observe that the coordinate directions method is faster than the other one. The reason is that the hypersphere directions method has to do more vector multiplications to find interception points and $m \times n$ more divisions during each walk step.

Table 4. Random walk by 10 million steps

n	m	time t_1(s)	time t_2(s)
10	20	6.104	13.761
20	40	10.701	24.502
30	60	17.541	40.455
40	80	27.494	61.484

In addition, we also compare the two hit-and-run methods on accuracy. The results in Table 5 show that the relative errors and standard deviations of the coordinate directions method are smaller.

Table 5. Comparison about accuracy between two methods

		Simplified			Original		
Instance	Exact Vol v	Volume $\bar{v}$	Err $\frac{|\bar{v}-v|}{v}$	Std Dev σ	Volume $\bar{v}'$	Err $\frac{|\bar{v}-v|}{v}$	Std Dev σ'
cube_10	1024	1024.91	0.089 %	41.7534	1028.31	0.421 %	62.6198
cube_14	16384	16382.3	0.010 %	3.020	16324.6	0.363 %	1145.76
cube_20	1.04858e+6	1.04551e+6	0.293 %	49092.6	1.04426e+6	0.412 %	81699.9
rh_8_25	785.989	786.240	0.032 %	23.5826	791.594	0.713 %	50.5415
rh_10_20	13882.7	13876.3	0.046 %	473.224	13994.4	0.805 %	963.197
rh_10_25	5729.52	5736.83	0.128 %	193.715	5765.18	0.622 %	368.887
rh_10_30	2015.58	2013.08	0.124 %	62.1032	2041.60	1.291 %	124.204

3.4   The Advantage of Reutilization of Sample Points

In Table 6, we demonstrate the effectiveness of reutilization technique. Values of n_1 are the number of sample points without this technique. Since our method is a randomized algorithm, the number of sample points with this technique is not a constant. So we list average values in column n_2. With this technique, the requirement of sample points is significantly reduced.

Table 6. Reutilize Sample Points

Instance	n_1	n_2	n_2/n_1
cube_10	2016000	535105.41	26.5 %
cube_15	5856000	1721280.3	29.4 %
cube_20	12249600	3789370.7	30.9 %
rh_8_25	1040000	181091.13	17.4 %
rh_10_30	2016000	304211.03	15.1 %
cross_7	809600	78428.755	9.69 %
fm_6	5856000	955656.79	16.3 %

4   Related Works

To our knowledge, there are two implementations of volume estimation methods in literature. Liu et al. [16] developed a tool to estimate volume of convex body with a direct Monte-Carlo method. Suffered from the curse of dimensionality, it can hardly solve problems as the dimension reaches 5. The recent work [17] is an implementation of the $O^*(n^4)$ volume algorithm in [9]. The algorithm is targeted for convex bodies, and only the computational results for instances within 10 dimensions are reported. The authors also report that they could not experiment with other convex bodies than cubes, since the oracle describing the convex bodies took too long to run.

5   Conclusion

In this paper, we propose an efficient volume estimation algorithm for convex polytopes which is based on Multiphase Monte Carlo algorithm. With simplified hit-and-run method and the technique of reutilizing sample points, we considerably improve the existing algorithm for volume estimation. Our tool, `PolyVest`, can efficiently handle instances with dozens of dimensions with high accuracy, while the exact volume computation algorithms often fail on instances with over 10 dimensions. In fact, the complexity of our method (excluding rounding procedure) is $O^*(mn^3)$ and it is measured in terms of basic operations instead of oracle queries. Therefore, our method requires much less computational overhead than the theoretical algorithms.

Appendix

A   About the Number of Sample Points

From Formula (1),

$$\frac{vol(P)}{vol(B(0,1))} = \prod_{i=0}^{l-1} \alpha_i = \prod_{i=0}^{l-1} \frac{step\_size}{c_i} = \frac{step\_size^l}{\prod_{i=0}^{l-1} c_i},$$

which shows that to obtain confidence interval of $vol(P)$, we only have to focus on $\prod_{i=0}^{l-1} c_i$. For a fixed P, $\{\alpha_i\}$ are fixed numbers. Let $c = \prod_{i=1}^{l} c_i$ and $\mathbb{D}(l, P)$ denote the distribution of c. With statistical results of substantial expriments on concentric balls, we observe that, when $step\_size$ is sufficiently large, the distribution of c_i is unbiased and its standard deviation is smaller than twice of the standard deviation of binomial distribution in dimensions below 80. Though such observation sometimes not holds when we sample on convex bodies other than balls, we still use this to approximate the distribution of c_i. Consider random variables X_i following binomial distribution $\mathbb{B}(step\_size, 1/\alpha_i)$, we have

$$E(c) = E(c_1)\dots E(c_l) = E(X_1)\dots E(X_l) = step\_size^l \prod_{i=1}^{l} \frac{1}{\alpha_i},$$

$$D(c) = E((c_1\dots c_l)^2) - E(c)^2 = \prod_{i=1}^{l}(D(c_i) + E(c_i)^2) - E(c)^2$$

$$= \prod_{i=1}^{l}(4D(X_i) + E(X_i)^2) - E(c)^2$$

$$= \prod_{i=1}^{l} \frac{step\_size^2}{\alpha_i^2}(1 + \frac{4\alpha_i}{step\_size}(1 - \frac{1}{\alpha_i})) - E(c)^2$$

$$= E(c)^2(\beta - 1),$$

where $\beta = \prod_{i=1}^{l}(1 + \frac{4\alpha_i}{step\_size} - \frac{4}{step\_size})$.

Suppose $\{\xi_1, \ldots, \xi_t\}$ is a sequence of i.i.d. random variables following $\mathbb{D}(l, P)$. Notice $D(c)$, the variance of $\mathbb{D}(l, P)$, is finite because $\beta - 1 \to 0$ as $t \to \infty$. According to **central limit theorem**, we have

$$\frac{\sum_{i=1}^{t}\xi_i - tE(c)}{\sqrt{t}D(c)} \xrightarrow{d} N(0, 1).$$

So we obtain the approximation of 95 % confidence interval of c, $[E(c) - \sigma\sqrt{D(c)}, E(c) + \sigma\sqrt{D(c)}]$, where $\sigma = 1.96$. And

$$Pr(\frac{vol(B(0,1))step\_size^l}{E(c) + \sigma\sqrt{D(c)}} \leq \overline{vol(P)} \leq \frac{vol(B(0,1))step\_size^l}{E(c) - \sigma\sqrt{D(c)}}) \approx 0.95.$$

Let $\epsilon \in [0, 1]$ denote the ratio of confidence interval's range to exact value of $vol(P)$, that is

$$\frac{vol(B(0,1))step\_size^l}{E(c) + \sigma\sqrt{D(c)}} - \frac{vol(B(0,1))step\_size^l}{E(c) - \sigma\sqrt{D(c)}} \leq vol(P) \cdot \epsilon \quad (5)$$

$$\Longleftrightarrow \frac{1}{E(c) - \sigma\sqrt{D(c)}} - \frac{1}{E(c) + \sigma\sqrt{D(c)}} \leq \frac{\epsilon}{E(c)} \quad (6)$$

$$\Longleftrightarrow \frac{1}{1 - \sigma\sqrt{\beta - 1}} - \frac{1}{1 + \sigma\sqrt{\beta - 1}} \leq \epsilon \quad (7)$$

$$\Longleftrightarrow 4\sigma^2(\beta - 1) \leq \epsilon^2(1 + \sigma^2 - \sigma^2\beta)^2 \quad (8)$$

$$\Longleftrightarrow \epsilon^2\sigma^2\beta^2 - 2\epsilon^2(1 + \sigma^2)\beta - 4\beta + (\frac{1}{\sigma} + \sigma)^2 + 4 \geq 0. \quad (9)$$

Solve inequality (9), we get $\beta_1(\epsilon, \sigma)$, $\beta_2(\epsilon, \sigma)$ that $\beta \leq \beta_1$ and $\beta \geq \beta_2$ (ignore $\beta \geq \beta_2$ because $1 - \sigma\sqrt{\beta_2 - 1} < 0$). $\beta \leq (1 + \frac{4}{step\_size})^l$, since $1 \leq \alpha_i \leq 2$.

$$(1 + \frac{4}{step\_size})^l \leq \beta_1 \Longleftrightarrow step\_size \geq \frac{4}{\beta_1^{1/l} - 1}, \quad (10)$$

(10) is a sufficient condition of $\beta \leq \beta_1$. Furthermore, $4/(l\beta_1^{1/l} - l)$ is nearly a constant as ϵ and σ are fixed. For example, $4/(l\beta_1^{1/l} - l) \approx 1569.2 \leq 1600$ when $\epsilon = 0.2$, $\sigma = 1.96$. So $step\_size = 1600l$ keeps the range of 95 % confidence interval of $vol(P)$ less than 20 % of the exact value of $vol(P)$.

References

1. Dyer, M., Frieze, A.: On the complexity of computing the volume of a polyhedron. SIAM J. Comput. **17**(5), 967–974 (1988)
2. Khachiyan, L.G.: On the complexity of computing the volume of a polytope. Izvestia Akad. Nauk SSSR Tekhn. Kibernet **3**, 216–217 (1988)

3. Khachiyan, L.G.: The problem of computing the volume of polytopes is NP-hard. Uspekhi Mat. Nauk. **44**, 179–180 (1989). In Russian; translation in. Russian Math. Surv. **44**(3), 199–200

4. Büeler, B., Enge, A., Fukuda, K.: Exact volume computation for polytopes: a practical study. In: Kalai, G., Ziegler, G.M. (eds.) Polytopes—Combinatorics and Computation, pp. 131–154. Birkhäuser Verlag, Birkhäuser Basel (2000)

5. Dyer, M., Frieze, A., Kannan, R.: A random polynomial time algorithm for approximating the volume of convex bodies. In: 21st Annual ACM Symposium on Theory of Computing, pp. 375–381 (1989)

6. Lovász, L., Simonovits M.: Mixing rate of Markov chains, an isoperimetric inequality, and computing a the volume. In: 31st Annual Symposium on Foundations of Computer Science, Vol. I, II, pp. 346–354 (1990)

7. Kannan, R., Lovász, L., Simonovits, M.: Random walks and an $O^*(n^5)$ volume algorithm for convex bodies. Random Struct. Algorithms **11**(1), 1–50 (1997)

8. Lovász, L.: Hit-and-Run mixes fast. Math. Prog. **86**(3), 443–461 (1999)

9. Lovász, L., Vempala, S.: Simulated annealing in convex bodies and an $O^*(n^4)$ volume algorithm. J. Comput. Syst. Sci. **72**(2), 392–417 (2006)

10. http://www.math.u-bordeaux1.fr/aenge/?category=software&page=vinci

11. Grötschel, M., Lovász, L., Schrijver, A.: Geometric Algorithms and Combinatorial Optimization. Springer, Heidelberg (1993)

12. Goldfarb, D., Todd, M.J.: Modifications and implementation of the ellipsoid algorithm for linear programming. Math. Program. **23**(1), 1–19 (1982)

13. Smith, R.L.: Efficient Monte-Carlo procedures for generating points uniformly distributed over bounded regions. Oper. Res. **32**, 1296–1308 (1984)

14. Berbee, H.C.P., Boender, C.G.E., Ran, A.R., Scheffer, C.L., Smith, R.L., Telgen, J.: Hit-and-run algorithms for the identification of nonredundant linear inequalities. Math. Program. **37**(2), 184–207 (1987)

15. Belisle, C.J.P., Romeijn, H.E., Smith, R.L.: Hit-and-run algorithms for generating multivariate distributions. Math. Oper. Res. **18**(2), 255–266 (1993)

16. Liu, S., Zhang, J., Zhu, B.: Volume computation using a direct Monte Carlo method. In: Lin, G. (ed.) COCOON 2007. LNCS, vol. 4598, pp. 198–209. Springer, Heidelberg (2007)

17. Lovász, L., Deák, I.: Computational results of an $O^*(n^4)$ volume algorithm. Eur. J. Oper. Res. **216**, 152–161 (2012)

Social Models and Algorithms for Optimization of Contact Immunity of Oral Polio Vaccine

Chengwei Guo, Chenglong Ma$^{(\boxtimes)}$, and Shengyu Zhang

Department of Computer Science and Engineering,
The Chinese University of Hong Kong, Shatin, Hong Kong S.A.R., China
`cwguo9@gmail.com`, {`clma,syzhang`}`@cse.cuhk.edu.hk`

Abstract. Oral polio vaccine (OPV) can produce contact immunity and help protect more individuals than the vaccinated from polio. To better capture the utilization of OPV's contact immunity, we model the community as a social network, and formulate the task of maximizing the contact immunity effect as an optimization problem on graphs, which is to find a sequence of vertices to be "vaccinated" to maximize the total number of "infected" vertices. Furthermore, we consider the restriction imported by immune deficient individuals, and study related problems. We present polynomial-time algorithms for these problems on trees, and show the intractability of problems on general graphs.

Keywords: Epidemic model · Social network · Graph theory · Parameterized complexity

1 Introduction

Polio, a common name for poliomyelitis, is an acute, viral, and highly infectious disease, transmitted by person-to-person spread mainly through the faecal-oral route or by a common vehicle, such as contaminated water or food, and multiplies in the intestine [2]. Individuals infected by polio can exhibit a range of symptoms if the virus enters the blood circulation [5]. When poliovirus enters the central nervous system, it can infect and destroy motor neurons, leading to muscle weakness and acute flaccid paralysis. Polio mainly affects children under 5 years of age, which is the reason that polio was called infantile paralysis. The paralysis caused by polio is usually in legs and irreversible, which makes many polio survivors disabled for life [9]. In fact, before the use of vaccine, polio was the most common cause of permanent disability.

This paper studies epidemics of polio, which started to appear in the late 19th century and became one of the most dreaded childhood diseases in the 20th century. Like most diseases caused by virus infection, there was hardly any cure for polio. In 1949, Jonas Salk made an effective polio vaccine [7] and the Global Polio Eradication Initiative was launched in 1988, since when polio cases have decreased by over 99 %. In this initial victory of the battle against polio, polio vaccine plays a crucial role. There are 2 safe and effective vaccines for

© Springer International Publishing Switzerland 2015
J. Wang and C. Yap (Eds.): FAW 2015, LNCS 9130, pp. 66–77, 2015.
DOI: 10.1007/978-3-319-19647-3_7

polio, the inactivated polio vaccine (IPV) which is injected and the oral polio vaccine (OPV) which is given by mouth. IPV consists of inactivated poliovirus, while OPV consists of live, attenuated poliovirus. Therefore IPV carries no risk of vaccine-associated polio paralysis, but induces very low levels of immunity in the intestine; OPV also produces a local immune response in the intestine and can limit the replication of the wild poliovirus inside the intestine [1], but the live attenuated virus in OPV can cause paralysis, in extremely rare cases [8]. When a person immunized with IPV is infected with wild poliovirus, the virus can still multiply inside the intestines and be shed in the faeces, risking continued circulation, which does not happen in the case of OPV. There are more advantages of OPV over IPV in terms of expenses and length of immunity [6].

In addition, there is yet another fact that makes OPV even more important in combating polio. For several weeks after vaccination of OPV, the attenuated virus replicates in the intestine, and is excreted in the faeces. Then the virus can be transmitted to others in close contact, making them immuned. This means that immunization with OPV can result in the immunization of people who have not been directly vaccinated, especially in areas where hygiene and sanitation are poor [4]. We call this phenomenon *contact immunity*.

Although human benefit from vaccines made form attenuated virus, for people with congenital or acquired immune deficiency, the attenuated virus in OPV can cause severe complications [3]. Immune deficient people are more common in recent years, as the Acquired Immune Deficiency Syndrome (AIDS) spreads, and application of immunosuppressor increases, in organ transplantations or cure for some autoimmune diseases. Therefore, when vaccinate OPV to a community, it is necessary to avoid vaccinating immune deficient individuals.

In this paper, we study how to take advantage of contact immunity, so that limited OPV can be applied to a community containing immune deficient patients, to give as much protection to the population as possible, while not infecting immune deficient individuals. By modeling the community as a social network, we formulate the task into an optimization problem on graphs, and propose efficient algorithms.

2 Preliminaries

2.1 Definitions and Notations

Graphs. Unless otherwise mentioned, a graph G in this paper is a simple, finite, and undirected graph with vertex set $V(G)$ and edge set $E(G)$. We usually use n to denote $|V(G)|$ and m to denote $|E(G)|$. A graph is *connected* if for any two vertices of the graph there exists a path connecting them. If a graph is disconnected, we refer to each maximal connected induced subgraph as a *connected component* of the graph.

The *r-ball of v* for the center $v \in V$ and the radius $r \in \mathbb{N}$, denoted by $B(v, r)$, is defined as $B(v, r) = \{x \mid d(v, x) \leq r, x \in V\}$, and $B(v, 0) = \{v\}$. When $u \in B(v, r)$, we also say that u is *covered* by $B(v, r)$. When $B(v_1, r_1) \cap B(v_2, r_2) \neq \emptyset$, we say that the 2 balls are *joint*.

Social Network. A *social network* is a social structure made up of a set of social actors, such as individuals, and a set of the dyadic ties between these actors. A social network can be modeled as a *social graph* $G = (V, E)$, where V is a finite set of vertices, and $E \subseteq V \times V$ is the set of edges connecting pairs of vertices. A vertex in G represents an individual in the social network, while an edge connecting u and v represents the relationship between individuals u and v. In this paper, we use the terms social network and social graph interchangably, and corresponding elements of social network and social graph interchangably for convenience and simplicity.

Parameterized Complexity. A *paramerized problem* Q is a subset of $\Sigma^* \times \mathbb{N}$ for some finite alphabet Σ. The second component is called the *parameter*. The problem Q is *fixed-parameter tractable* (FPT) if it admits an algorithm deciding whether $(I, k) \in Q$ in time $f(k) \cdot |I|^{O(1)}$, where $|I|$ is the size of I and f is a computable function depending only on k.

To prove the intractability of a parameterized problem Q', we usually present an FPT reduction from a known W[t]-hard problem Q to Q'. An *FPT reduction* from a problem Q to a problem Q' is a function that maps (I, k) to (I', k') such that (a) $(I, k) \in Q \Leftrightarrow (I', k') \in Q'$, (b) the function is computable in time $g(k) \cdot (|I| + k)^{O(1)}$ for some function g, and (c) $k' \leq h(k)$ for some function h.

2.2   Models

In this section, we introduce the epidemic model of OPV's contact immunity.

We model the social network of a community as a social graph $G = (V, E)$. In the context of contagion spreading, that two individuals are related means that virus can be transmitted directly from one to the other in daily-life contact. We model the relationship as an undirected edge $e = uv$, since the contact between u and v is basically undirected and symmetric.

There are some properties of OPV, or attenuated poliovirus, when it develops contact immunity. When an individual v is vaccinated with OPV, v gets immunity against polio, and gains infectivity at the same time. Because of the attenuated poliovirus spreaded by v, all individuals that are neighbors of v in the social graph get infected with high probability, and therefore get immuned. Note that OPV's are usually vaccinated multiple times to ensure immunity, and the vaccine used on an already vaccinated individual is called a *booster vaccine*. Booster vaccination leads to longer and stronger infectiousness, since the virus inhabits in intestine for longer time and reproduces more. Also, like the cases of most infectious viruses, the farther v is from infectious individuals, in the sense of either space or social network, the less possible he or she gets infected.

To model the spreading of attenuated poliovirus in a better way, we try to retain its properties, then reduce and simplify the real situation. There are 3 states of vertices in G. The *susceptible* are those who are susceptible to polio, whose set is denoted by P; the *infected* are those who get infected by attenuated poliovirus, whose set is denoted by I; and among the infected, the *infectious* are those who are active in spreading the attenuated virus, whose set is denote by

F, so $F \subseteq I$. Infectious individuals will infect their neighbors. When an individual turns infected, he or her will not be susceptible again; when an individual becomes infectious, he or her will stay infectious, in sufficiently long time.

We assume that, before the vaccination, all individuals in the social network are susceptible to polio, i.e., $P = V$, since we can just remove from the social network those individuals who are already immune to polio, because they scarcely involve in the process of attenuated poliovirus spreading. Suppose we vaccinate people one by one, and there's enough time for contact immunity to take effect. Suppose there are k doses of OPV, denote the *sequence of k vaccinations* by $S_k = (s_1, s_2, \ldots, s_k)$, where $s_i \in V$ for $1 \le i \le k$. Note that the same individuals can appear in S for multiple times.

When s_1 is vaccinated, s_1 becomes infected and infectious, and the neighbor set of s_1, $N(s_1)$, become infected because of s_1. The induced subgraph of the closed neighbor set of s_1, $N[s_1]$, is a connected component, which we call an *infected component*, denoted by $IC(v)$, where v is any vertex in this component. The corresponding sets are updated once the status of vertices changes. In this case, vertices in $N[s_1]$ are removed from P and contained in I, and s_1 is contained in F. When s_i, $1 < i \le k$ is vaccinated, 2 different situations may occur. If $s_i \in I$, meaning that this is a booster vaccination. We simplify the effect of booster vaccination, such that it makes all vertices in $IC(s_i) - F$ infectious, and $\cup_{v \in IC(s_i)} N(v) - IC(s_i)$ infected, then $IC(s_i)$ is updated to be $\cup_{v \in IC(s_i)} N[v]$. If $s_i \notin I$, then s_i becomes infectious, $N[s_i]$ become infected, and $IC(s_i)$ is generated to be $N[s_i]$. Infected components grow indepenently and don't interfere each other. When all vacinations finish, all individuals in I get infected by attenuated poliovirus, and therefore immuned to polio in some level, while infectious individuals in F get strengthened immunity because of booster vaccines.

There are some variants of the model. Sometimes some individuals need the immunity to polio more than others. For example, children under 5 years old, or people living nearby water polluted by virus are more susceptible. In these cases, a *demand index (DI)* is introduced, $DI(v)$ quantifies how pressing v needs the immunity. v's demand is met if and only if v is in F after vaccinations, and the *benefit* of a vaccination is defined as the sum of $DI(v)$'s where v is newly added to F after the vaccination.

In other cases, some individuals must get strengthened immunity, while some individuals should avoid vaccination, like immune deficient individuals. We define the set of individuals who must get strengthened immunity as *target set*, denoted by S, and define the set of individuals who must not get vaccinated or indirectly strengthened as *restriction set*, denoted by R. It should be ensured that $S \subseteq F$ and $R \cap F = \emptyset$ after vaccinations.

As in many papers that study models of propagation in social network, it's a very common method to simplify the complicated social graph, which is usually a general graph, into a tree. In this paper, we study problems both on general graphs and on trees.

With these models, we want to optimize the effect of contact immunity of OPV on the community, with limited doses of OPV, by making a plan of vaccinations. We propose problems as follows.

2.3 Problems

Five parameterized problems are studied in this paper, where the number k of vaccinations is considered as the parameter.

Problem 1 (MAXIMUM CONTACT IMMUNITY (M-CI)). For an undirected graph $G = (V, E)$, and $k \in \mathbb{Z}^+$, find a sequence of k vaccinations S_k on G to maximize $|I|$.

Problem 2 (MAXIMUM BENEFIT OF CONTACT IMMUNITY (MB-CI)). For a vertex weighted graph $G = (V, E; \omega)$, where $\omega : V \to \mathbb{R}^+ \cup \{0\}$, and $k \in \mathbb{Z}^+$, find a sequence of k vaccinations S_k on G to maximize the sum of benefit of vaccinations.

Problem 3 (SPECIFIC TARGETING CONTACT IMMUNITY (ST-CI)). For an undirected graph $G = (V, E)$, a set of targets $S \subseteq V$, and $k \in \mathbb{Z}^+$, find a sequence of k vaccinations S_k on G, such that $S \subseteq F$.

Problem 4 (MAXIMUM BENEFIT OF RESTRICTED CONTACT IMMUNITY (MB-RCI)). For a vertex weighted graph $G = (V, E; \omega)$, where $\omega : V \to \mathbb{R}^+ \cup \{0\}$, a set of restricted vertices $R \subseteq V$, and $k \in \mathbb{Z}^+$, find a sequence of k vaccinations S_k on G to maximize the sum of benefit of vaccinations, and $R \cap F = \emptyset$.

Problem 5 (SPECIFIC TARGETING RESTRICTED CONTACT IMMUNITY (ST-RCI)). For an undirected graph $G = (V, E)$, a set of restricted vertices $R \subseteq V$, a set of targets $S \subseteq V$, and $k \in \mathbb{Z}^+$, find a sequence of k vaccinations S_k on G, such that $S \subseteq F$, and $R \cap F = \emptyset$.

2.4 Overview

Our results for above problems are listed in Table 1.

Table 1. Results of computational complexities

	M-CI	MB-CI	ST-CI	MB-RCI	ST-RCI
Tree	P	P	P	P	P
General graph	Unknown	Unknown	Unknown	W[2]-hard	W[2]-hard

3 Polynomial-Time Algorithms

For a sequence of k vaccinations in a graph G, let C be the set of vaccinees of k vaccinations. The effect of these k vaccinations can be considered as a collection of $|C|$ balls: for each ball $B(v, r)$, the center $v \in C$ and the radius r equals the number of vaccinations on v or $IC(v)$. Moreover, for any collection of balls where every pair of balls has even depth of intersection, we can infer a sequence of vaccinations producing those balls. Thus, we have the following lemma.

Lemma 1. *A collection of balls is equivalent to a sequence of vaccinations.*

To find the optimum solutions of problems with no restriction set, say M-CI, MB-CI and ST-CI, we have the following lemma.

Lemma 2. *There is an optimum solution for* M-CI *(MB-CI, ST-CI) having the property that no mergence of infected components happens, i.e., all balls in this solution are disjoint.*

Proof. Suppose that in a optimum solution of M-CI (MB-CI, ST-CI), there exist two balls, $B(v_1, r_1)$ and $B(v_2, r_2)$, covering a vertex u, so $u \in B(v_1, r_1) \cap B(v_2, r_2)$. Therefore $d(v_1, u) \leq r_1$ and $d(v_2, u) \leq r_2$. By connecting the shortest path between v_1 and u with the shortest path between u and v_2, we have a path P between v_1 and v_2, whose length is at most $r_1 + r_2$. We can find a vertex w on P satisfying that $d(v_1, w) \leq r_2$ and $d(v_2, w) \leq r_1$. Since $B(v_1, r_1) \cup B(v_2, r_2) \subseteq B(w, r_1 + r_2)$, all vertices covered by $B(v_1, r_1)$ or $B(v_2, r_2)$ can also be covered by a single ball $B(w, r_1 + r_2)$. Therefore we can replace these two joint balls by the new ball, which also yields an optimum solution. By repeating this procedure, we can obtain an optimum solution without any joint balls. □

The above lemma implies that our algorithm for M-CI can be simplified into finding a collection of disjoint balls, the sum of whose radiuses is equal to k, to cover a maximum number of vertices in G. We now present a polynomial-time algorithm for M-CI on trees.

Theorem 1. M-CI *and* MB-CI *can be solved in* $O(k^2 n^2)$ *time when input graph is a tree.*

Proof. Given an instance $I = (G, k)$, where G is a tree. We make G into a rooted tree by arbitrarily choosing a vertex r as the root.

For every $v \in V(G)$, we denote by $Tr(v)$ the set of vertices in the subtree rooted by v, and define $C_l(v) = \{x \mid d(v, x) = l, x \in V(G) \cap Tr(v)\}$, for $l \geq 0$. Therefore $B(v, l) \cap Tr(v) = \sum_{i=0}^{l} C_i(v)$.

For any vertex $v \in V(G)$ and any non-negative integers l and t, we define the following notations.

$T_0(v, t) :=$ maximum number of vertices covered by any ball in the subproblem on $Tr(v)$ with the parameter t, in the case that v is not covered by any ball;

$T_1(v, l, t) :=$ maximum number of vertices covered by any ball in the subproblem on $Tr(v)$ with the parameter t, in the case that v is covered, and any vertices outside the subtree whose distance from v is at most l can also be covered by these balls;

$M(v, t) := \max\{T_0(v, t), T_1(v, 0, t)\}$;

$N(v, u, l, t) := \max_{k_0 + k_1 + \ldots + k_p = t}\{T_1(u, l+1, k_0) - |B(u, l-1) \cap Tr(u)| + \sum_{j=1}^{p} M(w_j, k_j)\}$, where $u \in C_1(v)$, $k_0 \geq l+1$, and $\{w_1, \ldots, w_p\} = C_{l+1}(v) \backslash C_l(u)$.

We use dynamic programming to compute $\{T_0(v,t), T_1(v,l,t)\}_{(v,l,t)}$, where $v \in V(G), t = 0, 1, \ldots, k, l = 0, 1, \ldots, t$:

$$T_0(v,t) = \max_{\substack{k_1+\ldots+k_d=t \\ \{v_1,\ldots,v_d\}=C_1(v)}} \sum_{i=1}^{d} M(v_i, k_i) \tag{1}$$

$$T_1(v,l,t) = |B(v,l) \cap Tr(v)| \tag{2}$$

$$+ \max \left\{ \max_{\substack{k_1+\ldots+k_d=t-l \\ \{v_1,\ldots,v_d\}=C_{l+1}(v)}} \sum_{i=1}^{d} M(v_i, k_i), \max_{\substack{i=1,\ldots,d \\ \{v_1,\ldots,v_d\}=C_1(v)}} N(v, v_i, l, t) \right\}$$

In Eq. 1, we distribute all t vaccinations among subtrees rooted by v's children $\{v_1, \ldots, v_d\} = C_1(v)$, since v is not covered. The number $M(v_i, k_i)$ denotes the maximum number of vertices covered in $Tr(v_i)$ after k_i vaccinations satisfying that the vertex v is not covered by any balls whose centers are in $Tr(v_i)$. However, if we enumerate all possible allcations of t vaccinations to d subtrees such that $k_1 + \ldots + k_d = t$, there will be $O(t^d)$ combinations, making the running time of the algorithm superpolynomial. To reduce the time of computing $T_0(v,t)$, we need another dynamic programming to compute the sequence

$$P(i,j) = \max_{k_1+\ldots+k_i=j} \sum_{l=1}^{i} M(v_l, k_l), 1 \leq i \leq d, 0 \leq j \leq t,$$

according to the fact that

$$P(i,j) = \max_{a+b=j} \{P(i-1,a) + M(v_i, b)\},$$

supposing $P(0,j) = P(i,0) = 0$. Therefore we get $T_0(v,t) = P(d,t)$, and the time for computing $T_0(v,t)$ is $O(d \cdot t^2)$. Similar methods are used multiple times in this paper when we allocate a sum of vaccinations to subtrees and get the maximum sum of some functions on subtrees.

In Eq. 2, it is easy to see that any vertex $u \in Tr(v)$ with $d(u,v) \leq l$ must be covered, whose set is $B(v,l) \cap Tr(v)$. There are two cases when v is covered by a ball centered in $Tr(v)$: (i) vertex v is vaccinated for l times, (ii) there exists a child $v_i \in C_1(v)$ such that there is a ball centered in $Tr(v_i)$ that covers v and all vertices with distances $\leq l$ to v. The number $N(v, v_i, l, t)$ denotes the maximum number of vertices in $Tr(v) \backslash B(v,l)$ that can be covered, with other vaccinations allocated properly to subtrees other than $Tr(v_i)$.

Finally, we use $\max\{T_0(r,k), \max_{l=0}^{k} T_1(r,l,k)\}$ to denote the maximum number of vertices covered, i.e., the maximum number of people infected by the attenuated poliovirus. In order to get the sequence of vaccinations, we attach a sequence to every T_0, T_1, M, N, P to keep track of the current sequence of vaccinations and maintain sequences during the dynamic programming.

The running time of the algorithm can be calculated as follows:

For fixed v, the time for computing $\{T_0(v,t)\}_{(t=0,\dots,k)}$ is $O(k^2 \cdot |C_1(v)|) = O(k^2 \cdot d(v))$;

For fixed v and l, the computing time of $\{T_1(v,l,t)\}_{(t=0,\dots,k)}$ is $O(n + (k - l)^2) \cdot |N_{l+1}(v)| + |N_1(v)| \cdot k^2 \cdot |N_{l+1}(v)| = O(n + k^2 \cdot |N_1(v)| \cdot |N_{l+1}(v)|)$;

Thus, for fixed v, the computing time of $\{T_1(v,l,t)\}_{(l=0,\dots,t,t=0,\dots,k)}$ is $O(\sum_{l=0}^{k}(n + k^2 \cdot |N_1(v)| \cdot |N_{l+1}(v)|)) = O(kn + k^2 n \cdot |N_{(v)}1|) = O(k^2 n \cdot d(v))$.

Consequently, the total running time is $O(k^2 n^2)$. □

Note that we can take a simple reduction from ST-CI to MB-CI (or from ST-RCI to MB-RCI) by assigning weight 1 to vertices in target set S and weight 0 to other vertices. Moreover, we can easily reduce MB-CI to MB-RCI (or ST-CI to ST-RCI) by setting the restriction set R to be an empty set. All these reductions take polynomial time and work whenever input graph is a tree or a general graph. We skip the details here. Thus, we have the following lemma.

Lemma 3. ST-CI $\leq_p$ MB-CI, ST-RCI $\leq_p$ MB-RCI, MB-CI $\leq_p$ MB-RCI, and ST-CI $\leq_p$ ST-RCI.

By Theorem 1 and Lemma 3, ST-CI on trees is also solvable in polynomial time.

Furthermore, Lemma 2 is not available for problems with restriction set, since we cannot easily replace two joint balls with one single ball when the new single ball may cover vertices in the restriction set. Therefore, the optimum solution of ST-RCI and MB-RCI allows joint balls. Although there are more difficulties introduced, we can also design dynamic programming algorithms to solve ST-RCI and MB-RCI on trees in polynomial time. By applying a dynamic programming for MB-RCI on trees, we have the following theorem.

Theorem 2. MB-RCI and ST-RCI can be solved in $O(kn^5)$ time when input graph is a tree.

Proof. Given an instance $I = (G, k)$, where G is a tree. We arbitrarily choose a vertex r as the root, and the graph G becomes a rooted tree. Denote the distance between v and the nearest restricted vertex in G by $r(v)$, for each vertex $v \in V(G)$. $r(v) = 0$ if v is a restrected vertex. Then the radius of a ball whose center is v should be at most $r(v)$.

For every $v \in V(G), a, b \in \mathbb{Z}, a \geq 0, b \leq a$, consider the subtree rooted by v, we define the function $S(v, a, b)$, which can be computed in linear time:

If $b \geq 0$, $S(v, a, b)$ is defined as the sum of weights of vertices u in subtree, such that $b < d(v, u) \leq a$;

otherwise, $S(v, a, b)$ is defined as (sum of weights of vertices u in subtree, such that $0 \leq d(v, u) \leq a$) + (sum of weights of vertices w outside subtree, such that $d(v, w) \leq \min\{a, |b| - 1\}$).

For every $v \in V(G), 0 \leq t \leq k, -n \leq l, h \leq t$, we define $T(v, t, l, h) :=$ the maximum benefit of subproblem on subtree rooted by v with the parameter t, such that:

If $l < 0$, the vertex v is not covered by any ball in the subtree, and the distance between v and the nearest vertex which is covered by some ball in the subtree is exactly $|l|$;

If $l \geq 0$, the vertex v is covered by some ball in the subtree, and the ball spreads outside the subtree by length l;

If $h < 0$, the vertex v is not covered by any ball outside subtree, and the distance between v and the nearest vertex which is covered by some ball outside the subtree is exact $|h|$;

If $h \geq 0$, the vertex v is covered by some ball outside the subtree, and the ball spreads inside subtree by length h.

Now we use dynamic programming to compute $\{T(v,t,l,h)\}_{(v,t,l,h)}$ where $v \in V(G), t \in [0,k], l, h \in [-n, t]$. Assume that the set of children of v is $\{v_1, \ldots, v_d\}$.

(1) $l > r(v)$

$$T(v,t,l,h) = 0$$

(2) $l \leq 0$

$$T(v,t,l,h) = \max_{k_1+\ldots+k_d=t} \sum_{i=1}^{d} T(v_i, k_i, l+1, h-1)$$

(3) $0 < l \leq r(v)$

$$T(v,t,l,h) = \max\{$$
$$S(v,l,h) + \max_{\substack{l_1,\ldots,l_d \leq l,\, k_1,\ldots,k_d \geq 0 \\ 1 \leq j \leq d \quad \Sigma k_j = t-l}} \Sigma_{j=1}^{d} T(v_j, k_j, l_j, \max\{l,h\}-1),$$
$$\max_{\substack{i=1,\ldots,d \\ l_i=l+1}} \max_{\substack{l_1,\ldots,l_d \leq l+1,\, k_1,\ldots,k_d \geq 0, k_i \geq l_i \\ 1 \leq j \leq d \quad \Sigma k_j = t}} \Sigma_{j=1}^{d} T(v_j, k_j, l_j, \max\{l,h\}-1)$$
$$\}$$

$\max_{l=-n}^{k}\{T(r,k,l,-n)\}$ is the maximum sum of benefit of vaccinations. And the corresponding sequence of vaccinations can be got from dynamic programming similar to the algorithm for M-CI on trees. The running time of this algorithm is $O(kn^5)$.

By Lemma 3, ST-RCI on trees is also solvable in polynomial time. □

4   Intractability

To give a complete picture of complexity of these problems, we show the intractability of ST-RCI and MB-RCI on general graphs. It is still open whether M-CI, MB-CI, and ST-CI are NP-hard when inputs are general graphs.

Theorem 3. ST-RCI *and* MB-RCI *on general graphs are NP-hard and W[2]-hard.*

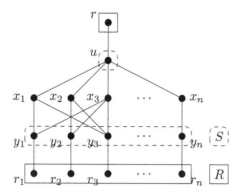

Fig. 1. Construction of ST-RCI instance from k-DOMINATING SET instance.

Proof. We prove it by constructing a polynomial reduction from k-DOMINATING SET to ST-RCI. Given an instance $I = (G, k)$ of DOMINATING SET where $G = (V, E)$ and $V = \{v_1, \ldots, v_n\}$, we construct an instance $I' = (G', S, R, k+1)$ of ST-RCI where $G' = (V', E')$ as following (Fig. 1):

$V' := \{r\} \cup \{u\} \cup \{x_1, \ldots, x_n\} \cup \{y_1, \ldots, y_n\} \cup \{r_1, \ldots, r_n\}$;

$E' := \{(r, u)\} \cup \{(u, x_i) \mid 1 \leq i \leq n\} \cup \{(x_i, y_j) \mid (v_i, v_j) \in E\} \cup \{(x_i, y_i) \mid 1 \leq i \leq n\} \cup \{(y_i, r_i) \mid 1 \leq i \leq n\}$;

$S := \{u\} \cup \{y_1, \ldots, y_n\}$;

$R := \{r\} \cup \{r_1, \ldots, r_n\}$.

Due to the restriction set R, there must be only one vaccination on each vertex of $\{u\} \cup \{y_i, \ldots, y_n\}$.

Suppose that G has a k-dominating set $\{v_{i_1}, \ldots, v_{i_k}\}$. It is easy to see that in graph G', the vertices $\{x_{i_1}, \ldots, x_{i_k}\}$ dominate all vertices of $\{y_1, \ldots, y_n\}$. We apply k vaccinations on these vertices $\{x_{i_1}, \ldots, x_{i_k}\}$ one by one. After these vaccinations, the vertices $\{x_{i_1}, \ldots, x_{i_k}\} \cup \{u\} \cup \{y_1, \ldots, y_n\}$ are merged into one infected component. In last step we perform a vaccination on this infected component. The above $k+1$ vaccinations form a solution of I'.

On the other side, suppose that I' has a $(k+1)$-size solution. Since vertex u is in targeting set, we may assume that the l-th vaccination is originated from u, where $1 \leq l \leq k+1$. We also assume that the first $l-1$ vaccinations are originated from vertices $\{x_{i_1}, \ldots, x_{i_p}\} \cup \{y_{j_1}, \ldots, y_{j_q}\}$, where $p + q = l - 1$. Let $\{y_{a_1}, \ldots, y_{a_s}\}$ be a subset of vertices in $\{y_1, \ldots, y_n\}$ that are dominated by $\{x_{i_1}, \ldots, x_{i_p}\}$. Then after the first $l-1$ steps, the vertices $\{x_{i_1}, \ldots, x_{i_p}\} \cup \{u\} \cup \{y_{k_1}, \ldots, y_{k_s}\}$ are merged into one infected component, and the l-th vaccination is performed on this infected component. Let $\{y_{b_1}, \ldots, y_{b_t}\} = \{y_1, \ldots, y_n\} - (\{y_{j_1}, \ldots, y_{j_q}\} \cup \{y_{a_1}, \ldots, y_{a_s}\})$ be the set of remaining specific vertices in G' after l vaccinations, where $t + q + s = n$. Note that in the next $(k + 1 - l)$ steps we can only perform vaccinations on vertices in $\{y_{b_1}, \ldots, y_{b_t}\}$, since other components become restricted, implying that $t \leq k + 1 - l$. It is clear that vertices in $\{x_{i_1}, \ldots, x_{i_p}\} \cup \{x_{j_1}, \ldots, x_{j_q}\} \cup \{x_{b_1}, \ldots, x_{b_t}\}$ dominate all vertices in $\{y_{a_1}, \ldots, y_{a_s}\} \cup \{y_{j_1}, \ldots, y_{j_q}\} \cup \{y_{b_1}, \ldots, y_{b_t}\} = \{y_1, \ldots, y_n\}$, and the total size is $p + q + t = l - 1 + t \leq k$. Thus, the original graph G has a dominating set of size at most k.

We have completed the proof of NP-hardness. Note that the above reduction is indeed an FPT reduction, and k-DOMINATING SET is W[2]-hard in the literature. Therefore, ST-RCI on general graphs is W[2]-hard. and thus is very unlikely to be FPT.

By Lemma 3, MB-RCI on general graphs is also NP-hard and W[2]-hard. □

5   Conclusion

In this paper, we have overviewed the history of people fighting polio and introduced oral polio vaccines (OPV). The contact immunity is an important property of OPV that is very important in helping eliminate polio thoroughly. And we have modeled the contact immunity of OPV into models on social graphs, and proposed 5 problems, including MAXIMUM CONTACT IMMUNITY, MAXIMUM BENEFIT OF CONTACT IMMUNITY, SPECIFIC TARGETING CONTACT IMMUNITY, MAXIMUM BENEFIT OF RESTRICTED CONTACT IMMUNITY, and SPECIFIC TARGETING RESTRICTED CONTACT IMMUNITY. We have studied these problems both on general graphs and on trees.

We have designed polynomial-time algorithms based on dynamic programming for all 5 problems on trees, and have proved the intractability for MAXIMUM BENEFIT OF RESTRICTED CONTACT IMMUNITY and SPECIFIC TARGETING RESTRICTED CONTACT IMMUNITY on general graphs. With these algorithms, we can possibly help improving the effect of OPV in certain circumstances, especially when the supply of vaccines is limited, or the community contains a proportion of immune deficient individuals.

However, we still have some future work to do. For problems MAXIMUM CONTACT IMMUNITY, MAXIMUM BENEFIT OF CONTACT IMMUNITY, and SPECIFIC TARGETING CONTACT IMMUNITY on general graphs, we haven't found polynomial-time algorithms, neither have we proved their intractabilities.

Furthermore, there may be variant models. For example, we can introduce IPV into the model. As IPV contains no live virus, it is basically safe for even immune deficient people. Moreover, when applying POV to an epidemic area of polio, the normal poliovirus and attenuated poliovirus may compete when transmitting in the social network. Such variant models can also induce problems that have practical significance.

References

1. Pediatrics Committee on Infectious Diseases: Poliomyelitis prevention: recommendations for use of inactivated poliovirus vaccine and live oral poliovirus vaccine. **99**, 300–305 (1997)
2. Cohen, J.I.: Enteroviruses and reoviruses. In: Isselbacher, K.J., Braunwald, E., Wilson, J.D., Martin, J.B., Fauci, A.S., Kasper, D.L. (eds.) Harrison's Principles of Internal Medicine. 16th edn., vol. 2, p. 1144. McGraw-Hill Professional (2004)

3. Kroger, A.T., Sumaya, C.V., Pickering, L.K., Atkinson, W.L.: General recommendations on immunization: recommendations of the Advisory Committee on Immunizations Practices (ACIP). Morb. Mortal. Wkly Rep. **60**, 1–60 (2011)
4. Plotkin, S.A., Orenstein, W., Offit, P.A.: Poliovirus vaccine-inactivated and poliovirus vaccine-live. In: Vaccines. 6th edn., pp. 573–645. Elsevier Saunders, Philadelphia (2012)
5. Ray, C.G., Ryan, K.J.: Enteroviruses. In: Sherris Medical Microbiology: An Introduction to Infectious Diseases. 4th edn., pp. 535–537. McGraw-Hill Medica, New York (2003)
6. Robertson, S.: Module 6: Poliomyelitis. In: The Immunological Basis for Immunization Series. World Health Organization, Geneva (1993)
7. Rosen, F.S.: Isolation of poliovirus-John Enders and the Nobel Prize. N. Engl. J. Med. **351**, 1481–1483 (2004)
8. Shimizu, H., Thorley, B., Paladin, F.J., Brussen, K.A., Stambos, V., Yuen, L., Utama, A., Tano, Y., Arita, M., Yoshida, H., Yoneyama, T., Benegas, A., Roesel, S., Pallansch, M., Kew, O., Miyamura, T.: Circulation of type 1 vaccine-derived poliovirus in the Philippines in 2001. J. Virol. **78**, 13512–13521 (2004)
9. World Health Organization, Fact Sheet No.114: Poliomyelitis. In: WHO Fact Sheets (2014)

The Directed Dominating Set Problem: Generalized Leaf Removal and Belief Propagation

Yusupjan Habibulla, Jin-Hua Zhao, and Hai-Jun Zhou$^{(\boxtimes)}$

State Key Laboratory of Theoretical Physics, Institute of Theoretical Physics,
Chinese Academy of Sciences, Beijing 100190, China
zhouhj@itp.ac.cn

Abstract. A minimum dominating set for a digraph (directed graph) is a smallest set of vertices such that each vertex either belongs to this set or has at least one parent vertex in this set. We solve this hard combinatorial optimization problem approximately by a local algorithm of generalized leaf removal and by a message-passing algorithm of belief propagation. These algorithms can construct near-optimal dominating sets or even exact minimum dominating sets for random digraphs and also for real-world digraph instances. We further develop a core percolation theory and a replica-symmetric spin glass theory for this problem. Our algorithmic and theoretical results may facilitate applications of dominating sets to various network problems involving directed interactions.

Keywords: Directed graph · Dominating vertices · Graph observation · Core percolation · Message passing

1   Introduction

The construction of a minimum dominating set (MDS) for a general digraph (directed graph) [1,2] is a fundamental nondeterministic polynomial-hard (NP-hard) combinatorial optimization problem [3]. A digraph $D = \{V, A\}$ is formed by a set $V \equiv \{1, 2, .., N\}$ of N vertices and a set $A \equiv \{(i, j) : i, j \in V\}$ of M arcs (directed edges), each arc (i, j) pointing from a parent vertex (predecessor) i to a child vertex (successor) j. The arc density α is defined simply as $\alpha \equiv M/N$. Each vertex i of digraph D brings a constraint requiring that either i belongs to a vertex set Γ or at least one of its predecessors belongs to Γ. A dominating set Γ is therefore a vertex set which satisfies all the N vertex constraints, and the dominating set problem can be regarded as a special case of the more general hitting set problem [4,5].

A dominating set containing the smallest number of vertices is a MDS, which might not necessarily be unique for a digraph D. As a MDS is a smallest set of vertices which has directed edges to all the other vertices of a given digraph, it is conceptually and practically important for analyzing, monitoring, and controlling many directed interaction processes in complex networked systems, such as

© Springer International Publishing Switzerland 2015
J. Wang and C. Yap (Eds.): FAW 2015, LNCS 9130, pp. 78–88, 2015.
DOI: 10.1007/978-3-319-19647-3_8

infectious disease spreading [6], genetic regulation [7,8], chemical reaction and metabolic regulation [9], and power generation and transportation [10]. Previous heuristic algorithms on the directed MDS problem all came from the computer science/applied mathematics communities [2] and they are based on vertices' local properties such as in- and out-degrees [6,11,12]. In the present work we study the directed MDS problem through statistical mechanical approaches.

In the next section we introduce a generalized leaf-removal (GLR) process to simplify an input digraph D. If GLR reduces the original digraph D into an empty one, it then succeeds in constructing an exact MDS. If a core is left behind, we implement a hybrid algorithm combining GLR with an impact-based greedy process to search for near-optimal dominating sets (see Fig. 3 and Table 1). We also study the GLR-induced core percolation by a mean field theory (see Fig. 2). In Sect. 3 we introduce a spin glass model for the directed MDS problem and obtain a belief-propagation decimation (BPD) algorithm based on the replica-symmetric mean field theory. By comparing with ensemble-averaged theoretical results, we demonstrate that the message-passing BPD algorithm has excellent performance on random digraphs and real-world network instances, and it outperforms the local hybrid algorithm (Fig. 3 and Table 1).

This paper is a continuation of our earlier effort [13] which studied the undirected MDS problem. Since each undirected edge between two vertices i and j can be treated as two opposite-direction arcs (i, j) and (j, i), the methods of this paper are more general and they are applicable to graphs with both directed and undirected edges. The algorithmic and theoretical results presented here and in [13] may promote the application of dominating sets to various network problems involving directed and undirected interactions.

In the remainder of this paper, we denote by ∂i^+ the set of predecessors of a vertex i, and refer to the size of this set as the in-degree of i; similarly ∂i^- denotes the set of successors of vertex i and its size defines the out-degree of this vertex. With respective to a dominating set Γ, if vertex i belongs to this set, we say i is occupied, otherwise it is unoccupied (empty). If vertex i belongs to the dominating set Γ or at least one of its predecessors belongs to Γ, then we say i is observed, otherwise it is unobserved.

2 Generalized Leaf Removal and the Hybrid Algorithm

The leaf-removal process was initially applied in the vertex-cover problem [14]. It causes a core percolation phase transition in random undirected or directed graphs [15]. Here we consider a generalized leaf-removal process for the directed MDS problem. This GLR process iteratively deletes vertices and arcs from an input digraph D starting from all the N vertices being unoccupied (and unobserved) and the dominating set Γ being empty. The microscopic rules of digraph simplification are as follows:

Rule 1: If an unobserved vertex i has no predecessor in the current digraph D, it is added to set Γ and become occupied (see Fig. 1A). All the previously unobserved successors of i then become observed.

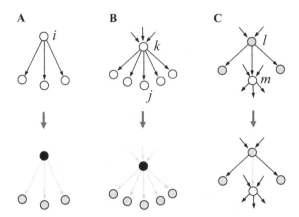

Fig. 1. The generalized leaf-removal process. White circles represent unobserved vertices, black circles are occupied vertices, and blue (gray) circles are observed but unoccupied vertices. Pink (light gray) arrows represent deleted arcs, while black arrows are arcs that are still present in the digraph. (A) vertex i has no predecessor, so it is occupied. (B) vertex j has only one predecessor k and no successor, so vertex k is occupied. (C) vertex l has only a single unobserved successor m, so the arc (l, m) is deleted (color figure online).

Rule 2: If an unobserved vertex j has only a single unoccupied predecessor (say vertex k) and no unobserved successor in the current digraph D, vertex k is added to set Γ and become occupied (Fig. 1B). All the previously unobserved successors of k (including j) then become observed.

Rule 3: If an unoccupied but observed vertex l has only a single unobserved successor (say m) in the current digraph D, occupying l is *not* better than occupying m, therefore the arc (l, m) is deleted from D (Fig. 1C). We emphasize that vertex m is still unobserved after this arc deletion. (Rule 3 is specific to the dominating set problem and it is absent in the conventional leaf-removal process [14, 15].)

The above-mentioned microscopic rules only involve the local structure of the digraph, they are simple to implement. Following the same line of reasoning in [13], we can prove that if all the vertices are observed after the GLR process, the constructed vertex set Γ must be a MDS for the original digraph D. If some vertices remain to be unobserved after the GLR process, this set of remaining vertices is unique and is independent of the particular order of the GLR process.

2.1   Core Percolation Transition

We apply GLR on a set of random Erdös-Rényi (ER) digraphs and random regular (RR) digraphs (see Fig. 2) and also on a set of real-world directed networks (see Table 1). To generate an ER digraph of size N and arc density α, we first select αN different pairs of vertices totally at random from the set of $N(N-1)/2$ possible pairs, and then create an arc of random direction between each selected

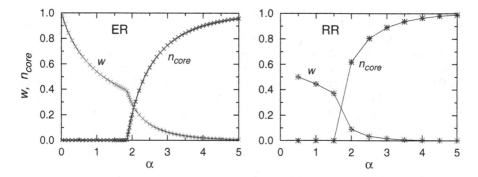

Fig. 2. GLR-induced core percolation transition in Erdös-Rényi (left panel) and regular random (right panel) digraphs. w is the fraction of occupied vertices, n_{core} is the fraction of remaining unobserved vertices. Cross symbols are results obtained on a single digraph with $N = 10^6$ vertices and $M = \alpha N$ arcs, lines (left panel) and plus symbols connected by lines (right panel) are mean-field theoretical results for $N = \infty$.

vertex pair. Similarly, to generate a RR digraph, we first generate an undirected RR graph with every vertex having the same integer number ($= 2\alpha$) of edges [13], and then randomly specify a direction for each undirected edge.

If the arc density α of an ER digraph is less than 1.852 and that of a RR digraph is less than 2.0, a MDS can be constructed by applying GLR alone. However, if $\alpha > 1.852$ for an ER digraph and $\alpha \geq 2.0$ for a RR digraph, GLR only constructs a partial dominating set for the digraph, and a fraction n_{core} of vertices remain to be unobserved after the termination of GLR. For ER digraphs n_{core} increases continuously from zero as α exceeds 1.852. The sub-digraph induced by all these unobserved vertices and all their predecessor vertices is referred to as the core of digraph D.

We develop a percolation theory to quantitatively understand the GLR dynamics on random digraphs. For theoretical simplicity we consider a GLR process carried out in discrete time steps $t = 0, 1, \ldots$. In each time step t, first Rule 1 is applied to all the eligible vertices, then Rule 2 is applied to all the eligible vertices, then Rule 3 is applied to all the eligible arcs, and finally all the newly occupied vertices and their attached arcs are all deleted from digraph D. The fraction w of occupied vertices during the whole GLR process and the fraction n_{core} of remaining unobserved vertices are quantitatively predicted by this mean-field theory (see the Appendix for technical details). These theoretical predictions are in complete agreement with simulation results on single digraph instances (Fig. 2). We believe that when there is no core ($n_{core} = 0$), the MDS relative size w as predicted by our theory is the exact ensemble-averaged result for finite-connectivity random digraphs.

2.2 The Hybrid Algorithm

The GLR process can not construct a MDS for the whole digraph D if it contains a core. For such a difficult case we combine GLR with a simple greedy

Table 1. Constructing dominating sets for several real-world network instances containing N vertices and $M = \alpha N$ arcs. For each graph, we list the number of unobserved vertices after the GLR process (Core), the size of the dominating set obtained by a single running of the greedy algorithm (Greedy), the hybrid algorithm (Hybrid), and the BPD algorithm at fixed re-weighting parameter $x = 8.0$ (BPD). Epinions1 [16] and WikiVote [17,18] are two social networks, Email [19] and WikiTalk [17,18] are two communication networks, HepPh and HepTh [20] are two research citation networks, Google and Stanford [21] are two webpage connection networks, and Gnutella31 [22] is a peer-to-peer network.

Network	N	M	α	Core	Greedy	Hybrid	BPD
Epinions1	75879	405740	5.347	348	37172	37128	37127
WikiVote	7115	100762	14.162	7	4786	4784	4784
Email	265214	364481	1.374	0	203980	203980	203980
WikiTalk	2394385	4659565	1.946	72	63617	63614	63614
HepPh	34546	420877	12.183	982	9628	9518	9512
HepTh	27770	352285	12.686	1900	7302	7213	7203
Google	875713	4322051	4.935	98473	315585	314201	313986
Stanford	281903	1992636	7.069	68947	90403	89388	89466
Gnutella31	62586	147892	2.363	26	12939	12784	12784

process to construct a dominating set that is not necessarily a MDS. We define the impact of an unoccupied vertex as the number of newly observed vertices caused by occupying this vertex [2,6,12]. For example, an unobserved vertex with three unobserved successors has impact 4, while an observed vertex with three unobserved successors has impact 3. Our hybrid algorithm has two modes, the default mode and the greedy mode. In the default mode, the digraph is iteratively simplified by occupying vertices according to the microscopic rules of GLR. If there are still unobserved vertices after this process, the algorithm first switches to the greedy mode, in which the digraph is simplified by occupying a vertex randomly chosen from the subset of highest-impact vertices, and then switches back to the default mode.

The hybrid algorithm can be regarded as an extension of the pure greedy algorithm which always works in the greedy mode. The simulation results obtained by the hybrid algorithm and the pure greedy algorithm are shown in Fig. 3 for random digraphs and in Table 1 for real-world network instances. The hybrid algorithm improves over the greedy algorithm considerably on random digraph instances when the arc density $\alpha \leq 10$. But when the relative size n_{core} of the core in the digraph is close to 1, the hybrid algorithm only slightly outperforms the pure greedy algorithm.

3   Spin Glass Model and Belief-Propagation

We now introduce a spin glass model for the directed MDS problem and solve it by the replica-symmetric mean field theory, which is based on the Bethe-Peierls

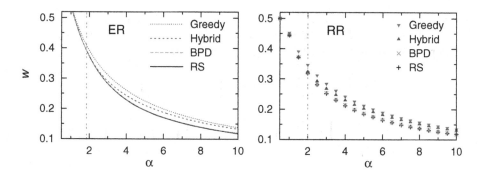

Fig. 3. Relative sizes w of dominating sets for Erdős-Rényi (left panel) and random regular (right panel) digraphs. We compare the mean sizes of 96 dominating sets obtained by the Greedy, the Hybrid, and the BPD algorithm on 96 digraph instances of size $N = 10^5$ and arc density α (fluctuations to the mean are of order 10^{-4} and are not shown). The MDS relative sizes predicted by the replica-symmetric theory are also shown. The re-weighting parameter is fixed to $x = 10.0$ for ER digraphs and to $x = 8.0$ for RR digraphs. The vertical dashed lines mark the core-percolation transition point $\alpha \approx 1.852$ for ER digraphs and $\alpha = 2.0$ for RR digraphs.

approximation [23,24] but can also be derived without any physical assumptions through partition function expansion [25,26]. We define a partition function $Z(x)$ for a given input digraph D as follows:

$$Z(x) = \sum_{\underline{c}} \prod_{i \in V} \left[e^{-xc_i} \left(1 - (1 - c_i) \prod_{j \in \partial i^+} (1 - c_j) \right) \right] . \tag{1}$$

The summation in this expression is over all the microscopic configurations $\underline{c} \equiv \{c_1, c_2, ..., c_N\}$ of the N vertices, with $c_i \in \{0, 1\}$ being the state of vertex i ($c_i = 0$, empty; $c_i = 1$, occupied). A configuration $\underline{c}$ has zero contribution to $Z(x)$ if it does not satisfy all the vertex constraints; if it does satisfy all these constraints and therefore is equivalent to a dominating set, it contributes a statistical weight $e^{-xW(\underline{c})}$, with $W(\underline{c}) \equiv \sum_{i \in V} c_i$ being the total number of occupied vertices. When the positive re-weighting parameter x is sufficiently large, $Z(x)$ will be overwhelmingly contributed by the MDS configurations.

We define on each arc (i, j) of digraph D a distribution function $q_{i \to j}^{c_i, c_j}$, which is the probability of vertex i being in state c_i and vertex j being in state c_j if all the other attached arcs of j are deleted and the constraint of j is relaxed, and another distribution function $q_{j \leftarrow i}^{c_j, c_i}$, which is the probability of i being in state c_i and j being in state c_j if all the other attached arcs of i are deleted and the constraint of i is relaxed. Assuming all the neighboring vertices of any vertex i are mutually independent of each other when the constraint of vertex i is relaxed (the Bethe-Peierls approximation), then when this constraint is present, the marginal probability $q_i^{c_i}$ of vertex i being in state c_i is estimated by

$$q_i^{c_i} = \frac{1}{z_i} e^{-xc_i} \left[\prod_{j \in \partial i^+} \sum_{c_j} q_{j \to i}^{c_j, c_i} - \delta_0^{c_i} \prod_{j \in \partial i^+} q_{j \to i}^{0,0} \right] \prod_{k \in \partial i^-} \sum_{c_k} q_{k \leftarrow i}^{c_k, c_i} , \tag{2}$$

where z_i is a normalization constant, and δ_m^n is the Kronecker symbol with $\delta_m^n = 1$ if $m = n$ and $\delta_m^n = 0$ if otherwise. Under the same approximation we can derive the following Belief-Propagation (BP) equations on each arc (i, j):

$$q_{i \to j}^{c_i, c_j} = \frac{1}{z_{i \to j}} e^{-x c_i} \left[\prod_{k \in \partial i+} \sum_{c_k} q_{k \to i}^{c_k, c_i} - \delta_0^{c_i} \prod_{k \in \partial i+} q_{k \to i}^{0,0} \right] \prod_{l \in \partial i- \backslash j} \sum_{c_l} q_{l \gets i}^{c_l, c_i}, \qquad (3a)$$

$$q_{j \gets i}^{c_j, c_i} = \frac{1}{z_{j \gets i}} e^{-x c_j} \left[\prod_{k \in \partial j+ \backslash i} \sum_{c_k} q_{k \to j}^{c_k, c_j} - \delta_0^{c_j + c_i} \prod_{k \in \partial j+ \backslash i} q_{k \to j}^{0,0} \right] \prod_{l \in \partial j-} \sum_{c_l} q_{l \gets j}^{c_l, c_j},$$

$$(3b)$$

where $z_{i \to j}$ and $z_{j \gets i}$ are also normalization constants, and $\partial j+ \backslash i$ is the vertex set obtained after removing i from $\partial j+$. We can easily verify that $q_{i \to j}^{c_i, 0} = q_{i \to j}^{c_i, 1}$ for $c_i = 0$ or 1, and that $q_{j \gets i}^{1,0} = q_{j \gets i}^{1,1}$.

We let Eqs. (2) and (3) guide our construction of a near-optimal dominating set Γ through a belief propagation decimation algorithm. This BPD algorithm is implemented in the same way as the BPD algorithm for undirected graphs [13], therefore its implementing details are omitted here (the source code is available upon request). Roughly speaking, at each iteration step of BPD we first iterate Eq. (3) for several rounds, then we estimate the occupation probabilities for all the unoccupied vertices using Eq. (2), and then we occupy those vertices whose estimated occupation probabilities are the highest. Such a BPD process is repeated on the input digraph until all the vertices are observed. The results of this message-passing algorithm are shown in Fig. 3 for random digraphs and in Table 1 for real-world networks.

If we can find a fixed point for the set of BP equations at a given value of the re-weighting parameter x, we can then compute the mean fraction w of occupied vertices as $w = (1/N) \sum_{i \in V} q_i^1$. The total free energy $F = -(1/x) \ln Z(x)$ can be evaluated as the total vertex contributions subtracting the total arc contributions:

$$F = -\sum_{i \in V} \frac{1}{x} \ln \left[\sum_{c_i} e^{-x c_i} \left[\prod_{j \in \partial i+} \sum_{c_j} q_{j \to i}^{c_j, c_i} - \delta_0^{c_i} \prod_{j \in \partial i+} q_{j \to i}^{0,0} \right] \prod_{k \in \partial i-} \sum_{c_k} q_{k \gets i}^{c_k, c_i} \right]$$

$$+ \sum_{(i,j) \in A} \frac{1}{x} \ln \left[\sum_{c_i, c_j} q_{i \to j}^{c_i, c_j} q_{j \gets i}^{c_j, c_i} \right]. \qquad (4)$$

The entropy density s of the system is then estimated through $s = x(w - F/N)$.

For a given ensemble of random digraphs, the ensemble-averaged occupation fraction w and entropy density s at each fixed value of x can also be obtained from Eqs. (2), (3) and (4) through population dynamics simulation [13]. Both w and s decrease with x, and s may change to be negative as x exceeds certain critical value. The value of w at this critical point of x is then taken as the ensemble-averaged MDS relative size w_0 (very likely it is only a lower bound to w_0). For example, at arc density $\alpha = 5$ the entropy density of ER digraphs decreases to zero at $x \approx 9.9$, at which point $w \approx 0.195$. These ensemble-averaged results for random ER and RR digraphs are also shown in Fig. 3. We notice that the BPD results and the replica-symmetric mean field results almost superimpose with each other, suggesting that dominating sets obtained by the BPD algorithm are extremely close to be optimal.

4   Conclusion

In this paper we studied the directed dominating set problem by a core percolation theory and a replica-symmetric mean field theory, and proposed a generalized leaf-removal local algorithm and a BPD message-passing algorithm to construct near-optimal dominating sets for single digraph instances. We expect these theoretical and algorithmic results to be useful for many future practical applications.

The spin glass model (1) was treated in this paper only at the replica-symmetric mean field level. It should be interesting to extend the theoretical investigations to the level of replica-symmetry-breaking [27] for a more complete understanding of this spin glass system. The replica-symmetry-breaking mean field theory can also lead to other message-passing algorithms that perform even better than the BPD algorithm [23] (the review paper [28] offers a demonstration of this point for the minimum vertex-cover problem).

Acknowledgments. This research is partially supported by the National Basic Research Program of China (grant number 2013CB932804) and by the National Natural Science Foundations of China (grant numbers 11121403 and 11225526). HJZ conceived research, JHZ and YH performed research, HJZ and JHZ wrote the paper. Correspondence should be addressed to HJZ (zhouhj@itp.ac.cn) or to JHZ (zhaojh@itp.ac.cn).

Appendix: Mean Field Equations for the GLR Process

The mean field theory for the directed GLR process is a simple extension of the same theory presented in [13] for undirected graphs. Therefore here we only list the main equations of this theory but do not give the derivation details. We denote by $P(k_+, k_-)$ the probability that a randomly chosen vertex of a digraph has in-degree k_+ and out-degree k_-. Similarly, the in- and out-degree joint probabilities of the predecessor vertex i and successor vertex j of a randomly chosen arc (i, j) of the digraph are denoted as $Q_+(k_+, k_-)$ and $Q_-(k_+, k_-)$, respectively. We assume that there is no structural correlation in the digraph, therefore

$$Q_+(k_+, k_-) = \frac{k_- P(k_+, k_-)}{\alpha}, \quad Q_-(k_+, k_-) = \frac{k_+ P(k_+, k_-)}{\alpha}, \qquad (5)$$

where $\alpha \equiv \sum_{k_+, k_-} k_+ P(k_+, k_-) = \sum_{k_+, k_-} k_- P(k_+, k_-)$ is the arc density.

Consider a randomly chosen arc (i, j) from vertex i to vertex j, suppose vertex i is always unobserved, then we denote by α_t the probability that vertex

j becomes an unobserved leaf vertex (i.e., it has no unobserved successor and has only a single predecessor) at the t-th GLR evolution step, and by $\gamma_{[0,t]}$ the probability that j has been observed at the end of the t-th GLR step. Similarly, suppose the successor vertex j of a randomly chosen arc (i,j) is always unobserved, we denote by $\beta_{[0,t]}$ the probability that the predecessor vertex i has been occupied at the end of the t-th GLR step, and by η_t the probability that at the end of the t-th GLR step vertex i becomes observed but unoccupied and having no other unoccupied successors except vertex j. These four set of probabilities are related by the following set of iterative equations:

$$\alpha_t = \delta_t^0 Q_-(1,0) + \sum_{k_+,\,k_-} Q_-(k_+,k_-)\left[\delta_t^1\left[(\eta_0)^{k_+-1}(\gamma_{[0,0]})^{k_-} - \delta_{k_+}^1\delta_{k_-}^0\right]+\right.$$

$$\left.(1-\delta_t^0-\delta_t^1)\left[\left(\sum_{t'=0}^{t-1}\eta_{t'}\right)^{k_+-1}(\gamma_{[0,t-1]})^{k_-} - \left(\sum_{t'=0}^{t-2}\eta_{t'}\right)^{k_+-1}(\gamma_{[0,t-2]})^{k_-}\right]\right],$$

$$\tag{6a}$$

$$\beta_{[0,t]} = 1 - \sum_{k_+,\,k_-} Q_+(k_+,k_-)\left[\delta_t^0(1-\delta_{k_+}^0)(1-\alpha_0)^{k_--1}+\right.$$

$$\left.(1-\delta_t^0)\left[1-\left(\sum_{t'=0}^{t-1}\eta_{t'}\right)^{k_+}\right]\left(1-\sum_{t'=0}^{t}\alpha_{t'}\right)^{k_--1}\right],\tag{6b}$$

$$\gamma_{[0,t]} = 1 - \sum_{k_+,\,k_-} Q_-(k_+,k_-)(1-\beta_{[0,t]})^{k_+-1}\left(1-\sum_{t'=0}^{t}\alpha_{t'}\right)^{k_-},\tag{6c}$$

$$\eta_t = \delta_t^0 \sum_{k_+,\,k_-} Q_+(k_+,k_-)\left(1-(1-\beta_{[0,0]})^{k_+}\right)(\gamma_{[0,0]})^{k_--1}+$$

$$(1-\delta_t^0) \sum_{k_+,\,k_-} Q_+(k_+,k_-)\left[\left(1-(1-\beta_{[0,t]})^{k_+}\right)(\gamma_{[0,t]})^{k_--1}\right.$$

$$\left.-\left(1-(1-\beta_{[0,t-1]})^{k_+}\right)(\gamma_{[0,t-1]})^{k_--1}\right].\tag{6d}$$

Let us define $\alpha_{cum} \equiv \sum_{t\geq 0}^{+\infty} \alpha_t$, $\beta_{cum} \equiv \beta_{[0,\infty]}$, $\gamma_{cum} \equiv \gamma_{[0,\infty]}$ and $\eta_{cum} \equiv \sum_{t\geq 0}^{\infty} \eta_t$ as the cumulative probabilities over the whole GLR process. From Eq. (6) we can verify that these four cumulative probabilities satisfy the following self-consistent equations:

$$\alpha_{cum} = \sum_{k_+,\,k_-} Q_-(k_+,\,k_-)(\eta_{cum})^{k_+-1}(\gamma_{cum})^{k_-},\tag{7a}$$

$$\beta_{cum} = 1 - \sum_{k_+,\,k_-} Q_+(k_+,\,k_-)\left[1-(\eta_{cum})^{k_+}\right](1-\alpha_{cum})^{k_--1},\tag{7b}$$

$$\gamma_{cum} = 1 - \sum_{k_+,\,k_-} Q_-(k_+,\,k_-)(1-\beta_{cum})^{k_+-1}(1-\alpha_{cum})^{k_-},\tag{7c}$$

$$\eta_{cum} = \sum_{k_+,\,k_-} Q_+(k_+,\,k_-)\left[1-(1-\beta_{cum})^{k_+}\right](\gamma_{cum})^{k_--1}.\tag{7d}$$

The fraction n_{core} of vertices that remain to be unobserved at the end of the GLR process is

$$n_{core} = \sum_{k_+, k_-} P(k_+, k_-) \big[(1 - \beta_{cum})^{k_+} - (\eta_{cum})^{k_+}\big](1 - \alpha_{cum})^{k_-}$$

$$- \sum_{k_+, k_-} P(k_+, k_-) k_+ (1 - \beta_{cum} - \eta_{cum})(\eta_{cum})^{k_+ - 1}(\gamma_{cum})^{k_-} . \quad (8)$$

The fraction w of vertices that are occupied during the whole GLR process is evaluated through

$$w = 1 - \sum_{k_+, k_-} P(k_+, k_-) \big[1 - (\eta_{cum})^{k_+}\big](1 - \alpha_{cum})^{k_-}$$

$$- P(1,0)\eta_0 - \sum_{t \geq 1} \sum_{k_+, k_-} P(k_+, k_-) k_+ \eta_t \Big(\sum_{t'=0}^{t-1} \eta_{t'}\Big)^{k_+ - 1} \Big(\sum_{t'=0}^{t-1} \gamma_{t'}\Big)^{k_-}$$

$$- \sum_{t \geq 1} \sum_{k_+, k_-} P(k_+, k_-) k_- \alpha_t \Big(\sum_{t'=0}^{t-1} \gamma_{t'}\Big)^{k_- - 1} \Big[1 - \Big(1 - \sum_{t'=0}^{t-1} \beta_{t'}\Big)^{k_+}\Big] . \quad (9)$$

References

1. Fu, Y.: Dominating set and converse dominating set of a directed graph. Amer. Math. Monthly **75**, 861–863 (1968)
2. Haynes, T.W., Hedetniemi, S.T., Slater, P.J.: Fundamentals of Domination in Graphs. Marcel Dekker, New York (1998)
3. Garey, M., Johnson, D.S.: Computers and Intractability: A Guide to the Theory of NP-Completeness. Freeman, San Francisco (1979)
4. Mézard, M., Tarzia, M.: Statistical mechanics of the hitting set problem. Phys. Rev. E **76**, 041124 (2007)
5. Gutin, G., Jones, M., Yeo, A.: Kernels for below-upper-bound parameterizations of the hitting set and directed dominating set problems. Theor. Comput. Sci. **412**, 5744–5751 (2011)
6. Takaguchi, T., Hasegawa, T., Yoshida, Y.: Suppressing epidemics on networks by exploiting observer nodes. Phys. Rev. E **90**, 012807 (2014)
7. Wuchty, S.: Controllability in protein interaction networks. Proc. Natl. Acad. Sci. USA **111**, 7156–7160 (2014)
8. Wang, H., Zheng, H., Browne, F., Wang, C.: Minimum dominating sets in cell cycle specific protein interaction networks. In: Proceedings of International Conference on Bioinformatics and Biomedicine (BIBM 2014), pp. 25–30. IEEE (2014)
9. Liu, Y.Y., Slotine, J.J., Barabási, A.L.: Observability of complex systems. Proc. Natl. Acad. Sci. USA **110**, 2460–2465 (2013)
10. Yang, Y., Wang, J., Motter, A.E.: Network observability transitions. Phys. Rev. Lett. **109**, 258701 (2012)
11. Pang, C., Zhang, R., Zhang, Q., Wang, J.: Dominating sets in directed graphs. Infor. Sci. **180**, 3647–3652 (2010)
12. Molnár Jr., F., Sreenivasan, S., Szymanski, B.K., Korniss, K.: Minimum dominating sets in scale-free network ensembles. Sci. Rep. **3**, 1736 (2013)

13. Zhao, J.H., Habibulla, Y., Zhou, H.J.: Statistical mechanics of the minimum dominating set problem. J. Stat, Phys. (2015). doi:10.1007/s10955-015-1220-2

14. Bauer, M., Golinelli, O.: Core percolation in random graphs: a critical phenomena analysis. Eur. Phys. J. B **24**, 339–352 (2001)

15. Liu, Y.Y., Csóka, E., Zhou, H.J., Pósfai, M.: Core percolation on complex networks. Phys. Rev. Lett. **109**, 205703 (2012)

16. Richardson, M., Agrawal, R., Domingos, P.: Trust management for the semantic web. In: Fensel, D., Sycara, K., Mylopoulos, J. (eds.) ISWC 2003. LNCS, vol. 2870, pp. 351–368. Springer, Heidelberg (2003)

17. Leskovec, J., Huttenlocher, D., Kleinberg, J.: Signed networks in social media. In: Proceedings of the SIGCHI Conference on Human Factors in Computing Systems, pp. 1361–1370. ACM, New York (2010)

18. Leskovec, J., Huttenlocher, D., Kleinberg, J.: Predicting positive and negative links in online social networks. In: Proceedings of the 19th International Conference on World Wide Web, pp. 641–650. ACM, New York (2010)

19. Leskovec, J., Kleinberg, J., Faloutsos, C.: Graph evolution: densification and shrinking diameters. ACM Trans. Knowl. Disc. Data **1**, 2 (2007)

20. Leskovec, J., Kleinberg, J., Faloutsos, C.: Graphs over time: densification laws, shrinking diameters and possible explanations. In: Proceedings of the Eleventh ACM SIGKDD International Conference on Knowledge Discovery in Data Mining, pp. 177–187. ACM, New York (2005)

21. Leskovec, J., Lang, K.J., Dasgupta, A., Mahoney, M.W.: Community structure in large networks: natural cluster sizes and the absence of large well-defined clusters. Internet Math. **6**, 29–123 (2009)

22. Ripeanu, M., Foster, I., Iamnitchi, A.: Mapping the gnutella network: properties of large-scale peer-to-peer systems and implications for system design. IEEE Internet Comput. **6**, 50–57 (2002)

23. Mézard, M., Montanari, A.: Information, Physics, and Computation. Oxford University Press, New York (2009)

24. Kschischang, F.R., Frey, B.J., Loeliger, H.A.: Factor graphs and the sum-product algorithm. IEEE Trans. Inf. Theory **47**, 498–519 (2001)

25. Xiao, J.Q., Zhou, H.J.: Partition function loop series for a general graphical model: free-energy corrections and message-passing equations. J. Phys. A: Math. Theor. **44**, 425001 (2011)

26. Zhou, H.J., Wang, C.: Region graph partition function expansion and approximate free energy landscapes: theory and some numerical results. J. Stat. Phys. **148**, 513–547 (2012)

27. Mézard, M., Parisi, G.: The bethe lattice spin glass revisited. Eur. Phys. J. B **20**, 217–233 (2001)

28. Zhao, J.H., Zhou, H.J.: Statistical physics of hard combinatorial optimization: vertex cover problem. Chin. Phys. B **23**, 078901 (2014)

A Linear Time Algorithm for Ordered Partition

Yijie Han$^{(\boxtimes)}$

School of Computing and Engineering,
University of Missouri at Kansas City,
Kansas City, MO 64110, USA
hanyij@umkc.edu

Abstract. We present a deterministic linear time and space algorithm for ordered partition of a set T of n integers into $n^{1/2}$ sets $T_0 \leq T_1 \leq \cdots \leq T_{n^{1/2}-1}$, where $|T_i| = \theta(n^{1/2})$ and $T_i \leq T_{i+1}$ means that $\max T_i \leq \min T_{i+1}$.

Keywords: Algorithms · Sorting · Integer sorting · Linear time algorithm · Ordered partition · Hashing · Perfect hash functions

1 Introduction

For a set T of n input integers we seek to partition them into $n^{1/2}$ sets $T_0, T_1, ..., T_{n^{1/2}} - 1$ such that $|T_i| = \theta(n^{1/2})$, $T_i \leq T_{i+1}$, $0 \leq i < n^{1/2} - 1$. Where $T_i \leq T_{i+1}$ means $\max T_i \leq \min T_{i+1}$. We call this ordered partition. We show that we can do this in deterministic optimal time, i.e. in $O(n)$ time.

This result, when applied iteratively for $O(\log \log n)$ iterations, partitions the n integers into $O(n^{3/4})$ sets because every set is further partitioned into $O(n^{1/4})$ sets, into $O(n^{7/8})$ sets, and so on, eventually partitions n integers into n ordered sets, i.e., having them sorted. The time for these iterations is $O(n \log \log n)$ and the space complexity is linear, i.e. $O(n)$. This complexity result for integer sorting was known [6] and is the current best result for deterministic integer sorting. However, ordered partition itself is an interesting topic for study and the result for ordered partition can be applicable in the design of algorithms for other problems. As an example in [7] our ordered partition algorithm is extended for obtaining an improved randomized integer sorting algorithm.

The problem of ordered partition was noticed in [11] and the linear time complexity was conceived there. However, the deterministic linear time ordered partition algorithm presented here has a nice structure and the mechanism for the design of our algorithm is particularly worth noting. In particular the conversion of the randomized signature sorting to the deterministic version shown in Sect. 3.2 and the mechanism shown there were not explained clearly in [11]. Besides, our ordered partition algorithm is an optimal algorithm and it has been extended to obtain a better randomized algorithm [7] for integer sorting. Therefore we present them here for the sake of encouraging future research toward an optimal algorithm for integer sorting.

© Springer International Publishing Switzerland 2015
J. Wang and C. Yap (Eds.): FAW 2015, LNCS 9130, pp. 89–103, 2015.
DOI: 10.1007/978-3-319-19647-3_9

2 Overview

Integers are sorted into their destination partition by moving them toward their destination partition. One way of moving integers is to use the ranks computed for them to move them, say a rank r is obtained for an integer i then move i to position r. We will call such moving as moving by indexing. Note that because our algorithm is a linear time algorithm such move by indexing can happen only a constant number of times for each integer. Since a constant number of moving by indexing is not sufficient through our algorithm we need to use the packed moving by indexing, i.e. we pack a integers into a word with each integer having the rank of $\log n/(2a)$ bits (and thus the total number of bits for all ranks of packed integers in a word is $\log n/2$) and move these integers to the destination specified by these $\log n/2$ bits in one step.

Here we packed a integers into a word in order to move them by indexing. In order to have a integers packed in a word we need to reduce the input integers to smaller valued integers. This can be done in the randomized setting by using the signature sorting [1] which basically says:

Lemma 1 [1]: In the randomized setting, sorting of n $p \log n$ bits integers can be done with 2 passes of sorting n $p^{1/2} \log n$ bits integers.

However, because our algorithm is a deterministic algorithm we converted Lemma 1 to Lemma 2.

Lemma 2: Ordered partition of n $p \log n$ bits integers into $n^{1/2}$ partitions can be done with 2 passes of ordered partition of n $p^{1/2} \log n$ bits integers into $n^{1/2}$ partitions.

Lemma 2 is basically proved in Sect. 3.2 as Lemma 2′ and it is a deterministic algorithm. This is one of our main contributions.

By Lemma 2 we can assume that the integers we are dealing with has $p \log n$ bits while each word has $p^3 \log n$ bits and thus we can pack more than p integers into a word.

We do ordered partition of n integers into $n^{1/2}$ partitions by a constant number of passes of ordered partition of n integers into $n^{1/8}$ partitions, i.e. ordered partition of n integers into $n^{1/8}$ partitions, into $n^{1/8+(1/8)(7/8)}$ partitions, ..., into $n^{1/2}$ partitions.

We do ordered partition of n integers in set T into $n^{1/8}$ partitions via ordered partition of n integers in T by $n^{1/8}$ integers $s_0 \le s_1 \le \cdots \le s_{n^{1/8}-1}$ in S, i.e. partition T into $|S| + 1$ sets $T_0, T_1, ..., T_{|S|}$ such that $T_0 \le s_0 \le T_1 \le \cdots \le s_{|S|-1} \le T_{|S|}$. We call this as ordered partition of T by S.

The ordered partition is done by *arbitrary* partition T into $n^{1/2}$ sets $T(i)$'s, $0 \le i < n^{1/2}$, with $|T(i)| = n^{1/2}$ and pick arbitrary $n^{1/16}$ integers $s_0, s_1, ..., s_{n^{1/16}-1}$ in T to form set S. Then use step i to do ordered partition of $T(i)$ by S, $i = 0, 1, ..., n^{1/2} - 1$. After step $i - 1$ we have ordered partitioned $\cup_{j=0}^{i-2} T(j)$ by S into $T_0 \le s_0 \le T_1 \le s_1 \le \cdots \le s_{n^{1/16}-1} \le T_{n^{1/16}}$ and we maintain that $|T_j| \le 2n^{15/16}$ for all j. If after step i $2n^{15/16} < |T_j| \le 2n^{15/16} + n^{1/2}$ then we use selection [4] in $O(|T_j|)$ time to select the median m of T_j and (ordered) partition T_j into two sets. We add m to S. The next time a partition has to be ordered partitioned into two sets is after adding additional $n^{15/16}+n^{1/2}$ integers (from $n^{15/16}$ integers to $2n^{15/16}+n^{1/2}$ integers). Thus the overall time of ordered partitioning of one set (partition) to two sets (partitions) is linear. This idea is shown in [2]. Thus after finishing we have partitioned T into $O(n^{1/16})$ sets with each set of size $\leq 2n^{15/16}$. We can then do selections for these sets to have them become $n^{1/16}$ ordered partitioned sets with each set of size $n^{15/16}$. The time is linear.

Thus we now left with the problem of ordered partition of T with $|T| = n$ by S with $|S| = n^{1/8}$. This is a complicated part of our algorithm. We achieve this using the techniques of deterministic perfect hashing, packed moving by indexing or sorting, progressive enlarge the size of hashed integers for packed moving and the determination of progress when hashed value matches, etc. The basic idea is outline in the next two paragraphs.

Let $s_0 \leq s_1 \leq \cdots \leq s_{n^{1/8}-1}$ be the integers in S. As will be shown in Sect. 4 that we can compare all integers in T with $s_i, s_{2i}, s_{3i}, ...,$ $s_{\lfloor n^{1/8}/i \rfloor i}$ in $O(n \log(n^{1/8}/i)/\log n)$ time because integers can be hashed to $O(\log(n^{1/8}/i))$ bits (this is a perfect hash for $s_i, s_{2i}, s_{3i}, ..., s_{\lfloor n^{1/8}/i \rfloor i}$ but not perfect for the integers in T, but the hash value for every integer has $O(\log(n^{1/8}/i))$ bits) and we can pack $O(\log n/\log(n^{1/8}/i)))$ integers into one word (note that we can pack more integers into one word but the number of hashed bits will exceed $\log n/2$ which cannot be handled by our algorithm). Because of this packing we left with $O(n \log(n^{1/8}/i)/\log n)$ words and we can sort them using bucket sorting (by treating the hashed bits together as one integer in each word) in $O(n \log(n^{1/8}/i)/\log n)$ time. This sorting will let us compare all integers in T with $s_i, s_{2i}, s_{3i}, ..., s_{\lfloor n^{1/8}/i \rfloor i}$. As noted in Sect. 4 if an integer t in T is equal to s_{ai} for some a (we call match) then we made progress (we say advanced). Otherwise t is not equal to any s_{ai} (we call unmatch) and then we will compare these unmatched t's with $s_j, s_{2j}, s_{3j}, ..., s_{\lfloor n^{1/8}/j \rfloor j}$, where $n^{1/8}/j = (n^{1/8}/i)^2$ (i.e. we double the number of bits for hashed value for every integer). We keep doing this when unmatch happens till i (or j) becomes 1 where we have compared t with all integers in S and none of them is equal to t.

If for every integer t in T we did the last paragraph then if t does not match any integer in S we will lose our work because we will not know which partition t will fall into. To prevent this to happen when we compare t to integers in S we really do is to compare the most significant $\log n$ bits of t to the most significant $\log n$ bits of integers in S. And thus if the most significant $\log n$ bits of t does not match the most significant $\log n$ bits of any integer in S we can then throw away all bits of t except the most significant $\log n$ bits. We can then bucket sort t (because there are only $\log n$ bits for t left) into its destination partition. The complication of this scheme such as when the most significant $\log n$ bits match is handled in Sect. 4 where we used the concept of current segment for every integer. Initially the most significant $\log n$ bits of an integer is the current segment of the integer. When the current segment matches we then throw away the current segment and use the next $\log n$ bits as the current segment. The details is shown in Sect. 4.

3   Preparation

3.1   Perfect Hash Function

We will use a perfect hash function H that can hash a set S of integers in $\{0, 1, ..., m-1\}$ to integers in $\{0, 1, ..., |S|^2 - 1\}$. A perfect hash function is a hash function that has no collisions for the input set. Such a hash function for S can be found in $O(|S|^2 \log m)$ time [13]. In [8] we improved this and made the time independent of the number of bits in integers ($\log m$):

Lemma 3 [8]: A perfect hash function that hashes a set S of integers in $\{0, 1, ..., m-1\}$ to integers in $\{0, 1, ..., |S|^2 - 1\}$ can be found in $O(|S|^4 \log |S|)$ time. Thereafter every batch of $|S|$ integers can be hashed to $\{0, 1, ..., |S|^2 - 1\}$ in $O(|S|)$ time and space.   □

We note that the hash function used in this paper was first devised in [5] where it is showed that such a hash function can be found with a randomized algorithm in constant time. Raman [13] showed how to use derandomization to obtain a deterministic version of the hash function in $O(n^2 \log m)$ time for a set of n integers in $\{0, 1, ..., m-1\}$.

The hash function given in [5,13] for hashing integers in $\{0, 1, ..., 2^k - 1\}$ to $\{0, 1, ..., 2^l - 1\}$ is

$$h_a(x) = (ax \bmod 2^k) \operatorname{div} 2^{k-l}$$

with different values of a different hash functions are defined. Raman [13] showed how to obtain the value of a such that the hash function becomes a deterministic perfect hash function. Note that the computation of hash values for multiple integers packed in a word can be done in constant time.

3.2   Converting Ordered Partition with Integers of $p \log n$ Bits to that with Integers of $p^{1/3} \log n$ Bits

Besides ordered partition, we now introduce the problem of ordered partition of set T of n integers by a set S of $|S| < n$ integers. Let $s_0 \leq s_1 \leq \cdots \leq s_{|S|-1}$ be the $|S|$ integers in S that are already sorted. Ordered partition T by S is to ordered partition T into $|S| + 1$ sets $T_0 \leq T_1 \leq \cdots \leq T_{|S|}$ such that $T_0 \leq s_0 \leq T_1 \leq \cdots \leq s_{|S|-1} \leq T_{|S|}$, where $T_i \leq s_i$ means $\max T_i \leq s_i$ and $s_i \leq T_{i+1}$ means $s_i \leq \min T_{i+1}$.

In our application in the following sections, $|S| = n^{1/8}$. Thus sorting set S takes at most $O(|S| \log |S|) = O(n)$ time. That is the reason we can consider the sorting of S as free as it will not dominate the time complexity of our algorithm.

We use a perfect hash function H for set S. Although there are many perfect hash functions given in many papers, we use the hash function in [5,13] because it has the property of hashing multiple integers packed in one word in constant time. Notice that H hashes S into $\{0, 1, ..., |S|^2 - 1\}$ instead of $\{0, 1, ..., c|S| - 1\}$ for any constant c.

We shall use H to hash integers in T. Although H is perfect for S it is not perfect for T as T has n integers. We cannot afford to compute a perfect hash function for T as it requires nonlinear time. The property we will use is that for any two integers in S if their hash values are different then these two integers are different. For any integer $t \in T$ if the hash value of t is different than the hash value of an integer s in S then $t \neq s$.

We will count the bits in an integer from the low order bits to the high order bits starting at the least significant bit as the 0-th bit.

For sorting integers in $\{0, 1, ..., m - 1\}$ we assume that the word size (the number of bits in a word) is $\log m + \log n$. $\log n$ bits are needed here for indexing into n input numbers. When $m = n^k$ for a constant k n integers can be sorted in linear time using radix sort. Let $p = \log m / \log n$ and therefore we have $p \log n$ bits for each integer.

As will be shown, we can do ordered partition of T by S, where $|T| = n$ and $|S| \leq n$, (here integers in $T \cup S$ are taken from $\{0, 1, ..., m - 1\}$, i.e. each integer has $p \log n$ bits) in constant number of passes with each pass being an ordered partition of set T'_i by S'_i, $i = 1, 2, 3, ..., c$, where T'_i has n integers each having $p^{1/3} \log n$ bits and S'_i has $|S|$ integers each having $p^{1/3} \log n$ bits, c is a constant.

This is done as follows. We view each integer of $p \log n$ bits in T (S) as composed of $p^{1/3}/2$ segments with each segment containing consecutive $2p^{2/3} \log n$ bits. We use a perfect hash function H which hashes each segment of an integer in S into $2 \log |S|$ bits and thus each integer in S is hashed to $p^{1/3} \log n$ bits. Note that the hash function provided in [5,13] can hash all segments contained in a word in constant time (It requires [5,13] that for each segment g of $2p^{2/3} \log n$ bits another $2p^{2/3} \log n$ bits g_1 has to be reserved and two segments of bits $g_1 g$ is used for the hashing of segment g. This can be achieved by separating even indexed segments into a word and odd indexed segments into another word).

We then use H to hash (segments of) integers in T and thus each integer in T is also hashed to $p^{1/3} \log n$ bits. Note here H is not perfect with respect to the (segments of) integers in T and thus different (segments) of integers in T may be hashed to the same hash value.

We use T'_1 (S'_1) to denote this hashed set, i.e. each integer in T'_1 (S'_1) has only $p^{1/3} \log n$ bits. Now assume that integers in S'_1 are sorted. If integer s (t) in S (T) is hashed to integer s' (t') in S'_1 (T'_1) then we use $H(s) = s'$ $(H(t) = t')$ to denote this. If s is in S then we also have that $s = H^{-1}(s')$. Note that if t is an integer in T and s is an integer in S, then if $H(t) = H(s)$ t may not be equal to s as H is not perfect for $T \cup S$. But if $H(t) \neq H(s)$ then $t \neq s$ and this is the property we will make use of. We will also use $H(S)$ $(H(T))$ to denote the set of integers hashed from S (T).

In the first pass we do ordered partition of T'_1 by S'_1. Assume that this is done. Let t' be an integer in T'_1 and $s'_i \leq t' \leq s'_{i+1}$, s'_i and s'_{i+1} be the two integers in S'_1 with ranks i and $i + 1$. If $t' = s'_i$ then we compare t and $H^{-1}(s'_i)$. Otherwise let the most significant bit that s'_i (s'_{i+1}) and t' differs be the d_i-th $(d_{i+1}$-th) bit and consider the case that $d_i < d_{i+1}$ (the situation that $d_i > d_{i+1}$ can be treated similarly) in which we will compare t and $H^{-1}(s'_i)$. Note that the situation $d_i = d_{i+1}$ cannot happen. Let d_i-th bit of s'_i be in the g-th segment of s'_i (each segment of s'_i has $\leq 2 \log n$ bits and is obtained by hashing the corresponding

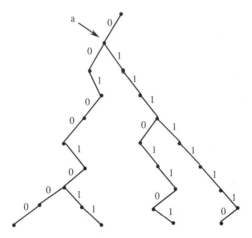

Fig. 1.

segment of $2p^{2/3}\log n$ bits in $H^{-1}(s'_i)$. Note here s'_i is not necessarily hashed from s_i because the rank order of s_i's are not kept in s'_i's after hashing, s_i need not be $H^{-1}(s'_i)$, i.e. s'_i is ranked i-th in S'_1 but $H^{-1}(s'_i)$ may not be ranked i-th in S). This says that in the more significant segments than the g-th segment both t' and s'_i are equal. Note that for every s'_j in S'_1 let the most significant bit that t' and s'_j differs be in the g'-th segment then $g' \geq g$ because $s'_i \leq t' \leq s'_{i+1}$.

If we look at the trie (binary tree built following the bits in the integers) for S we see that the most significant segment that t and $H^{-1}(s'_i)$ differs is the u-th segment with $u \geq g$. Note here that u may be not equal to g because H is not perfect for $T \cup S$. For all other s_k's if the most significant segment t and s_k differ is the u_1-th segment then $u_1 \geq u$ (u_1-th segment is at least as significant as u-th segment) and this comes from the property that H is perfect for S. Thus the further ordered partition for t by S can be restricted to the u-th segment of t with respect to the u-th segment of s_k's such that the most significant segment that s_k and t differ is the u-th segment.

Example 1: Figure 1. shows a trie for U (representing S) of 4 integers: $u_0 = 001001000, u_1 = 001001011, u_2 = 011101101, u_3 = 011111110$. Let set V (representing T) contain integers $v_0 = 000100111, v_1 = 010110011, v_2 = 011011101, v_3 = 010101011, v_4 = 011110101, v_5 = 011111111, v_6 = 001001010, v_7 = 011101011, v_8 = 011010100, v_9 = 011101111$. Let every 3 bits form a segment so that each integer has 3 segments. We will do ordered partition of V by U. Let $w_k = \min_{i=0}^{3}\{w \mid$ the most significant segment that $H(v_k)$ differs with $H(u_i)$ is the w-th segment $\}$. Let the u_i that achieves the w_k value be u_{v_k}.

Say that $w_0 = 0$ and u_{v_0} is u_0. Note that although the most significant segment that u_0 and v_0 differ is the 2nd segment, the most significant segment that $H(u_0)$ and $H(v_0)$ differ can be the 0th segment because H is not perfect on $T \cup S$. Now compare u_0 and v_0 we find that the most significant segment they differ is the 2nd segment. This says that v_0 "branches out" of the trie for U at the 2nd segment. Although the ordered partition for v_0 can be determined from the trie of U, this is not always the case for every integer in V. For example, say $w_3 = 1$ and u_{v_3} is u_0, then comparing u_0 and v_3 determines that v_3 branches out of u_0 at point a in Fig. 1, but we do not know the point where v_3 branches out of the trie for U. Because the 2nd segment of v_3 is different than the 2nd segment of any u_i's and thus the second segment of $H(v_3)$ can be equal to (or not equal to) any of that of $H(u_i)$. The further ordered partition of v_0 and v_3 by U can be restricted to the 2nd segment. □

In the trie for S we will say the g-th segment $s_{i,g}$ of $s_i \in S$ is nonempty if there is an $t \in T$ such that the most significant segment that t and s_i differ is in the g-th segment. If this happens then we will take the prefix $s_{i,prefix} = s_{i,p^{1/3}/2-1}s_{i,p^{1/3}/2-2}...s_{i,g+1}$, i.e. all segments more significant than the g-th segment of s_i. Because there are n integers in T and thus there are at most n nonempty segments. Thus we have at most n such $s_{i,prefix}$ prefix values. We will arbitrarily replace each such prefix value with a distinct number in $\{0, 1, ..., n-1\}$. Say $s_{i,prefix}$ is replaced by $R(s_{i,prefix}) \in \{0, 1, ..., n-1\}$ we then append the g-th segment of t to $R(s_{i,prefix})$ to form $t(1)$ of $\log n + 2p^{2/3} \log n$ bits, where $2p^{2/3} \log n$ bits come from the g-th segment of t and the other $\log n$ bits come from $R(s_{i,prefix})$. s_i is also transformed to $s_i(1)$ by appending the g-th segment of s_i to $R(s_{i,prefix})$. When all the n integers in T and s integers in S are thus transformed to form the sets $T(1)$ and $S(1)$ we left with the problem of ordered partition of $T(1)$ by $S(1)$ and here integers have only $\log n + 2p^{2/3} \log n$ bits.

Example 2: Follows Example 1 and Fig. 1. Because of v_0, v_1, v_3, the 2nd segment of u_0, u_1, u_2, u_3 are not empty. These segments are 001 and 011. The $s_{i,prefix}$ is ϵ. We may take $R(s_{i,prefix}) = 000$ for these two segments (case 0). Because of v_2, v_4, v_8, the 1st segments of u_2, u_3 are not empty. These segments are 101 and 111. The $s_{i,prefix}$ is 011 (the 2nd segment of u_2 and u_3). We may take $R(s_{i,prefix}) = 001$ for them (case 1). Because of v_5, the 0th segment of u_3 is not empty. This segment is 110. The $s_{i,prefix}$ is 011111. We may take $R(s_{i,prefix}) = 010$ for it (case 2). Because of v_6, the 0th segments of u_0 and u_1 are not empty. These segments are 000 and 011. The $s_{i,prefix}$ is 001001. We may take $R(s_{i,prefix}) = 011$ for these segments (case 3). Because of v_7 and v_9, the 0th segment of u_2 is not empty. This segment is 101. The $s_{i,prefix}$ is 0111101. We may take $R(s_{i,prefix}) = 100$ (case 4).

Thus V' contains 000000 (from v_0), 000010 (from v_1), 000010 (from v_3), 001011 (from v_2), 001110 (from v_4), 001010(from v_8), 010111 (from v_5), 011010 (from v_6), 100011 (from v_7), 100111 (from v_9).

U' contains 000001, 000011 (from case 0), 001101, 001111 (from case 1), 010110 (from case 2), 011000, 011011 (from case 3), 100101 (from case 4). □

Thus after one pass of ordered partitioning of a set of n $p^{1/3} \log n$ bit integers (T_1') by a set of $|S|$ $p^{1/3} \log n$ bit integers (S_1'), the original problem of ordered partitioning of T by S is transformed to the problem of ordered partitioning of $T(1)$ by $S(1)$.

If we iterate this process by executing one more pass of ordered partitioning of a set of n $p^{1/3} \log n$ bit integers (call it T_2') by a set of $|S|$ $p^{1/3} \log n$ bit integers (call it S_2') then the ordered partition of $T(1)$ by $S(1)$ will be transformed to the problem of ordered partitioning of $T(2)$ by $S(2)$ where integers in $T(2)$ and $S(2)$ have $2 \log n + 4p^{1/3} \log n$ bits. This basically demonstrates the idea of converting the ordered partition of $p \log n$ bit integers to constant passes of the ordered partition of $p^{1/3} \log n$ bit integers. This idea is not completely new. It can be traced back to [1] where it was named (randomized) signature sorting and used for randomized integer sorting. Here we converted it for deterministic ordered partition.

The principle explained in this subsection give us the follow lemma.

Lemma 2′: Ordered partition of set T by set S with integers of $p \log n$ bits can be converted to ordered partition of set T_i' by S_i' with integers of $p^{1/c_1} \log n$ bits, $i = 1, 2, ..., c_2$, where c_1 and $c_2 = O(c_1)$ are constants. $\qquad \square$

3.3 Nonconservative Integer Sorting

An integer sorting algorithm sorting integers in $\{0, 1, ..., m-1\}$ is a nonconservative algorithm [12] if the word size (the number of bits in a word) is $p(\log m + \log n)$, where p is the nonconservative advantage.

Lemma 4 [9,10]: n integers can be sorted in linear time with a nonconservative sorting algorithm with nonconservative advantage $\log n$. $\qquad \square$

4   Ordered Partition of T by S

Now let us consider the problem of the ordered partition of T by $S = \{s_0, s_1, ..., s_{|S|-1}\}$ where $s_i \leq s_{i+1}$ and each integer in T or S has $p \log n$ bits and the word size is $p^3 \log n$ bits. We here consider the case that $|T| = n$ and $|S| = n^{1/8}$. In Sect. 2 it is explained why we pick the size of S to be $n^{1/8}$.

First consider the case where $p \geq \log n$. Here we have $p^2 > \log n$ nonconservative advantage and therefore we can simply sort $T \cup S$ using Lemma 4. Thus we assume that $p < \log n$.

Because $|S| = n^{1/8}$ when we finish ordered partition T will be partitioned into $T_0 \leq T_1 \leq \cdots \leq T_{n^{1/8}}$ with $T_0 \leq s_0 \leq T_1 \leq \cdots \leq s_{|S|-1} \leq T_{|S|}$. Let us have a virtual view of the ordered partitioning of T by S. We can view as each integer t in T using $(1/8) \log n$ binary search steps to find the s_i such $s_i \leq t \leq s_{i+1}$ (i.e. in which set T_i it belongs to). The reason it takes $(1/8) \log n$ binary search steps is because we do a binary search on $n^{1/8}$ ordered numbers in S.

As we are partitioning integers in T by integers in S we say that an integer t in T is advanced k steps if t is in the status of having performed $k \log n/p$ binary search steps. Notice the difference when we use word "step" and the phrase "binary search step". Thus each integer needs to advance $p/8$ steps to finish in a set T_i.

Let $v(k) = n^{1/8}/2^{k\log n/p}$ and let $S(k) = \{s_{v(k)}, s_{2v(k)}, s_{3v(k)}, ...\}$, $k = 1, ..., p/8$. Thus $|S(k)| = 2^{k\log n/p}$. Also let $S[k, i] = \{s_{iv(k)}, s_{iv(k)+1}, s_{iv(k)+2}, ..., s_{(i+1)v(k)-1}\}$. If integer t in T is advanced k steps, then we know the i such that $\min S[k, i] \leq t \leq \max S[k, i]$. We will say that t falls into $S[k, i]$. Let $F(S[k, i])$ be the set of integers that fall into $S[k, i]$. When integers in T fall into different $S[k, i]$'s we will continue ordered partition for an $F(S[k, i])$ by $S[k, i]$ at a time. That is, the original ordered partition of T by S becomes multiple ordered partition problems.

Initially we pack p integers into a word. Because each word has size $p^3 \log n$ bits we have p nonconservative advantage after we packed p integers into a word (i.e. we used $p^2 \log n$ bits by packing and now $(word\ size)/(p^2 \log n) = p^3 \log n/(p^2 \log n) = p$).

The basic thoughts of our algorithm is as follows: As will be seen that advance n integers by a steps will result in time complexity $O(na/p)$ because when we try to advance a steps we will compare integers in T to integers in $S(a)$ and not to integers in S (the time $O(na/p)$ will be understood when we explain our algorithm). However, it is not that we can directly advance $p/8$ steps and get $O(n)$ time. There is a problem here. If we form one segment for each integer in T (as we did form $p^{1/3}/2$ segments in the previous section) then if the hash value of an integer t in T does not match the hash value of any integer in $S(a)$ then we lose our work because it does not provide any information as to which set T_i t will belong to. If we make multiple segments for each integer t in T then after hashing and comparing t to s_i's we can eliminate all segments except one for t as we did in the previous section. However, it will be understood that using multiple segments will increase the number of bits for the hashed value and this will result in a nonlinear time algorithm for us (the exact details of this will be understood when our algorithm is explained and understood). What we will do is to compare the most significant $\log n$ bits of the integers in T and in S. If the most significant $\log n$ bits of $t \in T$ is not equal to the most significant $\log n$ bits of any integer in S (we call this case as unmatch) then we win because we need to keep only the most significant $\log n$ bits of t and throw away the remaining bits of t as in the trie for S t branches out within the most significant $\log n$ bits. However, if the value of the most significant $\log n$ bits of t is equal to the value of the most significant $\log n$ bits of an integer in S (we call this case as match) then we can only eliminate the most significant $\log n$ bits of t. As t has $p \log n$ bits it will take $O(p)$ steps to eliminate all bits of t and this will result in a nonlinear time algorithm.

What we do is take the most significant $\log n$ bits of every integer and call it the current segment of the integer. The current segment of $a \in T \cup S$ is denoted by $c(a)$. We first advance 1 step for the current segments of T with respect to the current segments of S in time $O(n/p)$. If within this step we find a match for $c(t)$ for $t \in T$ then we eliminate $\log n$ bits of t and make the next $\log n$ bits as the current segment. Note that for $p \log n$ bits this will result in $O(n)$ time as removing $\log n$ bits takes $O(n/p)$ time. If we find unmatch for $c(t)$ then we advance 2 steps with time $O(2n/p)$. If we then find a match for $c(t)$ we then eliminate $\log n$ bits and make the next $\log n$ bits as the current segment. If unmatch we then advance 4 steps in time $O(4n/p)$. We keep doing this then if in the 1st, 2nd, ..., $(a-1)$-th passes we find unmatch for $c(t)$ and in a-th pass we first find a match for $c(t)$ then $c(t)$ matched the $c(s)$ for an $s \in S(2^{a-1})$ and did not match the $c(s)$ for any $s \in S(2^j)$ with $j < a-1$. Examining the structure of $S(2^{a-1})$ and $S(2^{a-2})$ will tell us that $c(t)$ has advanced 2^{a-2} steps (and therefore t advanced 2^{a-2} steps) because we can tell the two integers s_{k_1} and s_{k_2} with ranks k and $k+1$ in $S(2^{a-2})$ such that $s_{k_1} \le t \le s_{k_2}$ because $c(t)$ matches $c(s_i)$ for an s_i in $S(2^{a-1})$ while $S(2^{a-2}) \subset S(2^{a-1})$. The time we spent computing the 1st, 2nd, ..., a-th passes is $O(n2^a/p)$. Thus we advance 2^{a-2} steps by spending time $O(n2^a/p)$. Because we can advance only $p/8$ steps thus the time is linear. If for all passes from pass 1 to pass $\log(p/8)$ we do not find a match for $c(t)$ then we keep the most significant $\log n$ bits of t and remove other bits of t and the time we spent for all these passes is $O(n)$.

The way for us to find whether $c(t)$ matches any $c(s_i)$ is that we first find whether $H(c(t))$ matches any of $H(c(s_i))$ (here we are comparing $H(c(t))$ to all $H(c(s_i))$'s). If not then unmatch, if $H(c(t)) = H(c(s_i))$ then we compare $c(t)$ and $c(s_i)$ (here we are comparing $c(t)$ to a single $c(s_i)$) to find out whether $c(t) = c(s_i)$. Here we need to do this comparison because H is not perfect for the current segment of $T \cup S$. The reason we first compare $H(c(t))$ and all $H(c(s_i))$'s (we actually sort them) is because hashed values contain less number of bits and therefore can be compared more efficiently.

In our algorithm we need to prepend $\log n$ more significant bits (their value is equal to 0 for the first pass) to the $\log n$ bits (the current segment) for every integer we have mentioned above. This is because, say, for $c(t)$ we find unmatch for the first $a-1$ passes and find a match in the a-th pass ($c(t) = c(s_i)$), we then need to remove the most significant $\log n$ bits from t. As we did in the previous section we need to take the $s_{i,prefix}$ and replace it by a distinct integer $R(s_{i,prefix})$ in $\{0, 1, ..., n-1\}$ and then transform t by removing the segments of t that are not less significant than the current segment of t and replace it with $R(s_{i,prefix})$. This is significant as we go down the road as we can repeatedly removing $\log n$ bits (should be $2\log n$ bits after prepending $R(s_{i,prefix})$) from t. The $R(s_{i,prefix})$ value essentially tells on which part of the trie for S we are currently doing ordered partitioning for t. Note that we will do ordered partitioning for all integers fall into an $S[k, i]$ at a time. The prepending of $\log n$ bits is to tell apart of the different $s_{i,prefix}$ values for integers fall into the same $S[k, i]$.

If we have advanced $t \in T$ k steps, then we know the i such that $\min S[k, i] \le t \le \max S[k, i]$ and we need to continue to do ordered partition for t by $S[k, i]$. However, different integers may advance at different speed. If integer t is advanced k steps we will use $A(t) = k$ to denote about this. Let $T^{(k)}$ be the set of integers in T that have advanced k steps. $T^{(k)} = \cup_i F(S[k, i])$, where $F(S[k, i])$ are the integers in T that fall into $S[k, i]$.

Consider the case where $F(S[m, m_1])$ is selected for advancement. We will try to advance integers in $F(S[m, m_1])$ for 1 step. Let $T' = F(S[m, m_1])$.

Advance for One Step (T')

$U = T'; U' = \phi;$

while $|U| \geq n^{1/2}$ do

begin

1. Try to advance integers in U for 1 step by comparing them with $S[m, m_1] \cap S(m+1)$.
2. Let U_{match} be the set of matched integers and $U_{unmatch}$ be the set of unmatched integers.
3. Let U'_{match} be the subset of U_{match} such that the current segment is the least significant segment (i.e. $t \in U'_{match}$ is equal to some s_i in S). Integers in U'_{match} are done.
4. $U = U_{match} - U'_{match}; U' = U' \cup U_{unmatch};$
5. For $t \in U$ if $c(t)$ matches $c(s_i)$ remove the $c(t)$ from t and make the next $\log n$ bits as the current segment for t. Prepend $R(s_{i,prefix})$ to $c(t)$.

end

/* Now $|U| < n^{1/2}$. */

Use brute force to sort integers in U;

$T' = U';$

The technique to have **Advance for One Step** done in $O(n/p)$ time is the same as that used in Steps 3.1. and 3.2. in algorithm **Partition** (set $k = 1$ there) shown later and therefore we do not repeat them here.

After we have done algorithm **Advance for One Step** we know that integers in T' will advance at least 1 step later. We then try to advance integers in T' for 2^j steps, $j = 1, 2, 3, \ldots$. As we do this matched integers will be distributed to $T^{(m+2^{j-1})}$'s. The number of unmatched integers in T' will decrease. When $|T'| < n^{1/2}$ we will use brute force to sort all integers in T'. Thus for integers remain in $F(S[m, m_1])$ they will be sorted by brute force at most once. There are no more than $|S|^2 \leq n^{1/4}$ $S[k, i]$'s. Thus the total number of integers sorted by brute force is no more than $n^{1/2} \cdot n^{1/4} = O(n^{3/4})$. The reason we use brute force to sort them is that the (approximate) ratio of $|T|$ versus $|S|$ as n to $n^{1/8}$ may not hold any longer.

Now our algorithm is shown as follows.

Partition(T, S)

/* Partition T by S, where integers are in $\{0, 1, \ldots, 2^{p \log n} - 1\}$ and word size is $p^3 \log n$ bits. */

Initialization: Let current segment of each integer in T or S be the most significant $\log n$ bits of the integer.

For $a \in T \cup S$ let $c(a)$ be its current segment. Label all integers as undone. $T^{(0)} = T, T^{(i)} = \phi$ for $i \neq 0$. /* $T^{(i)}$ is the set of integers in T that has advanced i steps. */

1. for$(m = 0; m < p/8; m = m + 1)$ do Steps 2–4.

2. For all $S[m,j]$'s such that there are less than $n^{1/2}$ integers in $F(S[m,j])$ sort these integers and mark these integers done (advanced $p/8$ steps). For every m_1 such that there are at least $n^{1/2}$ integers in $F(S[m,m_1])$, let $T^{(m)} = T^{(m)} - F(S[m,m_1])$ and $T' = F(S[m,m_1])$ and let $S' = S[m,m_1]$, do steps 3 through 4. If $T^{(m)}$ is empty do next iteration of Step 1.

3. Call **Advance for One Step** (T'); $k = 2$.

4. Let $S'(k) = S' \cap S(m+k)$. /* $|S'(k)| = 2^{k \log n/p}$. */

   4.1 Use a hash function H to hash the current segment of each integer in $S'(k)$ into $2k \log n/p$ bits. H must be a perfect hash function for the current segments of $S'(k)$. Since $|S'(k)| = 2^{k \log n/p}$ such a hash function can be found [5,13]. The current segment of each integer in T' is also hashed by this hash function into $2k \log n/p$ bits (here note that there might be collisions as H is not perfect for the current segments of T'). Pack every $p/(8k)$ integers in T' into a word to form a set T'' of words. We also form a set S'' of $n^{1/4}$ words for $S'(k)$. The i-th word W_i in S'' has $p/(8k)$ integers $s_{i_0}, s_{i_1}, ..., s_{i_{p/(8k)-1}}$ in $S'(k)$ packed in it. Here $s_{i_0}, s_{i_1}, ..., s_{i_{p/(8k)-1}}$ must satisfy $H(s_{i_0})H(s_{i_1})...H(s_{i_{p/(8k)-1}}) = i$ (here $H(s_{i_0})H(s_{i_1})...H(s_{i_{p/(8k)-1}})$ is (the value of) the concatenation of the bits in $H(s_{i_j})$'s). Note that because we hash m integers in M to m integers in $M' = \{0, 1, ..., m^2\}$ there are some integers $a' \in M'$ such that no integer a in M is hashed to a'. Thus for some i and j we find that s_{i_j} does not exist in $S'(k)$. If this happens we leave the spot for $s_{i,j}$ in W_i blank, i.e. in this spot no integer in $S'(k)$ is packed into W_i. Note that hashing for integers in T' happens after packing them into words (due to time consideration) and hashing for integers in $S'(k)$ happens before packing as we have to satisfy $H(s_{i_0})H(s_{i_1})...H(s_{i_{p/(8k)-1}}) = i$.

   4.2. Since each word of T'' has $p/(8k)$ integers in T' packed in it it has only a total of $\log n/4$ hashed bits because each integer has hashed to $2k \log n/p$ bits. The $n^{1/4}$ words in S'' for packed integers in $S'(k)$ also have $\log n/4$ hashed bits in each of them. We treat these $\log n/4$ bits as one INTEGER and sort T'' and S'' together using the INTEGER of them as the sorting key. This sorting can be done by bucket sort (because INTEGER is the sorting key) in linear time in terms of the number of words or in $O(|T'|k/p)$ time (because we packed $p/(8k)$ integers in one word). This sorting will bring all words with same INTEGER key together. Because S'' has word with INTEGER keys with any value in $\{0, 1, ..., n^{1/4}\}$ any word w in T'' can find w_1 in S'' with the same INTEGER key value and thus w will be sorted together with w_1. Thus now each $c(t_i)$ with $t_i \in T'$ can find that either there is no $c(s_j)$ with $s_j \in S'(k)$ such that $H(c(t_i)) = H(c(s_j))$ (in this case $c(t_i) \neq c(s_j)$ for any $s_j \in S'(k)$) or there is an s_j such that $H(c(t_i)) = H(c(s_j))$. If $H(c(t_i)) = H(c(s_j))$ then we compare $c(t_i)$ with $c(s_j)$ (we need to do this comparison because H is not perfect for $T' \cup S'(k)$). If they are equal (match) then if the current segment is not the least significant segment then we eliminate the current segment of t_i (in this case we find all segments that are not less significant than the current segment of t_i are equal to the corresponding segments of s_j) and mark the next $\log n$ bits as its current segment. Also eliminate the current segment of s_j and use the next $\log n$ bits of s_j as its current segment (the idea here is similar to the mechanism in the previous section where we eliminates the preceding segments of t_i and s_j because they are equal). If the current segment before removing is the least significant segment then $t_i = s_j$ and we are done with t_i. Let D be the set of integers t_i in T' such that $c(t_i) = c(s_j)$ for some $s_j \in S'(k)$. Advance all integers in D $k/2$ steps (because they matched in $S'(k)$ and did not match in $S'(k/2)$) and let $T' = T' - D$, $T^{(m+k/2)} = T^{(m+k/2)} \cup D$. For integers in D we can also determine the $S[m + k/2, l]$ on which t_i falls into. For integers in D remove the current segment of these integers. For integers in D that has been distributed into $F(S[m + k/2, l])$'s we have to prepend $R(s_{j,prefix})$ to $c(t_i)$. T' is now the set of unmatched integers.

Note here we have to separate integers in D from integers not in D. For integers in D we have to further advance them later according to which $S[m + k/2, l]$ they fall into. This is done by labeling integers not in D with 0 and integers fall into $S[m + k/2, l]$ with $l + 1$. The labels form an integer LABEL (just as we formed INTEGER before) for integers packed in a word. We then use bucket sort to sort all words using LABEL as the key. This will bring all words with the same LABEL value together. For every p words with the same LABEL value we do a transpose (i.e. move the i-th integer (LABEL) in these words into a new word). This transposition would take $O(p \log p)$ time for p words if we do it directly. However we can first pack $\log p$ words into one word (because word size is $p^3 \log n$) and then do the transposition. This will bring the time for transposition to $O(p)$ for p words. This will have integers separated.

4.3. If $|T'| < n^{1/2}$ then use brute force to sort all integers in T' and mark them as done (advanced to step $p/8$). Goto Step 2.

4.4. If $m + k = p/8$ then mark all integers in T' as done (advanced to step $p/8$) keep the current segment of them and remove all other segments of them. Let $E(m, m_1) = T'$. Goto Step 2.

4.5. If $p/8 < m + 2k$ let $k = p/8 - m$ else let $k = 2k$ and goto Step 4.

5. We have to sort integers (of $2 \log n$ bits) in every nonempty set $E(m, m_1)$. Here each integer in $E(m, m_1)$ has its current segment differ from the current segment of any $s \in S[m, m_1]$. We mix them together and sort them by radix sort. Then we go from the smallest integer to the largest integer one by one and separate them by their index (m, m_1). This will have every set $E(m, m_1)$ sorted. This case corresponds to unmatch for all passes and we end up with $2 \log n$ bits for each of these integers.

5   Ordered Partition

We have explained in Sect. 2 how the ordered partition is done using the results in Sects. 3 and 4 (i.e. how ordered partition is done by ordered partition of T by S). Therefore we have our main theorem:

Main Theorem: Ordered partition of a set T of n integers into $n^{1/2}$ sets $T_0 \leq T_1 \leq \cdots \leq T_{n^{1/2}-1}$, where $|T_i| = \theta(n^{1/2})$ and $T_i \leq T_{i+1}$ means that $\max T_i \leq \min T_{i+1}$, can be computed in linear time and space.   □

6   Randomization and Nonconservativeness

Although our ordered partition algorithm is deterministic and conservative (i.e. use word of size $\log m + \log n$ for integers in $\{0, 1, ..., m-1\}$), it may be used in the randomized setting or the nonconservative setting. Here we consider the situation under these settings.

In randomized setting n integers with word size $\Omega(\log^2 n \log \log n)$ can be sorted in linear time [3] and thus ordered partition is not needed. Han and Thorup have presented a randomized integer sorting algorithm with time $O(n(\log \log n)^{1/2})$ [11] and it represents a tradeoff between ordered partition and sorting.

In [7] the schemes in this paper is extended and combined with other ideas to obtain a randomized linear time algorithm for sorting integers with word size $\Omega(\log^2 n)$, i.e. integers larger than $2^{\log^2 n}$, thus improve the results in [3].

7   Conclusion

We hope that our ordered partition algorithm will help in the search for a linear time algorithm for integer sorting. We tend to believe that integers can be sorted in linear time, as least in the randomized setting. There are still obstacles to overcome before we can achieve this goal.

Acknowledgement. Reviewers have given us very helpful comments and suggestions which helped us improve the presentation of this paper significantly. We very much appreciate their careful reviewing work.

References

1. Andersson, A., Hagerup, T., Nilsson, S., Raman, R.: Sorting in linear time? J. Comput. Syst. Sci. **57**, 74–93 (1998)
2. Andersson, A.: Faster deterministic sorting and searching in linear space. In: Proceedings of 1996 IEEE Symposium of Foundations of Computer Science FOCS 1996, pp. 135–141 (1996)
3. Belazzougui, D., Brodal, G.S., Nielsen, J.S.: Expected linear time sorting for word size $\Omega(\log^2 n \ \log \log n)$. In: Ravi, R., Gørtz, I.L. (eds.) SWAT 2014. LNCS, vol. 8503, pp. 26–37. Springer, Heidelberg (2014)
4. Blum, M., Floyd, R.W., Pratt, V.R., Rivest, R.L., Tarjan, R.E.: Time bounds for selection. J. Comput. Syst. Sci. **7**(4), 448–461 (1972)
5. Dietzfelbinger, M., Hagerup, T., Katajainen, J., Penttonen, M.: A reliable randomized algorithm for the closest-pair problem. J. Algorithms **25**, 19–51 (1997)

6. Han, Y.: Deterministic sorting in $O(n \log \log n)$ time and linear space. J. Algorithms **50**, 96–105 (2004)

7. Han, Y.: Optimal randomized integer sorting for integers of word size $\Omega(\log^2 n)$. Manuscript

8. Han, Y.: Construct a perfect hash function in time independent of the size of integers. To appear in FCS 2015

9. Han, Y., Shen, X.: Conservative algorithms for parallel and sequential integer sorting. In: Li, M., Du, D.-Z. (eds.) COCOON 1995. LNCS, vol. 959, pp. 324–333. Springer, Heidelberg (1995)

10. Han, Y., Shen, X.: Parallel integer sorting is more efficient than parallel comparison sorting on exclusive write PRAMs. SIAM J. Comput. **31**(6), 1852–1878 (2002)

11. Han, Y., Thorup, M.: Integer sorting in $O(n\sqrt{\log \log n})$ expected time and linear space, In: Proceedings of the 43rd IEEE Symposium on Foundations of Computer Science (FOCS 2002), pp. 135–144 (2002)

12. Kirkpatrick, D., Reisch, S.: Upper bounds for sorting integers on random access machines. Theor. Comput. Sci. **28**, 263–276 (1984)

13. Raman, R.: Priority queues: small, monotone and trans-dichotomous. In: Díaz, J. (ed.) ESA 1996. LNCS, vol. 1136, pp. 121–137. Springer, Heidelberg (1996)

Machine Scheduling with a Maintenance Interval and Job Delivery Coordination

Jueliang Hu[1], Taibo Luo[2,3], Xiaotong Su[1], Jianming Dong[1], Weitian Tong[3], Randy Goebel[3], Yinfeng Xu[2,4], and Guohui Lin[1,3(✉)]

[1] Department of Mathematics, Zhejiang Sci-Tech University, Hangzhou 310018, Zhejiang, China
[2] Business School, Sichuan University, Chengdu 610065, Sichuan, China
[3] Department of Computing Science, University of Alberta, Edmonton, AB T6G 2E8, Canada
guohui@ualberta.ca
[4] State Key Lab for Manufacturing Systems Engineering, Xi'an 710049, Shaanxi, China

Abstract. We investigate a scheduling problem with job delivery coordination in which the machine has a maintenance time interval. The goal is to minimize the makespan. In the problem, each job needs to be processed on the machine non-preemptively for a certain time, and then transported to a distribution center; transportation is by one vehicle with a limited physical capacity, and it takes constant time to deliver a shipment to the distribution center and return back to the machine. We present a 2-approximation algorithm for the problem, and show that the performance ratio is tight.

Keywords: Scheduling · Machine maintenance · Job delivery · Bin-packing · Approximation algorithm · Worst-case performance analysis

1 Introduction

We consider a scheduling problem that arises from supply chain management research at the operational level, with the goal to show that decision makers at different stages of a supply chain can make coordinated decisions at the detailed scheduling level, and achieve substantial efficiencies. This problem integrates production and delivery whereby the jobs are first processed in a manufacturing center, and then delivered to a distribution center. We use a machine to model the manufacturing center, which has a preventive maintenance time interval when it is unavailable for processing any jobs. Job delivery is performed by a single vehicle with a limited physical load capacity, between the manufacturing center and the distribution center. The goal is to minimize the makespan. A special case of this problem was first considered by Wang and Cheng [9], in which the jobs have uniform size.

J. Hu and T. Luo—Co-first authors.

© Springer International Publishing Switzerland 2015
J. Wang and C. Yap (Eds.): FAW 2015, LNCS 9130, pp. 104–114, 2015.
DOI: 10.1007/978-3-319-19647-3_10

Our target scheduling problem is formally described as follows. We are given a set of jobs $\mathcal{J} = \{J_1, J_2, \ldots, J_n\}$, each of which needs to be processed in a manufacturing center (the machine) and then delivered to a distribution center (the customer). Each job J_i requires a non-preemptive processing time of p_i in the manufacturing center; when transported by the only vehicle to the distribution center, it occupies a fraction s_i of physical space on the vehicle. The vehicle has a normalized space capacity of 1, is initially at the manufacturing center, and needs to return to the manufacturing center after all jobs are delivered. It takes the vehicle T units of time to deliver a shipment and return back to the manufacturing center. The manufacturing center, modeled as a single machine, has a known maintenance time interval $[s, t]$, where $0 \leq s \leq t$, during which no jobs can be processed. The problem objective is to minimize the makespan, that is, the time the vehicle returning to the manufacturing center after all jobs are delivered.

Using the notation of Lee et al. [7] and following Wang and Cheng [9], the problem under study is denoted as $(1, h(1) \mid non\text{-}pmtn, D, s_i \mid C_{\max})$. In this three-field notation, the first field denotes the machine environment, the second denotes the job characteristics, and the last denotes the performance measure to be optimized. In our case, "1" says that there is only a single machine to process the jobs and "$h(1)$" indicates that there is a hole (*i.e.* a maintenance interval) in the machine, "*non-pmtn*" states that each job needs a continuous processing or, if interrupted by the unavailable machine maintenance interval, it has to restart the processing after the machine becomes available,[1] "D" indicates the delivery requirement that jobs must be delivered to the distribution center after the processing is completed in the manufacturing center, "s_i" is the normalized physical size of job J_i on the single vehicle, and lastly, "$C_{\max}$" denotes the makespan, which is the time the vehicle returning to the manufacturing center after all jobs are delivered.

For the special case where all jobs have the same size, that is $s_i = \frac{1}{K}$ for some positive integer K, Wang and Cheng showed that $(1, h(1) \mid non\text{-}pmtn, D, s_i = \frac{1}{K} \mid C_{\max})$ is NP-hard, and presented a $\frac{3}{2}$-approximation algorithm based on the *shortest processing time* (SPT) rule [9]. Essentially, the SPT rule sorts the jobs into a non-decreasing order of the processing time and the machine processes the jobs in this order. The intuition is to let the machine finish processing as many jobs as possible at any given time point, to optimally supply the transportation vehicle.

While the SPT rule alone works well in this special uniform-size case, it can be very bad in the general case where the jobs have different sizes. Indeed, for another extremely special case where all jobs have zero processing time, the problem $(1, h(1) \mid non\text{-}pmtn, D, s_i \mid C_{\max})$ reduces to minimizing the number of shipments, or the classic *bin-packing* problem, which is NP-hard and APX-complete [2].

In this paper, we show that the *next-fit* (NF) algorithm [5] designed for the bin-packing problem can be employed for packing the jobs into a favorable

---

[1] In the literature, *non-resumable* (specified as "*nr-a*") has been used.

number of *batches*, where each batch is a shipment to be delivered by the single vehicle. This is followed by applying the SPT rule to sequence the batches with delivery coordination. We show that this algorithm has a worst-case performance guarantee of 2, and this ratio is tight. In the next section, we present the performance analysis in detail. We conclude the paper in the last section.

2   The Algorithm D-NF-SPT

In our target scheduling problem $(1, h(1) \mid non\text{-}pmtn, D, s_i \mid C_{\max})$ we assume the non-trivial case where $\sum_{i=1}^{n} p_i > s \geq \min_{i=1}^{n} p_i$, *i.e.* the machine maintenance interval does affect the schedule, since otherwise the problem reduces to the problem $(1 \mid D, s_i \mid C_{\max})$, which has been extensively investigated in the literature [1,4,6,8,10], and it admits a (best possible) 1.5-approximation algorithm [8]. In the 1.5-approximation algorithm, when the total size of the jobs is greater than 1 but less than or equal to 2, the jobs are packed by the NF algorithm; otherwise, the jobs are packed by the *modified first-fit decreasing* (MFFD) algorithm [3]. The resultant batches are then processed and delivered in the SPT order.

Recall that there are n jobs, and each job J_i, for $i = 1, 2, \ldots, n$, needs to be processed non-preemptively for p_i units of time on the machine, and then transported to the distribution center by a single vehicle. The machine has a known maintenance time interval $[s, t]$, during which no jobs can be processed. The job J_i has a physical size $s_i \in (0, 1]$, representing its fractional space requirement on the vehicle during the transportation. A shipment (*i.e.*, a batch, used interchangeably) can contain multiple jobs, as long as the total size of the jobs in the shipment is no greater than 1. The vehicle takes constant time T to deliver a shipment to the distribution center and return back to the machine. For ease of presentation, we use $\Delta = t - s$ to denote the length of the machine maintenance. As mentioned in the introduction, we have the following lemma due to the hardness results of the bin-packing problem.

Lemma 1. *The problem* $(1, h(1) \mid non\text{-}pmtn, D, s_i \mid C_{\max})$ *is NP-hard and APX-hard.*   □

Let π denote a feasible schedule, in which the jobs are transported in k shipments denoted as $B_1, B_2, \ldots, B_k$ in order. We extend the notation to use $p(B_j)$ ($s(B_j)$, respectively) to denote the total processing time (size, respectively) of the jobs of B_j, for every j. Note that in general the jobs of B_j are not necessarily processed before all the jobs of B_{j+1} on the machine. Let α denote the smallest batch index such that $\sum_{j=1}^{\alpha} p(B_j) > s$. Clearly, $1 \leq \alpha \leq k$. We can assume, without loss of generality, that for every $j < \alpha$ the jobs of B_j are processed before all the jobs of B_{j+1} on the machine, and furthermore they are processed on the machine consecutively in an arbitrary order. We use δ to denote the length of the machine idling period due to the pending maintenance. It follows that the machine finishes processing all the jobs at time $\sum_{j=1}^{k} p(B_j) + \Delta + \delta$. On the other hand, the total transportation time for this schedule is kT. We assume that the vehicle does not idle if there are shipments ready to be transported.

Our algorithm D-NF-SPT can be described as follows (see Fig. 1). First (the D-step), all jobs are sorted into a non-increasing order of the ratio $\frac{s_i}{p_i}$, which we also call the *density*. Next (the NF-step), in this order, the jobs are formed into shipments (batches) by their physical sizes using the next-fit (NF) bin-packing algorithm. The NF algorithm assigns the job at the head of the order to the last (largest indexed) shipment if the job fits in, or else to a newly created shipment for the job. This way, every shipment contains a number of consecutive jobs. The achieved batch sequence is denoted as $\langle B_1', B_2', \ldots, B_k' \rangle$. The processing times of the shipments are then calculated, and the shipments are sorted into a non-decreasing order of the processing time (the SPT-step). The final batch sequence is denoted as $\langle B_1, B_2, \ldots, B_k \rangle$. According to this shipment order, a maximum number of batches are processed before time s; the other batches are processed starting time t (when the maintenance ends). For each shipment, its jobs are processed consecutively on the machine in an arbitrary order; and a shipment is transported to the distribution center after all its jobs are finished and the vehicle is available. We denote the achieved schedule as π, that is $\pi = \langle B_1, B_2, \ldots, B_k \rangle$ with $p(B_1) \leq p(B_2) \leq \ldots \leq p(B_k)$. Let B_α denote the first shipment processed after time t;

$$\delta = s - \sum_{j=1}^{\alpha-1} p(B_j) \tag{1}$$

denotes the length of the machine idle time before the maintenance (see Fig. 2 for the configuration of π).

Algorithm D-NF-SPT:

Step 1. (The D-step) Sort the jobs into a non-increasing order of the ratio s_i/p_i;

Step 2. (The NF-step) Pack the jobs by size into a sequence of batches using the algorithm NF:

2.1. Place the current job into the last batch if it fits in;

2.2. Or else create a new batch for the current job;

2.3. The achieved batch sequence is denoted as $\langle B_1', B_2', \ldots, B_k' \rangle$;

Step 3. (The SPT-step) Sort the job batches into a non-decreasing order of the processing time:

3.1. The achieved batch sequence is denoted as $\langle B_1, B_2, \ldots, B_k \rangle$;

Step 4. Process the jobs in this batch order and deliver a finished batch as early as possible:

4.1. Let α denote the smallest batch index such that $\sum_{j=1}^{\alpha} p(B_j) > s$;

4.2. Batches $B_1, B_2, \ldots, B_{\alpha-1}$ are processed before time s;

4.3. Batches $B_\alpha, B_{\alpha+1}, \ldots, B_k$ are processed starting time t.

Fig. 1. A high-level description of the algorithm D-NF-SPT.

Fig. 2. A visual configuration of the schedule π produced by the D-NF-SPT algorithm.

We next prove some structural properties for the schedule π, and estimate its makespan denoted as $C_{\max}$. For ease of presentation, the finish processing time of the batch B_j on the machine is denoted as C_j, and let D_j denote the time at which the vehicle delivers the batch B_j to the distribution center and returns back to the machine. Clearly, $D_j - C_j \geq T$, for every j.

Lemma 2. *For the schedule π produced by the algorithm D-NF-SPT for the problem $(1, h(1) \mid non\text{-}pmtn, D, s_i \mid C_{\max})$, the makespan is*

$$C_{\max} = \begin{cases} \sum_{j=1}^{\alpha} p(B_j) + \Delta + \delta + (k - \alpha + 1)T, & \text{if } C_\alpha > D_{\alpha-1}, \ C_k < D_{k-1}; \\ p(B_1) + kT, & \text{if } C_\alpha \leq D_{\alpha-1}, \ C_k < D_{k-1}; \\ \sum_{j=1}^{k} p(B_j) + \Delta + \delta + T, & \text{if } C_k \geq D_{k-1}. \end{cases}$$

Proof. Recall that the machine processes the jobs of $B_1 \cup B_2 \cup \ldots \cup B_{\alpha-1}$ continuously before time s, and processes the jobs of $B_\alpha \cup B_{\alpha+1} \cup \ldots \cup B_k$ continuously after time t. Thus for the last job batch B_k, $C_k = \sum_{j=1}^{k} p(B_j) + \Delta + \delta$.

If the batch B_k has finished the processing while the vehicle is not ready for transporting it, *i.e.* $C_k < D_{k-1}$, we conclude that the vehicle does not idle during the time interval $[C_\alpha, C_k]$, where $C_\alpha = \sum_{j=1}^{\alpha} p(B_j) + \Delta + \delta$. This can be proven by a simple contradiction, as otherwise there would be a batch B_j for some $j > \alpha$, such that $C_j > D_{j-1}$. Then clearly $p(B_j) = C_j - C_{j-1} > D_{j-1} - C_{j-1} \geq T$. It follows that all the succeeding batches have a processing time greater than T. This indicates that for every successive batch, including B_k, the vehicle has to idle for a while before delivering it.

Using the same argument, if the vehicle idles inside the time interval $[C_1, C_\alpha]$ (note that the vehicle has to wait for the first batch B_1 to finish), then there must be $C_\alpha > D_{\alpha-1}$ and thus the vehicle must have delivered all the batches $B_1, B_2, \ldots, B_{\alpha-1}$ at time C_α. In this case, the makespan is $C_{\max} = \sum_{j=1}^{\alpha} p(B_j) + \Delta + \delta + (k - \alpha + 1)T$. If the vehicle does not idle before time C_α, that is, $\sum_{j=2}^{\alpha} p(B_j) + \Delta + \delta \leq (\alpha - 1)T$, then the makespan is $C_{\max} = p(B_1) + kT$.

If the job batch B_k has finished the processing and the vehicle is ready for transporting it, *i.e.* $C_k \geq D_{k-1}$, then the makespan is the finishing time of the batch B_k plus one shipment delivery time of the vehicle, which is $C_{\max} = \sum_{j=1}^{k} p(B_j) + \Delta + \delta + T$. $\square$

From the proof of Lemma 2, we have the following corollary.

Corollary 1. *For the schedule π produced by the algorithm D-NF-SPT for the problem $(1, h(1) \mid non\text{-}pmtn, D, s_i \mid C_{\max})$, the vehicle idles inside the time interval $[C_1, C_\alpha]$ if and only if $C_\alpha > D_{\alpha-1}$.* $\square$

Consider the associated instance I of the bin-packing problem to pack all the jobs of $\mathcal{J} = \{J_1, J_2, \ldots, J_n\}$ by their size into the minimum number of batches (of capacity 1); let k^o denote this minimum number of batches. It is known that $k \leq 2k^o - 1$ [5], where k is the number of batches by the algorithm NF. The algorithm NF is one of the simplest approximation algorithms designed for the bin-packing problem, but not the best in terms of approximation ratio. Nevertheless, there are important properties of the packing result achieved by the algorithm NF, stated in the next two lemmas.

Lemma 3. *Consider the job batch sequence $\langle B'_1, B'_2, \ldots, B'_k \rangle$ produced by the algorithm D-NF-SPT in Step 2. Let $\mathcal{J}'$ be any subset of jobs, and assume all its jobs can be packed into k' batches. For any k_1, if $\sum_{j=1}^{k_1} s(B'_j) \leq s(\mathcal{J}')$, then $k_1 \leq 2k' - 1$.*

Proof. From the execution of the NF algorithm, we know that every two adjacent batches, B'_j and B'_{j+1}, have a total size strictly greater than 1. If k_1 is odd, then $\sum_{j=1}^{k_1} s(B'_j) > \frac{k_1-1}{2}$; otherwise, $\sum_{j=1}^{k_1} s(B'_j) > \frac{k_1}{2}$. On the other hand, every one of the k' batches has size at most 1, and thus $s(\mathcal{J}') \leq k'$. Putting together, we have

$$k' \geq s(\mathcal{J}') \geq \sum_{j=1}^{k_1} s(B'_j) > \frac{k_1 - 1}{2}.$$

That is, $k' > \frac{k_1-1}{2} + 1 = \frac{k_1+1}{2}$. This proves the lemma. □

Lemma 4. *Consider the job batch sequence $\langle B'_1, B'_2, \ldots, B'_k \rangle$ produced by the algorithm D-NF-SPT in Step 2. Let $\mathcal{J}'$ be any subset of jobs. For any k_1, if $\sum_{j=1}^{k_1} p(B'_j) > p(\mathcal{J}')$, then $\sum_{j=1}^{k_1} s(B'_j) > s(\mathcal{J}')$.*

Proof. Recall that in Step 1 of the algorithm D-NF-SPT, all the jobs of $\mathcal{J}$ are sorted by non-increasing density $\frac{s_i}{p_i}$. Assume to the contrary that $\sum_{j=1}^{k_1} p(B'_j) > p(\mathcal{J}')$ and $\sum_{j=1}^{k_1} s(B'_j) \leq s(\mathcal{J}')$. There must exist at least one job $J \in \mathcal{J}'$ but not in the first k_1 batches $B'_1, B'_2, \ldots, B'_{k_1}$, such that

$$\frac{s(J)}{p(J)} > \frac{\sum_{j=1}^{k_1} s(B'_j)}{\sum_{j=1}^{k_1} p(B'_j)} \geq \min_{J_i \in B'_{k_1}} \left\{ \frac{s_i}{p_i} \right\}.$$

However, this is a contradiction since such a job J must have been in one of the first k_1 batches $B'_1, B'_2, \ldots, B'_{k_1}$, from the execution of the NF algorithm. This proves the lemma. □

Let π^* denote an optimal schedule, in which there are k^* job batches B_1^*, B_2^*, $\ldots, B_{k^*}^*$, when finished, delivered in this order. We assume that the batch $B_{\alpha^*}^*$ is the first one in this order containing a job processed after time t; and use δ^* to denote the length of the machine idle time before the maintenance. It is important to note that we may assume without loss of generality that the jobs of B_j^*, for each $j < \alpha^*$, are processed continuously (in an arbitrary order), but no specific processing order for the jobs of $B_{\alpha^*}^*, B_{\alpha^*+1}^*, \ldots, B_{k^*}^*$.

The makespan of the optimal schedule π^* is denoted as $C^*_{\max}$. Again for ease of presentation, the finish processing time of the batch B^*_j on the machine is denoted as C^*_j, and let D^*_j denote the time at which the vehicle delivers the batch B^*_j to the distribution center and returns back to the machine. Clearly, $D^*_j - C^*_j \geq T$, for every j.

Lemma 5. *For the optimal schedule π^* for the problem $(1, h(1) \mid non\text{-}pmtn, D, s_i \mid C_{\max})$, the makespan is*

$$C^*_{\max} \geq \max \left\{ p(B^*_1) + k^*T, \ \sum_{j=1}^{k^*} p(B^*_j) + \Delta + \delta^* + T \right\}.$$

Proof. Since after the first batch B^*_1 is processed on the machine, the vehicle needs to deliver all the k^* batches; thus the makespan is at least $p(B^*_1) + k^*T$.

On the other hand, the finish processing time of the last batch $B^*_{k^*}$ is $C^*_{k^*} = \sum_{j=1}^{k^*} p(B^*_j) + \Delta + \delta^*$, and afterwards it has to be delivered; hence the makespan is at least $\sum_{j=1}^{k^*} p(B^*_j) + \Delta + \delta^* + T$. This completes the proof. □

Now we are ready to prove the main theorem.

Theorem 1. *The algorithm D-NF-SPT is an $O(n \log n)$-time 2-approximation for the problem $(1, h(1) \mid non\text{-}pmtn, D, s_i \mid C_{\max})$.*

Proof. First, if $\sum_{i=1}^{n} p_i \leq s$, i.e. all the jobs can be processed before the machine maintenance, the target problem reduces to the problem $(1 \mid D, s_i \mid C_{\max})$, which admits a 1.5-approximation algorithm [8]. We thus assume in the following that $\sum_{i=1}^{n} p_i > s$. Consequently, $1 \leq \alpha \leq k$ and $1 \leq \alpha^* \leq k^*$ (these four quantities are all well defined).

If in the schedule π produced by the algorithm D-NF-SPT, $C_k \geq D_{k-1}$, then by Lemma 2 the makespan is $C_{\max} = \sum_{j=1}^{k} p(B_j) + \Delta + \delta + T$. On the other hand, from Lemma 5 we have $C^*_{\max} \geq \sum_{j=1}^{k^*} p(B^*_j) + \Delta + \delta^* + T$. Clearly, $\delta \leq p(B_\alpha) \leq \sum_{j=1}^{k} p(B_j) = \sum_{j=1}^{k^*} p(B^*_j)$. It follows that

$$C_{\max} = \sum_{j=1}^{k} p(B_j) + \Delta + \delta + T \leq 2 \sum_{j=1}^{k^*} p(B^*_j) + \Delta + T \leq 2C^*_{\max}.$$

That is, the makespan of the schedule π is no more than twice of the optimum.

If $k^* = 1$ in the optimal schedule π^*, that is, all the jobs can form into a single batch, then we also have $k = 1$ in the schedule π, and consequently $C_{\max} = p(B_1) + \Delta + \delta + T$. As in the previous paragraph the makespan of the schedule π is no more than twice that of the optimum.

In the following we consider $C_k < D_{k-1}$, $k \geq 2$ and $k^* \geq 2$, and we separate the discussion into two cases. Note that in the following $D_0 = 0$, meaning at the beginning the vehicle is ready.

Case 1. $C_\alpha \le D_{\alpha-1}$.

From Corollary 1 and Lemma 2, we know that in the schedule π the vehicle does not idle inside the time interval $[C_1, C_\alpha]$ and the makespan is $C_{\max} = p(B_1) + kT$.

By letting $\mathcal{J}'$ be the whole set $\mathcal{J}$ of jobs in Lemma 3, we have $k \le 2k' - 1 \le 2k^* - 1$, since k' is the minimum number of batches for all the jobs of $\mathcal{J}$.

One can check that for every possible value of α, we always have $C_2 \le D_1$ because there is no vehicle idling inside the time interval $[C_1, C_\alpha]$, and thus $p(B_1) \le p(B_2) = C_2 - C_1 \le D_1 - C_1 = T$. It follows from Lemma 5 that

$$C_{\max} = p(B_1) + kT \le T + (2k^* - 1)T = 2k^*T \le 2C_{\max}^*.$$

Case 2. $C_\alpha > D_{\alpha-1}$.

Note that we have $C_k < D_{k-1}$, and thus $\alpha \le k - 1$. From Corollary 1 and Lemma 2, we know that in the schedule π, the vehicle idles inside the time interval $[C_1, C_\alpha]$ and the makespan is $C_{\max} = \sum_{j=1}^{\alpha} p(B_j) + \Delta + \delta + (k - \alpha + 1)T$.

If $\alpha > 2$ and the vehicle idles inside the time interval $[C_1, C_{\alpha-1}]$, then $p(B_{\alpha-1}) > T$. Consequently, $p(B_k) > T$ too, which contradicts $C_k < D_{k-1}$. In the remaining situation, either $\alpha = 1$ (*i.e.*, no jobs processed before the machine maintenance), or $2 \le \alpha \le k - 1$ and the vehicle idles only inside the time interval $[C_{\alpha-1}, C_\alpha]$. Thus we always have $p(B_\alpha) \le p(B_{\alpha+1}) \le T$ (again, as otherwise $C_k > D_{k-1}$, a contradiction).

Subcase 2.1. $\alpha^* = 1$. In this subcase, all the batches are finished after time t, and thus $C_{\max}^* \ge t + k^*T$. It follows from $p(B_\alpha) \le T$ and $k \le 2k^* - 1$ [5] that

$$C_{\max} = t + p(B_\alpha) + (k - \alpha + 1)T \le t + T + 2k^*T - \alpha T = t + 2k^*T \le 2C_{\max}^*.$$

Subcase 2.2. $\alpha^* \ge 2$. In this subcase, $C_{\max}^* \ge \max\{t + (k^* - \alpha^* + 1)T, p(B_1^*) + k^*T\}$.

Let $\mathcal{J}'$ denote the subset of jobs that are processed before time s in the optimal schedule π^*, and $\mathcal{J}'' = \mathcal{J} - \mathcal{J}'$. Clearly, $\sum_{j=1}^{\alpha} p(B_j) > p(\mathcal{J}')$ since not all the jobs of B_α can be processed before time s. On the other hand, the batch sequence $\langle B_1, B_2, \ldots, B_k \rangle$ is the rearrangement of the batch sequence $\langle B_1', B_2', \ldots, B_k' \rangle$ in the SPT order; therefore, $\sum_{j=1}^{\alpha} p(B_j) \le \sum_{j=1}^{\alpha} p(B_j')$. It follows that

$$\sum_{j=1}^{\alpha} p(B_j') > p(\mathcal{J}').$$

By Lemma 4 we have

$$\sum_{j=1}^{\alpha} s(B_j') > s(\mathcal{J}'),$$

and thus

$$\sum_{j=\alpha+1}^{k} s(B_j') < s(\mathcal{J}'').$$

From Lemma 3 and that the jobs of $\mathcal{J}''$ are in $k^* - \alpha^* + 1$ batches, we conclude that

$$k - \alpha \leq 2(k^* - \alpha^* + 1) - 1.$$

It follows from $p(B_\alpha) \leq T$ that

$$
\begin{aligned}
C_{\max} &= t + p(B_\alpha) + (k - \alpha + 1)T \\
&\leq t + T + 2(k^* - \alpha^* + 1)T \\
&= (t + (k^* - \alpha^* + 1)T) + (k^* - \alpha^* + 2)T \\
&\leq 2C_{\max}^*.
\end{aligned}
$$

That is, in the remaining situation we also have $C_{\max} \leq 2C_{\max}^*$. Hence the algorithm D-NF-SPT is a 2-approximation.

The running time of the algorithm D-NF-SPT in $O(n \log n)$, where n is the number of jobs, is clearly seen, because the job sorting by density and the later job batch sorting by processing time take an $O(n \log n)$-time, and the algorithm NF takes only an $O(n)$-time. This proves the theorem. □

2.1 A Tight Instance

In this instance I there are $2n$ jobs, $\mathcal{J} = \{J_1, J_2, \ldots, J_{2n}\}$, with n being even. The processing time and the size of the job J_i is (p_i, s_i), and here $J_{2i-1} = (i\epsilon, \frac{1}{2})$ and $J_{2i} = ((2i + 1)\epsilon^2, \epsilon)$ for every $i = 1, 2, \ldots, n$. The positive constant ϵ is small such that $\epsilon < \frac{1}{2n+1}$. The machine maintenance time interval is $[s, t]$ where $s = \frac{1}{2}n(n - 1)\epsilon + (n - 1)(n + 1)\epsilon^2 + \frac{1}{2}(2n - 1)\epsilon^2$ and $t = s + \frac{1}{2}(2n - 1)\epsilon^2$, i.e. $\Delta = \frac{1}{2}(2n - 1)\epsilon^2$. The one shipment delivery time is $T = 1$.

Clearly, $\frac{s_i}{p_i} = \frac{1}{(i+1)\epsilon}$ for every $i = 1, 2, \ldots, 2n$ and therefore the job order after Step 1 of the algorithm D-NF-SPT is $\langle J_1, J_2, \ldots, J_{2n} \rangle$. Using this job order, the algorithm NF packs the jobs into a sequence of n batches $B_j' = \{J_{2j-1}, J_{2j}\}$, $j = 1, 2, \ldots, n$. Clearly, $s(B_j') = \frac{1}{2} + \epsilon$ for all j, and $p(B_j') = j\epsilon + (2j + 1)\epsilon^2$. Therefore, $B_j = B_j'$ for every j, and the final batch order is $\langle B_1, B_2, \ldots, B_n \rangle$. Note that

$$s = \frac{1}{2}n(n - 1)\epsilon + (n - 1)(n + 1)\epsilon^2 + \frac{1}{2}(2n - 1)\epsilon^2 = \sum_{j=1}^{n-1} p(B_j) + \frac{1}{2}(2n - 1)\epsilon^2.$$

Since $p(B_j) = j\epsilon + (2j + 1)\epsilon^2 < T$ for every j, $C_{n-1} < s < t < C_{n-1} + p(B_n)$ and $(2n - 1)\epsilon^2 + p(B_n) < T$, for the achieved schedule π its makespan is

$$C_{\max} = p(B_1) + nT = \epsilon + 3\epsilon^2 + nT. \tag{2}$$

Consider a feasible schedule in which there are $\frac{n}{2} + 1$ batches where $B_j^* = \{J_{4j-3}, J_{4j-1}\}$ for each $j = 1, 2, \ldots, \frac{n}{2}$, and $B_{\frac{n}{2}+1}^* = \{J_2, J_4, \ldots, J_{2n}\}$. Clearly, $s(B_j^*) = 1$ for all $1 \leq j \leq \frac{n}{2}$, $s(B_{\frac{n}{2}+1}^*) = n\epsilon$, $p(B_j^*) = (4j - 1)\epsilon$ for all $1 \leq j \leq \frac{n}{2}$, and $p(B_{\frac{n}{2}+1}^*) = n(n + 2)\epsilon^2$. Since $\sum_{j=1}^{\frac{n}{2}-1} p(B_j^*) < s$, all the jobs of the batches

$B_1^*, B_2^*, \ldots, B_{\frac{n}{2}-1}^*$ are processed before time s in this feasible schedule. Due to $\sum_{j=1}^{\frac{n}{2}+1} p(B_j^*) + (2n-1)\epsilon^2 < \frac{n}{2} - 1$, no matter when the jobs of the batches $B_{\frac{n}{2}}^*$ and $B_{\frac{n}{2}+1}^*$ are processed, the vehicle has not delivered the batch $B_{\frac{n}{2}-1}^*$ and comes back to the machine. It follows that the makespan of this feasible schedule is at most $p(B_1^*) + (\frac{n}{2} + 1)T = 3\epsilon + (\frac{n}{2} + 1)T$. Therefore, the makespan of an optimal schedule for the instance I is also

$$C_{\max}^* \leq 3\epsilon + \left(\frac{n}{2} + 1\right) T. \qquad (3)$$

Consequently, putting Eqs. (2) and (3) together gives

$$\frac{C_{\max}}{C_{\max}^*} \geq \frac{\epsilon + 3\epsilon^2 + nT}{3\epsilon + (\frac{n}{2} + 1)T} \rightarrow 2, \text{ when } n \rightarrow +\infty.$$

3 Conclusions

We have investigated the single scheduling problem with job delivery coordination, in which the machine has an unavailable maintenance interval. A good schedule needs not only to well organize jobs into a smaller number of shipments to save delivery time, but also must wisely exploit the machine time period before maintenance. The first consideration is addressed by employing a good approximation algorithm for the bin-packing problem, where the item size is the job physical size and the bin has a size that is the vehicle capacity. Nevertheless, the second consideration implies that the machine should perhaps process first those *jobs* of shorter processing times. We realized that this is *not the same as* the machine processing first those *batches* of shorter processing time. We thus propose to sort the jobs in a non-increasing order of the density s_i/p_i, and call the *next-fit* (NF) bin-packing algorithm to pack the jobs into batches. Two key properties of the packing results achieved by the algorithm NF lead to the desired performance analysis.

We also showed that the worst-case performance ratio 2 of the algorithm D-NF-SPT is tight. It would be really interesting to see whether the problem admits a better approximation algorithm, for example, by distinguishing the machine availability before and after the maintenance. From the practical point of view, it is worth investigating the problem where the machine has multiple maintenance periods, whether they occur on a regular basis, or irregularly but known in advance.

Acknowledgements. Hu is supported by the National Natural Science Foundation of China (NNSF) Grants No. 11271324 and 11471286. Luo is supported by a sabbatical research grant of Lin; his work was mostly done during his visit to the UofA. Dong is supported by the Zhejiang Provincial Natural Science Foundation Grants No. LY13A010015 and the Science Foundation of Zhejiang Sci-Tech University (ZSTU) Grants No. 13062171-Y. Tong is supported by an Alberta Innovates Technology Futures (AITF) Graduate Student Scholarship. Goebel is supported by AITF and the Natural

Sciences and Engineering Research Council of Canada (NSERC). Xu is supported by the NNSF Grants No. 61221063 and 71371129, and the Program for Changjiang Scholars and Innovative Research Team in University Grant No. IRT1173. Lin is supported by NSERC and the Science Foundation of Zhejiang Sci-Tech University (ZSTU) Grants No. 14062170-Y; his work was mostly done during his sabbatical leave at the ZSTU.

References

1. Chang, Y.-C., Lee, C.-Y.: Machine scheduling with job delivery coordination. Eur. J. Oper. Res. **158**, 470–487 (2004)
2. Garey, M.R., Johnson, D.S.: Computers and Intractability: A Guide to the Theory of NP-Completeness. W. H. Freeman and Company, San Francisco (1979)
3. Garey, M.R., Johnson, D.S.: A 71/60 theorem for bin-packing. J. Complex. **1**, 65–106 (1985)
4. He, Y., Zhong, W., Gu, H.: Improved algorithms for two single machine scheduling problems. Theoret. Comput. Sci. **363**, 257–265 (2006)
5. Johnson, D.S.: Near-optimal allocation algorithms. Ph.D. thesis, Massachusetts Institute of Technology (1973)
6. Lee, C.-Y., Chen, Z.-L.: Machine scheduling with transportation considerations. J. Sched. **4**, 3–24 (2001)
7. Lee, C.-Y., Lei, L., Pinedo, M.: Current trends in deterministic scheduling. Ann. Oper. Res. **70**, 1–41 (1997)
8. Lu, L., Yuan, J.: Single machine scheduling with job delivery to minimize makespan. Asia-Pac. J. Oper. Res. **25**, 1–10 (2008)
9. Wang, X., Cheng, T.C.E.: Machine scheduling with an availability constraint and job delivery coordination. Naval Res. Logistics **54**, 11–20 (2007)
10. Zhong, W., Dósa, G., Tan, Z.: On the machine scheduling problem with job delivery coordination. Eur. J. Oper. Res. **182**, 1057–1072 (2007)

Lower and Upper Bounds for Random Mimimum Satisfiability Problem

Ping Huang[1][(⊠)] and Kaile Su[2]

[1] Key Laboratory of High Confidence Software Technologies, Ministry of Education,
School of Electronic Engineering and Computer Science,
Institute of Software, Peking University, Beijing 100871, China
huangping@pku.edu.cn
[2] Institute for Integrated and Intelligent Systems,
Griffith University, Nathan, Australia
k.su@griffith.edu.au

Abstract. Given a Boolean formula in conjunctive normal form with n variables and $m = rn$ clauses, if there exists a truth assignment satisfying $(1 - 2^{-k} - q(1 - 2^{-k}))m$ clauses, call the formula q-satisfiable. The Minimum Satisfiability Problem (MinSAT) is a special case of q-satisfiable, which asks for an assignment to minimize the number of satisfied clauses. When each clause contains k literals, it is called Min$k$SAT. If each clause is independently and randomly selected from all possible clauses over the n variables, it is called random MinSAT. In this paper, we give upper and lower bounds of r (the ratio of clauses to variables) for random k-CNF formula with q-satisfiable. The upper bound is proved by the first moment argument, while the proof of lower bound is the second moment with weighted scheme. Interestingly, our experimental results about MinSAT demonstrate that the lower and upper bounds are very tight. Moreover, these results give a partial explanation for the excellent performance of MinSatz, the state-of-the-art MinSAT solver, from the perspective of pruning effects. Finally, we give a conjecture about the relationship between the minimum number and the maximum number of satisfied clauses on random SAT instances.

Keywords: MinSAT · Upper bounds · Lower bounds

1 Introduction

Given a Boolean formula in conjunctive normal form(CNF), the satisfiability problem (SAT), which is a prototype of many NP-complete problems, asks for the existence of a satisfying assignment to the formula. In the past decades, SAT has been one of the most active and prolific research areas. Many problems, such as planning [21] and Pseudo-Boolean Constraints, can be translated into SAT [10]. Recently, the success of SAT research has led to exploring its optimization formalisms, such as the maximum satisfiability problem (MaxSAT) [4–6,8,12,15, 19,23] and the minimum satisfiability problem (MinSAT) [14,16,17]. MaxSAT

© Springer International Publishing Switzerland 2015
J. Wang and C. Yap (Eds.): FAW 2015, LNCS 9130, pp. 115–124, 2015.
DOI: 10.1007/978-3-319-19647-3_11

asks for a Boolean assignment to maximize the number of satisfied clauses, while MinSAT asks for a Boolean assignment to minimize the number of satisfied clauses.

MaxSAT is considered as one of the fundamental combinatorial optimization problems, with close ties to important problems like max cut or max clique, and with applications in scheduling, routing, etc. For MaxSAT, there is a long tradition of theoretical works, e.g. [6,12,15,19]. Moreover, Coppersmith et al. consider the phase transitions of random Max$k$SAT ($k \geq 2$) problem, where each clause contains two literals and is selected independently and randomly [9]. They demonstrate that, with increasing of r, i.e. the ratio between number of clauses and number of variables, the expected number of unsatisfied clauses under an optimal assignment quickly changes from $\Theta(1/n)$ to $\Theta(n)$. Furthermore, they provide the upper and lower bounds of the maximum number of satisfied clauses for MaxSAT. Xu et al. improve the upper bound of Max2SAT by the first moment argument via correcting error items [24]. Achlioptas et al. first studied the p-satisfiable problem while there exists a truth assignment satisfying a fraction of $1 - 2^{-k} + p2^{-k}$ of all clauses. They introduced weighting second moment method to prove the upper and lower bound of r(clauses/variables) [2]. Zhou et al. improve the lower bound by giving a different weight to the truth assignment if exactly one of k literals in a clause is satisfied [26].

The research of MaxSAT leads to increasing interest in its counterpart, Min-SAT, which is introduced by Kohli, Krishnamurti and Mirchandani in [14]. They show that MinSAT is NP-complete, even when the formula is a 2-CNF formula, i.e. each clause of which contains at most two literals, or a Horn formula, i.e. each clause of which contains at most one positive variable. They also analyze the performances of deterministic greedy and probabilistic greedy heuristics for MinSAT. A reduction from MinSAT to the minimum vertex cover (MinVC) problem is given in [20], to improve the approximation ratio of MinSAT to 2. A simple randomized 1.1037-approximation algorithm for Min2SAT, and a 1.2136-approximation algorithm for Min3SAT, are given by Avidor and Zwick in [1]. The first exact algorithm for MinSAT is by encoding MinSAT to MaxSAT and solving it with a MaxSAT solver [16]. A branch and bound algorithm for solving MinSAT is proposed by Li et al. in [17]. Their experiments show that solving problems like MaxClique and combinatorial auction problems, is faster by encoding them to MinSAT than reducing them to MaxSAT.

Compared with MaxSAT, there is a lack of knowledge about the bounds of random MinSAT. A k-CNF of q-satisfiable asks for a truth assignment satisfying a fraction of $1 - 2^{-k} - q(1 - 2^{-k})$ of all clauses. In this paper, we give upper and lower bounds of r for k-CNF to be q-satisfiable. The upper bound is obtained by the first moment method, while the proof of lower bound is weighted second moment used by [2].

We also present experimental results to demonstrate the tightness of these lower and upper bounds. For MinSAT, the experimental results explain the excellent performance of the state-of-the-art solver MinSatz. Moreover, we investigate the relationship between MaxSAT and MinSAT, and propose a conjecture about the expected sum of the maximum number and the minimum number of satisfied clauses for a random SAT instance.

This paper is organized as follows. In the next section, we review some basic definitions about SAT and MinSAT. Then the lower and upper bounds of k-CNF to be q-satisfiable are proved respectively. After that, experimental results are presented, as well as a discussion on the relationship between MaxSAT and MinSAT suggested by the experimental results. Finally, we conclude our work and point out future research directions.

2   Preliminaries

A Boolean formula in conjunctive normal form F is a set of clauses $\{C_1, C_2, ..., C_m\}$, where m is the number of clauses in F. A clause is a disjunction of literals, $x_{i_1} \vee x_{i_2} \vee \cdots \vee x_{i_k}$, where k is the length of the clause. A literal is either a Boolean variable x or its negation $\bar{x}$. The SAT problem is to determine the existence of an assignment satisfying all the clauses. If there is no assignment to satisfy all the clauses, a natural but more practical question is, how far or how close can one get to satisfiability? This is the optimization version of SAT, which is to find an assignment satisfying the most or the least number of clauses.

Definition 1. *(MinSAT) Given a SAT instance F, MinSAT asks for an assignment to all the variables such that the minimum number of clauses are satisfied. This number is called the value of the MinSAT instance.*

Definition 2. *(q-satisfiable) Given a k-CNF formula F, if there exists a truth assignment satisfying a fraction of $1 - 2^{-k} - q(1 - 2^{-k})$ $(0 < q < 1)$ of all clauses, it is q-satisfiable.*

For simplicity, we use x in lieu of $\lfloor x \rfloor$, the largest integer no more than x. The following standard asymptotic notations will be used in this paper.

$$\lim_{x \to \infty} \frac{f(x)}{g(x)} = 0 \ \Rightarrow \ f(x) = o(g(x))$$

$$\lim_{x \to \infty} \frac{f(x)}{g(x)} = 1 \ \Rightarrow \ f(x) \simeq g(x)$$

$$\lim_{x \to \infty} \sup \frac{f(x)}{g(x)} \leq M(M > 0) \ \Rightarrow \ f(x) = O(g(x))$$

Especially, while $M = 1$, $f(x) \lesssim g(x)$ means $f(x)$ is less than or equal to $g(x)$ asymptotically.

3   The Upper Bound

Given a k-CNF instance F to be q-satisfiable. Let P_r be the probabilistic distribution and let N denote the solutions number of the instance.

Theorem 1. *Let* $r > \frac{2\ln 2}{q^2(2^k-1)}$, *we have*

$$\lim_{n\to\infty} Pr[F \text{ is } q - satisfiable] = 0$$

Proof. The expected value of N, denoted as $E(N)$, is given by

$$E(N) = 2^n \sum_{i=0}^{\rho rn} \binom{rn}{i} \left(\frac{1}{2^k}\right)^{rn-i} \left(1 - \frac{1}{2^k}\right)^i$$

where $\rho = 1 - 2^{-k} - q(1 - 2^{-k})$.

The last term is maximized for $q \in (0,1)$, so $E(N)$ is upper bounded by

$$E(N) \le 2^n(\rho rn + 1) \binom{rn}{\rho rn} \left(\frac{1}{2^k}\right)^{rn-\rho rn} \left(1 - \frac{1}{2^k}\right)^{\rho rn}$$

According to Stirling's formula $n! \simeq (n/e)^n O(n)$, and

$$\binom{rn}{\rho rn} \simeq \left(\rho^{-\rho}(1-\rho)^{\rho-1}\right)^{rn},$$

We have

$$E(N) \le 2^n(\rho rn + 1) \left(\rho^{-\rho}(1-\rho)^{\rho-1} \left(\frac{1}{2^k}\right)^{1-\rho} \left(1 - \frac{1}{2^k}\right)^{\rho}\right)^{rn}$$

It is easy to prove that $E(N) < 0$ while $r > \frac{2\ln 2}{q^2(2^k-1)}$. By the Markov inequality $Pr(SAT) \le E(N)$, the upper bound is obtained.

4   The Lower Bound

Generally, the standard second moment method can be used to prove the lower bound for random problems, such as SAT, CSP. Unfortunately, it fails by using to optimization problem like MaxSAT. To cover this problem, Achlioptas provide a weighting scheme of the second moment to improve the lower bound of kSAT [3] and MaxSAT [2]. We following this line to prove the lower bound of q-satisfiable problem.

For any truth assignment $\sigma \in \{0,1\}^n$, let

$$H = H(\sigma, F) = Unsat(l, \sigma) - Sat(l, \sigma)$$
$$S = S(\sigma, F) = Sat(c, \sigma)$$

where $Unsat(l, \sigma)(Sat(l, \sigma))$ is the number of unsatisfied(satisfied) literal, $Sat(c, \sigma)$ is the number of satisfied clauses.

Let N defined as the number of solutions to q-satisfiable problem. Define

$$N = \Sigma_\sigma \gamma^{H(\sigma,F)} \eta^{S(\sigma,F)-s_0 rn}$$

where $s_0 = 1 - q - (1-q)2^{-k}$, $0 < \gamma, \eta < 1$.

Using the following two functions,

$$
\begin{aligned}
&f(\alpha, \gamma, \eta) \\
&= \eta^{-2s_0} E[\gamma^{H(\sigma,F)+S(\tau,F)} \eta^{H(\sigma,F)+S(\tau,F)}] \\
&= \eta^{2s_0} \left[\left(\alpha(\tfrac{\gamma^2+\gamma^{-2}}{2}) + 1 - \alpha \right)^k \right. \\
&\quad - 2(1-\eta)\left(\left(\alpha(\tfrac{\gamma^2+\gamma^{-2}}{2}) + 1 - \alpha \right)^k - \left(\tfrac{\alpha\gamma^{-2}+1-\alpha}{2} \right)^k \right) \\
&\quad \left. + (1-\eta)^2 \left(\left(\left(\alpha(\tfrac{\gamma^2+\gamma^{-2}}{2}) + 1 - \alpha \right)^k - 2\left(\tfrac{\alpha\gamma^{-2}+1-\alpha}{2} \right)^k \right) + 2^{-k}\left(\alpha\gamma^{-2} \right)^k \right) \right]
\end{aligned}
$$

Let

$$
g_r(\alpha, \gamma, \eta) = \frac{f(\alpha, \gamma, \eta)^r}{\alpha^\alpha (1-\alpha)^{1-\alpha}}
$$

We have

$$
E^2[N] = \frac{2^n}{\eta^{s_0 rn}} \left[(\tfrac{\gamma+\gamma^{-1}}{2})^k - (1-\eta)[\tfrac{\gamma+\gamma^{-1}}{2}]^k - (2\gamma)^{-k} \right]^{rn} = \left(2g_r\left(\tfrac{1}{2}, \gamma, \eta\right) \right)^n
$$

Theorem 2. *Given a k-CNF formula F, if* $r \le \frac{\ln 2}{(q+(1-q)\ln(1-q))2^k}(1-O(k2^{-k}))$,

$$
\lim_{n \to \infty} [F \text{ is } q - satisfiable] = 1.
$$

The proof is similar to the lower bound for MaxSAT in [2], so we ignore here.

5 Experimental Results

We conduct experiments to k-CNF formula to be q-satisfiable, the results are presented in Fig. 1. The upper bound (*upper*) is proved by the first moment

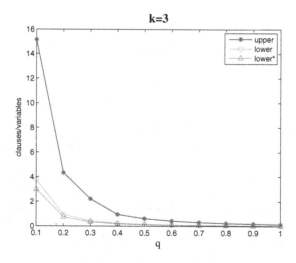

Fig. 1. The upper and lower bound for q-satisfiable

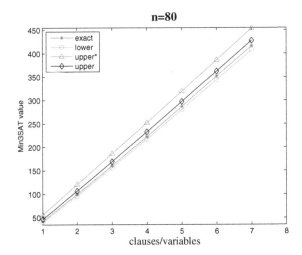

Fig. 2. The bounds of $n = 80$ for Min3SAT

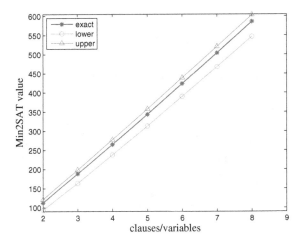

Fig. 3. The bounds of $n = 120$ for Min2SAT

method, while the proof of the lower bound (*lower*) is weighted scheme of the second moment method. We also give a poor lower bound (*lower**) by algorithm analysis, which was used to MaxSAT in [9].

From Fig. 1, we can see that the space between upper bond and lower bound become smaller as q increase. That is to say, the bounds is tighter while the number of satisfying clauses is less. Based on this, the lower and upper bounds of Min$k$SAT is presented in Fig. 2, which indicate that the bounds for MinSAT ($k = 3$) provided by this paper is very tight. However, the second moment is poor to upper bound for Min2SAT, so algorithm analysis is considered in Fig. 3.

Table 1. Comparing with *MinSATz* with branching number

| | r = 6 | | r = 7 | |
#n	*MinSatz* + LB	*MinSatz*	*Minsatz* + LB	*MinSatz*
120	6593	7331	14601	14764
130	24881	25605	319657	333949
140	109061	112213	946257	947779
150	49597	49744	552269	552442

In the second experiment, we apply our work to the state-of-the-art solver *MinSatz*. In *MinSatz*, UB is the largest number of falsified clauses while extending the current partial assignment to a complete one, and LB is the number of clauses falsified in the best assignment found so far. if LB < UB, MinSatz select a variable and instantiate it, otherwise, the solver backtracks. Besides, *MinSatz* introduces both clique partitioning algorithms and MaxSAT technology to improve the UB so as to prune the search tree quickly. However, we focus on the other side, and give LB an initial number so as to reduce the branching number. We conduct an experiment to test the performances before and after adding our work to *MinSatz* in Table 1. The branching number can be reduced while giving UB a initial value computed by this paper, but the improvements is not so obvious. Further analysis indicate that the genuine *MinSatz* is an excellent solver for MinSAT. However, if the MinSAT solver is not so excellent as *MinSatz*, such as a trivial MinSAT solver with no inference rules (*minsat*), the improvements will be more obvious. The results are presented in Table 2, which indicates that a MinSAT solver with our work outperforms the one without.

Li et al. have found from experiments that the relationship between MaxSAT and MinSAT is counter-intuitive [17,18]: for the same instances, the bigger the MaxSAT value is, the smaller the MinSAT value is, the opposite is also true. They focus these instances at the threshold ($c = 4.25$ for 3SAT). We follow this line, and conjecture that, the sum of the MaxSAT and MinSAT value is a constant value. In our experiments, the number of variables in random instances ranges from 40 to 140, and the density r considered are 6, 7 and 8. We compare the sum of exact MinSAT value and MaxSAT value ('sum' in Table 3) with our conjectured value ('conjecture' in Table 3). From experiments, the accuracy of our conjecture is found to be over 99 % ('accuracy' in Table 3). This indicates

Table 2. Application to MinSAT solver

| | r = 0.8 | | | r = 1 | | | r = 2 | | |
#n	*minsat*	*minsat* + LB	incr	*minsat*	*minsat* + LB	incr	*minsat*	*minsat* + LB	incr
20	0	0	0	1	1	0	1	1	0
30	0	0	0	1	1	0	100	85	15 %
40	500	420	16 %	600	500	16.67 %	1783	1458	18.22 %
50	1284	1076	16.2 %	2117	1916	9.49 %	4839	4279	11.57 %

Table 3. The relationship of MinSAT and MaxSAT

	$r = 6$			$r = 7$			$r = 8$		
#n	Sum	Conjecture	Accuracy	Sum	Conjecture	Accuracy	sum	Conjecture	Accuracy
40	357	360	99.17 %	418	420	99.52 %	478	480	99.58 %
60	535	540	99.07 %	627	630	99.52 %	716	720	99.44 %
80	715	720	99.31 %	834	840	99.29 %	954	960	99.37 %
100	892	900	99.11 %	1041	1050	99.14 %	1191	1200	99.25 %
120	1071	1080	99.17 %	1251	1260	99.29 %	1431	1440	99.37 %
140	1248	1260	99.05 %	1459	1470	99.25 %	1669	1680	99.35 %

that our conjecture is close to the exact value. For kSAT instances, the sum of MinSAT value and MaxSAT value is approximately $(2 - 2^{1-k})cn$. If the relationship between MinSAT and MaxSAT values is clear, this conjecture can be used to get a better upper bound for MaxSAT. In other words, give the exact value of MinSAT, the value of MaxSAT for the same instance can be approximated. Besides, this value is significantly better than the upper bound obtained by [9], see Table 4.

Table 4. Conjecture about the upper bound of MaxSAT. 'MinSAT value' ('MaxSAT value') is the exact value of MinSAT (MaxSAT), 'ub04' is the upper bound of MaxSAT obtained by [9], $ub^\star$ is our guess value.

#n	MinSAT value	MaxSAT value	$ub^\star$	ub04
40	142	**215**	**218**	229
60	213	**322**	**327**	344
80	283	**432**	**437**	459
100	353	**539**	**547**	574
120	423	**648**	**657**	689
140	493	**755**	**767**	804

6   Conclusions and Future Work

We have presented upper and lower bounds of the minimization versions of the SAT problem. For the upper bound, the first moment argument is used, while the lower bound is derived by weighting second moment. The experimental results confirm the correctness and accuracy of our work. Furthermore, we consider the relationship between MinSAT and MaxSAT, and an interesting conjecture is presented. As for future work, the bounds of other optimization problems such as Min-CUT of random graphs [7] and Min-CSP [13] of random CSPs [25] may be considered.

References

1. Avidor, A., Zwick, U.: Approximating MIN 2-SAT and MIN 3-SAT. Theor. Comput. Syst. **38**(3), 329–345 (2005)
2. Achlioptas D., Naor A., Peres, Y.: On the maximum satisfiability of random formulas. In: FOCS 2003, pp. 362–370 (2003)
3. Achlioptas, D., Peres, Y.: The threshold for random k-SAT is $2^k \log 2 - O(k)$. J. Am. Math. Soc. **17**(4), 947–973 (2004)
4. Anstegui, C., Bonet, M.L., Levy, J.: SAT-based MaxSAT algorithms. Artif. Intell. **196**, 77–105 (2013)
5. Anstegui, C., Malitsky, Y., Sellmann, M: MaxSAT by improved instance-specific algorithm configuration. In: AAAI 2014, pp. 2594–2600 (2014)
6. Bollobas, B., Borgs, C., Chayes, J.T., Kim, J.H., Wilson, D.B.: The scaling window of the 2-SAT transition. Random Struct. Algorithms **18**(3), 201–256 (2001)
7. Bollobas, B.: Random Graphs. Cambridge Studies in Advanced Mathematics, vol. 73, 2nd edn. Cambridge University Press, Cambridge (2001)
8. Cai, S.W., Luo, C., Thornton, J., Su, K.L.: Tailoring local search for partial MaxSAT. In: AAAI 2014, pp. 2623–2629 (2014)
9. Coppersmith, D., Gamarnik, D., Hajiaghayi, M.T., Sorkin, G.B.: Random MAX SAT random MAX CUT, and their phase transitions. Random Struct. Algorithms **24**, 502–545 (2004)
10. Eén, N., Sörensson, N.: Translating pseudo-boolean constraints into SAT. J. Satisfiability Boolean Model. Comput. **2**(2006), 21–26 (2006)
11. Gramm, J., Hirsch, E.A., Niedermeier, R., Rossmanith, P.: Worst-case upper bounds for MAX-2-SAT with an application to MAX-CUT. Discrete Appl. Math. **130**, 139–155 (2003)
12. Hirsch, E.A.: A new algorithm for MAX-2-SAT. In: Reichel, H., Tison, S. (eds.) STACS 2000. LNCS, vol. 1770, pp. 65–73. Springer, Heidelberg (2000)
13. Huang, P., Yin, M.H.: An upper (lower) bound for Max(Min) CSP. Sci. China Inf. Sci. **57**(7), 1–9 (2014)
14. Kohli, R., Krishnamurti, R., Mirchandani, P.: The minimum satisfiability problem. SIAM J. Discrete Math. **7**, 275–283 (1994)
15. Li, C.M., Many, F., Nouredine, N.O., Planes, J.: Resolution-based lower bounds in MaxSAT. Constraints **15**(4), 456–484 (2010)
16. Li, C.M., Manyà, F., Quan, Z., Zhu, Z.: Exact MinSAT solving. In: Strichman, O., Szeider, S. (eds.) SAT 2010. LNCS, vol. 6175, pp. 363–368. Springer, Heidelberg (2010)
17. Li, C.M., Zhu, Z., Many, F., Simon, L.: Minimum satisfiability and its applications. In: IJCAI 2011, pp. 605–610 (2011)
18. Li, C.M., Zhu, Z., Many, F., Simon, L.: Optimizing with minimum satisfiability. Artif. Intell. **190**, 32–44 (2012)
19. Lin, H., Su, K.L., Li, C.M.: Within problem learning for efficient lower bound computation in Max-SAT solving. In: AAAI 2008, pp. 351–356 (2008)
20. Marathe, M.V., Ravi, S.S.: On approximation algorithms for the minimum satisfiability problem. Inf. Process. Lett. **58**, 23–29 (1996)
21. Robinson, N., Gretton, C., Pham, D.N., Sattar, A.: A compact and efficient SAT encoding for planning. In: ICAPS 2008, pp. 296–303 (2008)
22. Spencer, J.H.: Ten Lectures on the Probabilistic Method. SIAM, Philadelphia (1994)

23. Whitley, D., Howe, A.E., Hains, D.: Greedy or Not? Best Improving versus first improving stochastic local search for MAXSAT. In: AAAI 2013 (2013)
24. Xu, X.L., Gao, Z.S., Xu, K.: A tighter upper bound for random MAX 2-SAT. Inf. Process. Lett. **111**(3), 115–119 (2011)
25. Xu, K., Li, W.: Exact phase transitions in random constraint satisfaction problems. J. Artif. Intell. Res. (JAIR) **12**, 93–103 (2000)
26. Zhou, G.Y., G Z.S., Liu J.: On the lower bounds of random Max 3 and 4-SAT. Manuscript

On Solving Systems of Diagonal Polynomial Equations Over Finite Fields

Gábor Ivanyos[1] and Miklos Santha[2,3](✉)

[1] Institute for Computer Science and Control, Hungarian Academy of Sciences,
Budapest, Hungary
Gabor.Ivanyos@sztaki.mta.hu
[2] LIAFA, Université Paris Diderot, CNRS, 75205 Paris, France
[3] Centre for Quantum Technologies, National University of Singapore,
Singapore, Singapore
miklos.santha@gmail.com

Abstract. We present a randomized algorithm to solve a system of diagonal polynomial equations over finite fields when the number of variables is greater than some fixed polynomial of the number of equations whose degree depends only on the degree of the polynomial equations. Our algorithm works in time polynomial in the number of equations and the logarithm of the size of the field, whenever the degree of the polynomial equations is constant. As a consequence we design polynomial time quantum algorithms for two algebraic hidden structure problems: for the hidden subgroup problem in certain semidirect product p-groups of constant nilpotency class, and for the multi-dimensional univariate hidden polynomial graph problem when the degree of the polynomials is constant.

Keywords: Algorithm · Polynomial equations · Finite fields · Chevalley–Warning theorem · Quantum computing

1 Introduction

Finding small solutions in some well defined sense for a system of integer linear equations is an important, well studied, and computationally hard problem. *Subset Sum*, which asks the solvability of a single equation in the binary domain is one of Karp's original 21 NP-complete problems [16].

The guarantees of many lattice based cryptographic system come from the average case hardness of *Short Integer Solution*, dating back to Ajtai's breakthrough work [1], where we try to find short nonzero vectors in a random integer lattice. Indeed, this problem has a remarkable worst case versus average case hardness property: solving it on the average is at least as hard as solving various lattice problems in the worst case, such as the decision version of the shortest vector problem, and finding short linearly independent vectors.

© Springer International Publishing Switzerland 2015
J. Wang and C. Yap (Eds.): FAW 2015, LNCS 9130, pp. 125–137, 2015.
DOI: 10.1007/978-3-319-19647-3_12

Turning back to binary solutions, deciding, if there exists a nonzero solution of the system of linear equations

$$a_{11}x_1 + \ldots + a_{1n}x_n = 0$$
$$\vdots \qquad \vdots \; \vdots \qquad (1)$$
$$a_{m1}x_1 + \ldots + a_{mn}x_n = 0$$

in the finite field $\mathbb{F}_p$, for some prime number p is easy when $p = 2$. However, by modifying the standard reduction of *Satisfiability* to *Subset Sum* [24] it can be shown that it is an NP-hard problem for $p \geq 3$.

The system (1) is equivalent to the system of equations

$$a_{11}x_1^{p-1} + \ldots + a_{1n}x_n^{p-1} = 0$$
$$\vdots \qquad \vdots \; \vdots \qquad (2)$$
$$a_{m1}x_1^{p-1} + \ldots + a_{mn}x_n^{p-1} = 0$$

where we look for a nonzero solution in the whole $\mathbb{F}_p^n$.

In this paper we will consider finding a nonzero solution for a system of diagonal polynomial equations similar to (2), but where more generally, the variables are raised to some power $2 \leq d$. We state formally this problem.

Definition 1. The *System of Diagonal Equation* problem SDE is parametrized by a finite field $\mathbb{F}$ and three positive integers n, m and d.

SDE$(\mathbb{F}, n, m, d)$

    Input: A system of polynomial equations over $\mathbb{F}$:

$$a_{11}x_1^d + \ldots + a_{1n}x_n^d = 0$$
$$\vdots \qquad \vdots \; \vdots \qquad (3)$$
$$a_{m1}x_1^d + \ldots + a_{mn}x_n^d = 0$$

    Output: A nonzero solution $(x_1, \ldots, x_n) \neq 0^n$.

For $j = 1, \ldots, n$, let us denote by v_j the vector $(a_{1j}, \ldots, a_{mj}) \in \mathbb{F}^m$. Then the system of Eq. (3) is the same as

$$\sum_{j=1}^n x_j^d v_j = 0. \qquad (4)$$

That is, solving SDE$(\mathbb{F}, n, m, d)$ is equivalent to the task of representing the zero vector as a nontrivial linear combinations of a subset of $\{v_1, \ldots, v_n\}$ with dth power coefficients. We present our algorithm actually as solving this vector problem. The special case $d = |\mathbb{F}| - 1$ is the vector zero sum problem where the goal is to find a non-empty subset of the given vectors with zero sum.

Under which conditions can we be sure that for system (3) there exists a nonzero solution? The elegant result of Chevalley [3] states that a system of homogeneous polynomial equations has a nonzero solution if the number of variables is greater than the sum of the degrees of the polynomials. In our case this means that when $n > dm$, the existence of a nonzero solution is assured. In addition, Warning has proven [26] that under similar condition the number of solutions is in fact a multiple of the characteristic of $\mathbb{F}$.

In general where little is known about the complexity of finding a nonzero solution for systems which satisfy the Chevalley condition. When $|\mathbb{F}| = 2$, Papadimitriou has shown [20] that this problem is in the complexity class Polynomial Parity Argument (PPA), the class of NP search problems where the existence of the solution is guaranteed by the fact that in every finite graph the number of vertices with odd degree is even. This implies that it can not be NP-hard unless NP = co-NP. Nonetheless finding efficiently a nonzero solution in general seems to be a very hard task.

Let us come back to our special system of Eq. (3). In the case $m = 1$, a nonzero solution can be found in polynomial time for the single equation which satisfies the Chevalley condition due to the remarkable work of van de Woestijne [25] where he proves the following.

Fact 1. *In deterministic polynomial time in d and $\log|\mathbb{F}|$ we can find a nontrivial solution for $a_1 x_1^d + \ldots + a_{d+1} x_{d+1}^d = 0$.*

In the case of more than one equation we don't know how to find a nonzero solution for Eq. (3) under just the Chevalley condition. However, if we relax the problem, and take much more variable than required for the existence of a nonzero solution, we are able to give a polynomial time solution. Using van de Woestijne's result for the one dimensional case, a simple recursion on m shows that if $n \geq (d+1)^m$ then $\mathrm{SDE}(\mathbb{F}_p, n, m, d)$ can be solved in deterministic polynomial time in n and $\log p$. The time complexity of this algorithm is therefore polynomial for any fixed m. The case when d is fixed and m grows appears to be more difficult. To our knowledge, the only existing result in this direction is the case $d = 2$ for which it was shown in [14] that there exists a randomized algorithm that, when $n = \Omega(m^2)$, solves $\mathrm{SDE}(\mathbb{F}_p, n, m, d)$ in polynomial time in n and $\log p$. In the main result of this paper we generalize this result by showing, for every constant d, the existence of a randomized algorithm that, for every n larger than some polynomial function of m, solves $\mathrm{SDE}(\mathbb{F}_p, n, m, d)$ in polynomial time in n and $\log p$.

Theorem 2. *Let d be constant. For $n > d^{d^2 \log d}(m+1)^{d \log d}$, the problem $\mathrm{SDE}(\mathbb{F}_p, n, m, d)$ can be solved by a randomized algorithm in polynomial time in n and $\log p$.*

The large number of variables that makes possible a polynomial time solution unfortunately also makes our algorithm most probably irrelevant for cryptographic applications. Nonetheless, it turns out the algorithm is widely applicable in quantum computing for solving efficiently various algebraic hidden structure problems. We explain now this connection.

Simply speaking, in a hidden structure problem we have to find some hidden object related to some explicitly given algebraic structure A. We have access to an oracle input, which is an unknown member f of a family of black-box functions which map A to some finite set S. The task is to identify the hidden object solely from the information one can obtain by querying the oracle f. This means that the only useful information we can obtain is the structure of the level sets $f^{-1}(s) = \{a \in A : f(a) = s\}$, $s \in S$, that is, we can only determine whether two elements in A are mapped to the same value or not. In these problems we say that the input f *hides* the hidden structure, the output of the problem. We define now the two problems for which we can apply our algorithm for SDE.

Definition 2. The *hidden subgroup problem* HSP is parametrized by a finite group G and a family $\mathcal{H}$ of subgroups of G.

　HSP$(G, \mathcal{H})$
　　Oracle input: A function f from G to some finite set S.
　　Promise: For some $H \in \mathcal{H}$, we have $f(x) = f(y) \Longleftrightarrow Hx = Hy$.
　　Output: H.

The *hidden polynomial graph problem* HPGP is parametrized by a finite field $\mathbb{F}_p$ and three positive integers n, m and d.

　HPGP$(\mathbb{F}_p, n, m, d)$.
　　Oracle input: A function f from $\mathbb{F}_p^n \times \mathbb{F}_p^m$ to a finite set S.
　　Promise: For some $Q : \mathbb{F}_p^n \to \mathbb{F}_p^m$, where $Q(x) = (Q_1(x), \ldots, Q_m(x))$,
　　　and $Q_i(x)$ is an n-variate degree d polynomial over $\mathbb{F}_p$ with zero constant
　　　term, we have $f(x, y) = f(x', y') \Longleftrightarrow y - Q(x) = y' - Q(x')$.
　　Output: Q.

While no classical algorithm can solve the HSP with polynomial query complexity even if the group G is abelian, one of the most powerful results of quantum computing is that it can be solved by a polynomial time quantum algorithm for any abelian G (see, e.g., [15]). Shor's factorization and discrete logarithm finding algorithms [23], and Kitaev's algorithm [17] for the abelian stabilizer problem are all special cases of this general solution.

Extending the quantum solution of the abelian HSP to non abelian groups is an active research area since these instances include several algorithmically important problems. For example, efficient solutions for the dihedral and the symmetric group would imply efficient solutions, respectively, for several lattice problems [21] and for graph isomorphism. While the non abelian HSP has been solved efficiently by quantum algorithms in various groups [2,8–11,18,19], finding a general solutions seems totally elusive.

A different type of extension was proposed by Childs, Schulman and Vazirani [4] who considered the problem where the hidden object is a polynomial. To recover it we have at our disposal an oracle whose level sets coincide with the level sets of the polynomial. Childs et al. [4] showed that the quantum query complexity of this problem is polynomial in the logarithm of the field size when the degree and the number of variables are constant. In [7] the first time efficient quantum algorithm was given for the case of multivariate quadratic polynomials over fields of constant characteristic.

The hidden polynomial graph problem HPGP was defined in [5] by Decker, Draisma and Wocjan. Here the hidden object is again a polynomial, but the oracle is more powerful than in [4] because it can also be queried on the graphs that are defined by the polynomial functions. They obtained a polynomial time quantum algorithm that correctly identifies the hidden polynomial when the degree and the number of variables are considered to be constant. In [7] this result was extended to polynomials of constant degree. The version of the HPGP we define here is more general than the one considered in [5] in the sense that we are dealing not only with a single polynomial but with a vector of several polynomials. The restriction on the constant terms of the polynomials are due to the fact that level sets of two polynomials are the same if they differ only in their constant terms, and therefore the value of the constant term can not be recovered.

It will be convenient for us to consider a slight variant of the hidden polynomial graph problem which we denote by HPGP'. The only difference between the two problems is that in the case of HPGP' the input is not given by an oracle function but by the ability to access random *level set states*, which are quantum states of the form

$$\sum_{x \in \mathbb{F}_p^n} |x\rangle |u + Q(x)\rangle,$$

where u is a random element of $\mathbb{F}_p^m$. Given an oracle input f for HPGP, a simple and efficient quantum algorithm can create such a random coset state. Therefore an efficient quantum algorithm for HPGP' immediately provides an efficient quantum algorithm for HPGP.

In [6] it was shown that HPGP'$(\mathbb{F}_p, 1, m, d)$ is solvable in quantum polynomial time when d and m are both constant. Part of the quantum algorithm repeatedly solved instances of SDE$(\mathbb{F}_p, n, m, d)$ under such conditions. We present here a modification of this method which works in polynomial time even if m is not constant.

Theorem 3. *Let d be constant. If* SDE$(\mathbb{F}_p, n, m, d)$ *is solvable in randomized polynomial time for some n, then* HPGP'$(\mathbb{F}_p, 1, m, d)$ *is solvable in quantum polynomial time.*

Using Theorem 2 it is possible to dispense in the result of [6] with the assumption that m is constant.

Corollary 1. *If d is constant then* HPGP'$(\mathbb{F}_p, 1, m, d)$ *is solvable in quantum polynomial time.*

Bacon, Childs and van Dam in [2] have considered the HSP in p-groups of the form $G = \mathbb{F}_p \ltimes \mathbb{F}_p^m$ when the hidden subgroup belongs to the family $\mathcal{H}$ of subgroups of order p which are not subgroups of the normal subgroup $0 \times \mathbb{F}_p^m$. They have found an efficient quantum algorithm for such groups as long as m is constant. In [7], based on arguments from [2] it was sketched how the HSP$(\mathbb{F}_p \ltimes \mathbb{F}_p^m, \mathcal{H})$ can be translated into a hidden polynomial graph problem. For the sake of completeness we state here and prove the exact statement about such a reduction.

Proposition 1. *Let d be the nilpotency class of a group G of the form $\mathbb{F}_p \ltimes \mathbb{F}_p^m$. There is a polynomial time quantum algorithm which reduces $\mathrm{HSP}(G, \mathcal{H})$ to $\mathrm{HPGP}'(\mathbb{F}_p, 1, m, d)$.*

Putting together Corollary 1 and Proposition 1, it is also possible to get rid of the assumption that m is constant in the result of [2].

Corollary 2. *If the nilpotency class of the group G of the form $\mathbb{F}_p \ltimes \mathbb{F}_p^m$ is constant then $\mathrm{HSP}(G, \mathcal{H})$ can be solved in quantum polynomial time.*

The special cases of Theorem 2 for $d = 2, 3$ will be shown in Sect. 2. The proof of Theorem 2 will be given in Sect. 3. The proofs of Theorem 3 and Proposition 1 are given in the full and improved version of the paper [13]. We remark that the proof of Theorem 2 extends to arbitrary finite fields (only minor notational changes are needed). Also, the method can be made deterministic using techniques similar to those used by van de Woestijne in [25]. Details of these can also be found in [13].

2   Warm-Up: The Quadratic and Cubic Cases

2.1   The Quadratic Case

Proposition 2. *The problem $\mathrm{SDE}(\mathbb{F}_p, (m+1)^2, m, 2)$ can be solved in randomized polynomial time.*

Proof. We assume that $p > 2$ and that we have a non-square ζ in $\mathbb{F}_p$ at hand. Such an element can be efficiently found by a random choice. Assuming GRH, even a deterministic polynomial time method exists for finding a non-square.

Our input is a set V of $(m+1)^2$ vectors in $\mathbb{F}_p^m$, and we want to represent the zero vector as a nontrivial linear combination of some vectors from V where all the coefficients are squares. The construction is based on the following. Pick any $m+1$ vectors $u_1, \ldots, u_{m+1}$ from $\mathbb{F}_p^m$. Since they are linearly dependent, it is easy to represent the zero vector as a proper linear combination $\sum_{i=1}^{m+1} \alpha_i u_i = 0$. Let $J_1 = \{i : \alpha_i^{\frac{p-1}{2}} = 1\}$ and $J_2 = \{i : \alpha_i^{\frac{p-1}{2}} = -1\}$. Using ζ, we can efficiently find in deterministic polynomial time in $\log p$ by the Shanks-Tonelli algorithm [22] field elements β_i such that $\alpha_i = \beta_i^2$ for $i \in J_1$ and $\alpha_i = \beta_i^2 \zeta$ for $i \in J_2$. Let $w_1 = \sum_{i \in J_1} \beta_i^2 v_i$ and $w_2 = \sum_{i \in J_2} \beta_i^2 v_i$. Then $w_1 = -\zeta w_2$. Notice that we are done if either of the sets J_1 or J_2 is empty.

What we have done so far, can be considered as a high-level version of the approach of [14]. The method of [14] then proceeds with recursion to $m - 1$. Unfortunately, that approach is appropriate only in the quadratic case. Here we use a completely different idea which will turn to be extensible to more general degrees.

From the vectors in V we form $m + 1$ pairwise disjoint sets of vectors of size $m + 1$. By the construction above, we compute $w_1(1), w_2(1), \ldots, w_1(m + 1)$, $w_2(m + 1)$, where

$$w_1(i) = -\zeta w_2(i), \tag{5}$$

for $i = 1, \ldots, m+1$. Moreover, these $2m$ vectors are represented as linear combinations with nonzero square coefficients of $2m$ pairwise disjoint nonempty subsets of the original vectors.

Now $w_1(1), \ldots, w_1(m+1)$ are linearly dependent and again we can find disjoint subsets J_1 and J_2 and scalars γ_i for $i \in J_1 \cup J_2$ such that for $w_{11} = \sum_{i \in J_1} \gamma_i^2 w_1(i)$ and $w_{12} = \sum_{i \in J_2} \gamma_i^2 w_1(i)$ we have $w_{11} = -\zeta w_{12}$. But then for $w_{21} = \sum_{i \in J_2} \gamma_i^2 w_2(i)$ and $w_{22} = \sum_{i \in J_2} \gamma_i^2 w_1(i)$, using Eq. (5) for all i, we similarly have $w_{21} = -\zeta w_{22}$. On the other hand, if we sum up Eq. (5) for $i \in J_1$, we get $w_{11} = -\zeta w_{21}$. Therefore $w_{11} = \zeta^2 w_{22}$ and $w_{12} = w_{21} = -\zeta w_{22}$. By Fact 1 we can find field elements $\delta_{11}, \delta_{22}, \delta_{12}$, not all zero, such that $\zeta^2 \delta_{11}^2 - 2\zeta \delta_{12}^2 + \delta_{22}^2 = 0$, and therefore $(\zeta^2 \delta_{11}^2 - 2\zeta \delta_{12}^2 + \delta_{22}^2) w_{22} = 0$. But $(\zeta^2 \delta_{11}^2 - 2\zeta \delta_{12}^2 + \delta_{22}^2) w_{22} = \delta_{11}^2 w_{11} + \delta_{12}^2 (w_{12} + w_{21}) + \delta_{22}^2 \zeta^2 w_{22}$. Then expanding $\delta_{11}^2 w_{11} + \delta_{12}^2 (w_{12} + w_{21}) + \delta_{22}^2 \zeta^2 w_{22} = 0$ gives a representation of the zero vector as a linear combination with square coefficients (squares of appropriate product of βs, γs and δs) of a subset of the original vectors.

2.2 The Cubic Case

Proposition 3. *Let $n = (9m+1)(3m+1)(m+1)$. Then $\mathrm{SDE}(\mathbb{F}_p, n, m, 3)$ can be solved in randomized polynomial time.*

Proof. We assume that $p-1$ is divisible by 3 since otherwise the problem is trivial. By a randomized polynomial time algorithm we can compute two elements ζ_2, ζ_3 from $\mathbb{F}_p$ such that $\zeta_1 = 1, \zeta_2, \zeta_3$ are a complete set of representatives of the cosets of the subgroup $\{x^3 : x \in \mathbb{F}_p^*\}$ of $\mathbb{F}_p^*$. Let V be our input set of n vectors in $\mathbb{F}_p^m$, now we want to represent the zero vector as a nontrivial linear combination of some vectors from V where all the coefficients are cubes.

As in the quadratic case, for any subset of $m+1$ vectors $u_1, \ldots, u_{m+1}$ from V, we can easily find a proper linear combination summing to zero, $\sum_{i=1}^{m+1} \alpha_i u_i = 0$. For $r = 1, 2, 3$, let J_r be the set of indices such that $0 \neq \alpha_i = \beta_i^3 \zeta_r$. We know that at least one of these three sets is non-empty. For each $\alpha_i \neq 0$ we can efficiently identify the coset of α_i and even find β_i. Let $w_r = \sum_{i \in J_r} \beta_i^3 v_i$. Then $\zeta_1 w_1 + \zeta_2 w_2 + \zeta_3 w_3 = 0$. Without loss of generality we can suppose that J_1 is non-empty since if J_r is non-empty for $r \in \{2, 3\}$, we can just multiply α_is simultaneously by ζ_1 / ζ_r.

From any subset of size $(3m+1)(m+1)$ of V we can form $3m+1$ groups of size $m+1$, and within each group we can do the procedure outlined above. This way we obtain, for $k = 1, \ldots, 3m+1$, and $r = 1, 2, 3$, pairwise disjoint subsets $J_r(k)$ of indices and vectors $w_r(k)$ such that

$$\zeta_1 w_1(k) + \zeta_2 w_2(k) + \zeta_3 w_3(k) = 0. \tag{6}$$

For $k = 1, \ldots, 3m+1$, we know that $J_1(k) \neq \emptyset$ and the vectors $w_r(k)$ are combinations of input vectors with indices form $J_r(k)$ having coefficients which are nonzero cubes. Let $W(k) \in F_p^{3m}$ denote the vector obtained by concatenating $w_1(k), w_2(k)$ and $w_3(k)$ (in this order). Then we can find three pairwise disjoint subsets M_1, M_2, M_3 of $\{1, \ldots, 3m+1\}$, and for each $k \in M_s$, a nonzero field element γ_k such that

$$\sum_{s=1}^{3} \zeta_s \sum_{k \in M_s} \gamma_k^3 W(k) = 0. \tag{7}$$

We can arrange that M_2 is non-empty. For $r, s \in \{1, 2, 3\}$, set $J_{rs} = \bigcup_{k \in M_s} J_r(k)$ and $w_{rs} = \sum_{k \in M_s} \gamma_k^3 w_r(k)$. Then w_{rs} is a linear combination of input vectors with indices from J_{rs} having coefficients that are nonzero cubes. The equality (7) just states that $\zeta_1 w_{r1} + \zeta_2 w_{r2} + \zeta_3 w_{r3} = 0$, for $r = 1, 2, 3$. Furthermore, summing up the equalities (6) for $k \in M_s$, we get $\zeta_1 w_{1s} + \zeta_2 w_{2s} + \zeta_3 w_{3s} = 0$, for $s = 1, 2, 3$.

Continuing this way, from $(9m+1)(3m+1)(m+1)$ input vectors we can make 27 linear combinations with cubic coefficients w_{rst}, for $r, s, t = 1, 2, 3$, having pairwise disjoint supports such that the support of w_{123} is non-empty and they satisfy the 27 equalities $\zeta_1 w_{1st} + \zeta_2 w_{2st} + \zeta_3 w_{3st} = 0$ ($s, t = 1, 2, 3$); $\zeta_1 w_{r1t} + \zeta_2 w_{r2t} + \zeta_3 w_{r3t} = 0$ ($r, t = 1, 2, 3$); $\zeta_1 w_{rs1} + \zeta_2 w_{rs2} + \zeta_3 w_{rs3} = 0$ ($r, s = 1, 2, 3$). From these we use the following 6 equalities: $\zeta_1 w_{123} + \zeta_2 w_{223} + \zeta_3 w_{323} = 0$; $\zeta_1 w_{132} + \zeta_2 w_{232} + \zeta_3 w_{332} = 0$; $\zeta_1 w_{213} + \zeta_2 w_{223} + \zeta_3 w_{233} = 0$; $\zeta_1 w_{312} + \zeta_2 w_{322} + \zeta_3 w_{332} = 0$; $\zeta_1 w_{231} + \zeta_2 w_{232} + \zeta_3 w_{233} = 0$; $\zeta_1 w_{321} + \zeta_2 w_{322} + \zeta_3 w_{323} = 0$. Adding these equalities with appropriate signs so that the terms with coefficients ζ_2 and ζ_3 cancel and dividing by ζ_1, we obtain $w_{123} + w_{231} + w_{312} - w_{132} - w_{213} - w_{321} = 0$. Observing that $-1 = (-1)^3$, this gives a representation of zero as a linear combination of the input vectors with coefficients that are cubes.

3   The General Case

In this section we prove Theorem 2. First we make the simple observation that it is sufficient to solve $\mathrm{SDE}(\mathbb{F}_p, n, m, d)$ in the case when d divides $p - 1$. If it is not the case, then let $d' = \gcd(d, p-1)$. Then from a nonzero solution of the system

$$\sum_{j=1}^{n} x_j^{d'} v_j = 0,$$

one can efficiently find a nonzero solution of the original equation. Indeed, the extended Euclidean algorithm efficiently finds a positive integer t such that $td = u(p-1) + d'$ for some integer u. Then for any nonzero $x \in \mathbb{F}_p$ we have $(x^t)^d = x^{d'} \mod p$, and therefore $(x_1^t, \ldots, x_n^t)$ is a solution of Eq. (4). From now on we suppose that d divides $p - 1$.

Our algorithm will distinguish two cases, according to the value of d. The first case is when -1 is not a dth power in $\mathbb{F}_p$. Then d is necessarily an even number, and we give a method which reduces to the problem HPGP with polynomials of degree $d/2$. Observe that in that case -1 is a $d/2$th power, and the algorithm proceeds with the method of the second case. The second case is when -1 is a dth power in $\mathbb{F}_p$, then our algorithm directly solves the problem. For both cases we will denote by $C(d, m)$ the number of vectors (variables) used by our algorithm. For $d = 1$, we can take $C(1, m) = m + 1$.

3.1   The Reduction When d is Even

We assume that $p - 1$ is divisible by d and that we have a non-square ζ in $\mathbb{F}_p$ at hand. We also assume that we can efficiently express the zero vector as a nontrivial linear combination with dth power coefficients of any given $t = C(d/2, m)$ vectors $u_1, \ldots, u_t \in \mathbb{F}_p^m$: $\sum_{i=1}^{t} \alpha_i^d u_i = 0$.

As in the quadratic case, let $J_1 = \{i : \alpha_i^{\frac{p-1}{2}} = 1\}$ and $J_2 = \{i : \alpha_i^{\frac{p-1}{2}} = -1\}$. Using ζ, we can efficiently find β_i such that $\alpha_i = \beta_i^2$ for $i \in J_1$ and $\alpha_i = \beta_i^2 \zeta$ for $i \in J_2$. Let $w_1 = \sum_{i \in J_1} \beta_i^2 v_i$ and $w_2 = \sum_{i \in J_2} \beta_i^2 v_i$. Then $w_1 = -\zeta^d w_2$. Note that we are done if either of the sets J_1 or J_2 is empty.

Suppose that we have $C(d/2, m)$ groups, each consisting of $C(d/2, m)$ vectors of length m. For each i, we can build vectors $w_1(i)$ and $w_2(i)$ in the ith group with the properties of w_1 and w_2 above. Then we can express the zero vector as a linear combination with nonzero dth power coefficients from a subset of the vectors $w_1(i)$. Like in the quadratic case, we find four vectors, a scalar multiple of each other, represented as nontrivial linear combinations with dth power coefficients of four pairwise disjoint subsets of the original variables.

We can iterate this process. In the ℓth iteration we start with $C(d/2, m)$ groups, each consisting of $C(d/2, m)^{\ell-1}$ vectors of length m. At the end of the ℓth iteration we can find a nonzero vector w and scalars $\lambda_1, \ldots, \lambda_{2^\ell}$ together with representations of the vectors $\lambda_1 w, \ldots, \lambda_{2^\ell} w$ as linear combination with nonzero dth power coefficients of ℓ pairwise disjoint subsets of the original vectors.

After $\lceil \log_2(d + 1) \rceil \leq \log d + 1$ iterations, starting from at most $C(d/2, m)^{\log d + 1}$ input vectors, we get a vector w and scalars $\lambda_1, \ldots, \lambda_{d+1}$, together with the representations of the vectors $w_1 = \lambda_1 w, \ldots, w_{d+1} = \lambda_{d+1} w$ as above.

By Fact 1 we can find field elements $z_1, \ldots, z_{d+1}$ such that $\sum_{i=1}^{d+1} \lambda_i z_i^d = 0$, which implies that $\sum_{i=1}^{d+1} z_i^d w_i = 0$. The representations of $w_1, \ldots, w_{d+1}$ give then the desired representation of the zero vector. Observe that we have also shown that in that case $C(d, m) \leq C(d/2, m)^{\log d + 1}$.

3.2   The Algorithm When $\sqrt[d]{-1} \in \mathbb{F}_p$

We assume that $p - 1$ is divisible by d, we have a dth root μ of -1 as well as $\zeta_2, \ldots, \zeta_d$ in $\mathbb{F}_p$ at hand such that $\zeta_1 = 1, \zeta_2, \ldots, \zeta_d$ are a complete set of representatives of the cosets of $\mathbb{F}_p^{*d}$ in $\mathbb{F}_p^*$. To construct such elements $\mu, \zeta_2, \ldots, \zeta_d$ we need ρth non-residues for any prime factor ρ of $2d$. Such non-residues can be found in time polynomial in $\log p$ and d by random choice or a deterministic search assuming GRH [12].

For $\ell = 1, \ldots, d$, put $B_\ell(d, m) = d^{\frac{\ell(\ell-1)}{2}}(m + 1)^\ell$. For any ℓ-tuple $\underline{a} = (a_1, \ldots, a_\ell) \in \{1, \ldots, d\}^\ell$, for $s \in \{1, \ldots, d\}$ and for $1 \leq j \leq \ell$, set $\underline{a}(j, s) = (a_1, \ldots, a_{j-1}, s, a_{j+1}, \ldots, a_\ell)$.

Claim. From $B = B_\ell(d, m)$ input vectors $v_1, \ldots, v_B$, in time polynomial in B and $\log p$, we can find d^ℓ pairwise disjoint subsets $J_{\underline{a}} \subseteq \{1, \ldots, B\}$ and field elements $\beta_1, \ldots, \beta_B$ such that $J_{(1,\ldots,\ell)} \neq \emptyset$, and if we set $w_{\underline{a}} = \sum_{i \in J_{\underline{a}}} \beta_i^d v_i$, then we have

$$\sum_{s=1}^d \zeta_s w_{\underline{a}(j,s)} = 0, \text{ for every } \underline{a} \in \{1, \ldots, d\}^\ell \text{ and } j = 1, \ldots, \ell.$$

Proof. We prove it by recursion on ℓ. If $\ell = 1$ then any $B_\ell(d, m) = m+1$ vectors from $\mathbb{F}_p^m$ are linearly dependent. Therefore there exist $\alpha_1, \ldots, \alpha_{m+1} \in \mathbb{F}_p$, not all zero, such that $\sum_{i=1}^{m+1} \alpha_i v_i = 0$. For $r = 1, \ldots, d$, let J_r be the set of indices i such that there exists $\beta_i \in \mathbb{F}_p^*$ with $\alpha_i = \zeta_r \beta_i^d$. For $i \in J_r$, such a β_i can be efficiently found. At least one of the sets J_r is non-empty. If J_1 is empty then we multiply the coefficients α_i simultaneously by ζ_1/ζ_r^{-1} where J_r is nonempty to arrange that J_1 becomes nonempty.

To describe the recursive step, assume that we are given $B_{\ell+1}(d, m) = d^\ell(m + 1)B$ vectors. Put $E = d^\ell(m + 1)$, and for convenience assume that the input vectors are denoted by v_{ki}, for $k = 1, \ldots, E$ and $i = 1, \ldots, B$. By the recursive hypothesis, for every $k \in \{1, \ldots, E\}$, there exist subsets $J_{\underline{a}}(k) \subseteq \{1, \ldots, B\}$ and field elements $\beta_i(k)$ such that $J_{(1,\ldots,\ell)}(k) \neq \emptyset$, and with $w_{\underline{a}}(k) = \sum_{i \in J_{\underline{a}}(k)} \beta_i(k)^d v_{ki}$, we have

$$\sum_{s=1}^d \zeta_s w_{\underline{a}(j,s)}(k) = 0, \tag{8}$$

for every $\underline{a} \in \{1, \ldots, d\}^\ell$ and $j = 1, \ldots, \ell$.

For every $k = 1, \ldots, E$, let $W(k)$ be the concatenation of the vectors $w_{\underline{a}}(k)$ in a fixed, say the lexicographic, order of $\{1, \ldots, d\}^\ell$. Then the $W(k)$'s are vectors of length $d^\ell m < E$. Therefore there exist field elements $\alpha_1, \ldots, \alpha_E$, not all zero, such that $\sum_{i=k}^E \alpha(k)W(k) = 0$. For a k such that $\alpha(k) \neq 0$, let $\alpha(k) = \zeta_r \gamma(k)^d$ for some $1 \leq r \leq d$ and $\gamma(k) \in \mathbb{F}_p^*$. The index r and $\gamma(k)$ can be computed efficiently. For $r = 1, \ldots, d$, let M_r be the set of k's such that $\alpha(k) = \zeta_r \gamma(k)^d$. We can arrange that $M_{\ell+1}$ is nonzero by simultaneously multiplying the $\alpha(k)$'s by $\zeta_{\ell+1}/\zeta_r$ for some r, if necessary. Observe that we have

$$\sum_{s=1}^d \zeta_s \sum_{k \in M_s} \gamma(k)^d W(k) = 0. \tag{9}$$

For $i \in \{1, \ldots, B\}$ and $k \in \{1, \ldots, E\}$ set $\beta'_{ki} = \gamma(k)\beta_i(k)$. We fix $\underline{a}' \in \{1, \ldots, d\}^{\ell+1}$, and we set $\underline{a} = (a'_1, \ldots a'_\ell)$ and $r = a'_{\ell+1}$. We define $J'_{\underline{a}'} = \{(k,i) : k \in M_r \text{ and } i \in J_{\underline{a}}(k)\}$ and $w'_{\underline{a}'} = \sum_{(k,i) \in J'_{\underline{a}'}} \beta'^d_{ki} v_{ki}$. Then $w'_{\underline{a}'} = \sum_{k \in M_r} \gamma_k^d w_{\underline{a}}(k)$. This equality, together with the Eq. (8) imply that for every $j = 1, \ldots, \ell$, we have

$$\sum_{s=1}^d \zeta_s w_{\underline{a}'(j,s)} = 0.$$

Eq. (9) for $j - \ell + 1$ gives $\sum_{s=1}^{d} \zeta_s \sum_{k \in M_s} \gamma(k)^d w_{\underline{a}}(k) = 0$. Expanding $w_{\underline{a}}(k)$ in the inner sum $\sum_{k \in M_s} \gamma(k)^d w_{\underline{a}}(k)$ gives that it equals $w_{\underline{a}'(\ell+1,s)}$. Thus also

$$\sum_{s=1}^{d} \zeta_s w_{\underline{a}'(\ell+1,s)} = 0,$$

finishing the proof of the claim.

We apply the procedure of the claim for $\ell = d$. From any $B = B_d(d, m) = d^{\frac{d(d-1)}{2}}(m+1)^d$ input vectors $v_1, \ldots, v_B$, we compute in time polynomial in $\log p$ and B subsets $J_{\underline{a}}$, with $J_{(12\ldots d)} \neq \emptyset$, as well as nonzero elements $\beta_1, \ldots, \beta_B \in \mathbb{F}_p$ such that with $w_{\underline{a}} = \sum_{i \in J_{\underline{a}}} \beta_i^d v_i$, we have

$$\sum_{s=1}^{d} \zeta_s w_{\underline{a}(j,s)} = 0, \tag{10}$$

for every $j = 1, \ldots, d$ and for every $\underline{a} \in \{1, \ldots, d\}^d$.

Permutative tuples $\underline{a} \in S_d$ are of special interest. By $\mathrm{sgn}(\underline{a})$ we denote the *sign* of such a permutation, which is 1 if $\underline{a}$ is even and -1 if $\underline{a}$ is odd. We show that

$$\sum_{\underline{a} \in S_d} \mathrm{sgn}(\underline{a}) w_{\underline{a}} = 0. \tag{11}$$

For $\underline{a} \in S_d$, let $j_{\underline{a}}$ be the position of 1 in $\underline{a}$ and for every $s \in \{1, \ldots, d\}$, we denote by $\underline{a}[s]$ the sequence obtained from a by replacing 1 with s. Notice that $\underline{a}[s] = \underline{a}(j_{\underline{a}}, s)$, therefore (10) implies

$$\sum_{\underline{a} \in S_d} \mathrm{sgn}(\underline{a}) \sum_{s=1}^{d} \zeta_s w_{\underline{a}[s]} = 0.$$

We claim that

$$\sum_{\underline{a} \in S_d} \mathrm{sgn}(\underline{a}) \sum_{s=2}^{d} \zeta_s w_{\underline{a}[s]} = 0.$$

To see this, observe that for $s > 1$ the tuple $\underline{a}[s]$ has entries from $\{2, \ldots, d\}$, where s occurs twice, while the others once. Any such sequence $\underline{a}'$ can come from exactly two permutations which differ by a transposition: these are obtained from $\underline{a}'$ by replacing one of the occurrences of s with 1. Then (11) is just the difference of the above two equalities.

For $i \in J_{\underline{a}}$, let $\gamma_i = 0$ if $\underline{a}$ is not a permutation, $\gamma_i = \beta_i$ if $\underline{a}$ is an even permutation and $\gamma_i = \mu\beta_i$ if $\underline{a}$ is an odd permutation. Then (11) gives $\sum_{i=1}^{B} \gamma_i^d v_i = 0$, the required representation of the zero vector. Observe that in that case $C(d, m) \leq d^{\frac{d(d-1)}{2}}(m+1)^d$. The bounds obtained in the two cases imply that $C(d, m) \leq d^{d^2 \log d}(m+1)^{d \log d}$ in general.

Acknowledgements. Research was supported in part by the Hungarian Scientific Research Fund (OTKA) Grant NK105645, the Singapore Ministry of Education and the National Research Foundation Tier 3 Grant MOE2012-T3-1-009, by the European Commission IST STREP project Quantum Algorithms (QALGO) 600700, and the French ANR Blanc Program Contract ANR-12-BS02-005.

References

1. Ajtai, M.: Generating hard instances of lattice problems. In: 28th Annual ACM Symposium on Theory of Computing (STOC), pp. 99–108 (1996)
2. Bacon, D., Childs, A.M., van Dam, W.: From optimal measurement to efficient quantum algorithms for the hidden subgroup problem over semidirect product groups. In: 46th IEEE Symposium on Foundations of Computer Science (FOCS), pp. 469–478 (2005)
3. Chevalley, C.: Démonstration d'une hypothèse de M. Artin. Abh. Math. Sem. Hamburg **11**, 73–75 (1936)
4. Childs, A.M., Schulman, L., Vazirani, U.: Quantum algorithms for hidden nonlinear structures. In: 48th IEEE Symposium on Foundations of Computer Science (FOCS), pp. 395–404 (2007)
5. Decker, T., Draisma, J., Wocjan, P.: Quantum algorithm for identifying hidden polynomial function graphs. Quantum Inf. Comput. **9**, 0215–0230 (2009)
6. Decker, T., Høyer, P., Ivanyos, G., Santha, M.: Polynomial time quantum algorithms for certain bivariate hidden polynomial problems. Quantum Inf. Comput. **14**, 790–806 (2014)
7. Decker, T., Ivanyos, G., Santha, M., Wocjan, P.: Hidden symmetry subgroup problems. SIAM J. Comput. **42**, 1987–2007 (2013)
8. Denney, A., Moore, C., Russell, A.: Finding conjugate stabilizer subgroups in $PSL(2; q)$ and related groups. Quantum Inf. Comput. **10**, 282–291 (2010)
9. Friedl, K., Ivanyos, G., Magniez, F., Santha, M., Sen, P.: Hidden translation and translating coset in quantum computing. SIAM J. Comput. **43**, 1–24 (2014)
10. Grigni, M., Schulman, L., Vazirani M., Vazirani, U.: Quantum mechanical algorithms for the nonabelian hidden subgroup problem. In: 33rd ACM Symposium on Theory of Computing (STOC), pp. 68–74 (2001)
11. Hallgren, S., Russell, A., Ta-Shma, A.: Normal subgroup reconstruction and quantum computation using group representations. SIAM J. Comput. **32**, 916–934 (2003)
12. Huang, M-D.A: Riemann hypothesis and finding roots over finite fields. In: 17th Annual ACM Symposium on Theory of Computing (STOC), pp. 121–130 (1985)
13. Ivanyos, G., Santha, M.: On solving systems of diagonal polynomial equations over finite fields. arXiv:1503.09016 [cs.CC]
14. Ivanyos, G., Sanselme, L., Santha, M.: An efficient quantum algorithm for the hidden subgroup problem in nil-2 groups. Algoritmica **62**, 480–498 (2012)
15. Jozsa, R.: Quantum factoring, discrete logarithms, and the hidden subgroup problem. Comput. Sci. Engin. **3**, 34–43 (2001)
16. Karp, R.: Reducibility among combinatorial problems. In: Miller, R., Thatcher, J.W., Bohlinger, J.D. (eds.) Complexity of Computer Computations. The IBM Research Symposia Series, pp. 85–103. Springer, New York (1972)
17. Kitaev, A.Y.: Quantum measurements and the Abelian Stabilizer Problem (1995). arXiv:quant-ph/9511026v1

18. Kuperberg, G.: A subexponential-time quantum algorithm for the dihedral hidden subgroup problem. SIAM J. Comput. **35**, 170–188 (2005)
19. Moore, C., Rockmore, D., Russell, A., Schulman, L.: The power of basis selection in Fourier sampling: hidden subgroup problems in affine groups. In: 15th Annual ACM-SIAM Symposium on Discrete Algorithms, pp. 1113–1122 (2004)
20. Papadimitriou, C.: On the complexity of the parity argument and other inefficient proofs of existence. J. Comput. Syst. Sci. **48**, 498–532 (1994)
21. Regev, O.: Quantum computation and lattice problems. SIAM J. Comput. **33**, 738–760 (2004)
22. Shanks., D.: Five number-theoretic algorithms. In: 2nd Manitoba Conference on Numerical Mathematics, pp. 51–70 (1972)
23. Shor, P.: Algorithms for quantum computation: discrete logarithm and factoring. SIAM J. Comput. **26**, 1484–1509 (1997)
24. Sipser, M.: Introduction to the Theory of Computation. PWS Publishing Company, Boston (1997)
25. van de Woestijne, C.E.: Deterministic equation solving over finite fields. Ph.D. thesis, Universiteit Leiden (2006)
26. Warning, E.: Bemerkung zur vorstehenden Arbeit von Herrn Chevalley. Abh. Math. Sem. Hamburg **11**, 76–83 (1936)

Pattern Backtracking Algorithm
for the Workflow Satisfiability Problem
with User-Independent Constraints

Daniel Karapetyan[1], Andrei Gagarin[2], and Gregory Gutin[2(✉)]

[1] University of Nottingham, Nottingham, UK
Daniel.Karapetyan@gmail.com
[2] Royal Holloway, University of London, Surrey, UK
{Andrei.Gagarin,G.Gutin}@rhul.ac.uk

Abstract. The workflow satisfiability problem (WSP) asks whether there exists an assignment of authorised users to the steps in a workflow specification, subject to certain constraints on the assignment. (Such an assignment is called valid.) The problem is NP-hard even when restricted to the large class of user-independent constraints. Since the number of steps k is relatively small in practice, it is natural to consider a parametrisation of the WSP by k. We propose a new fixed-parameter algorithm to solve the WSP with user-independent constraints. The assignments in our method are partitioned into equivalence classes such that the number of classes is exponential in k only. We show that one can decide, in polynomial time, whether there is a valid assignment in an equivalence class. By exploiting this property, our algorithm reduces the search space to the space of equivalence classes, which it browses within a backtracking framework, hence emerging as an efficient yet relatively simple-to-implement or generalise solution method. We empirically evaluate our algorithm against the state-of-the-art methods and show that it clearly wins the competition on the whole range of our test problems and significantly extends the domain of practically solvable instances of the WSP.

1 Introduction

In the *workflow satisfiability problem* (WSP), we aim at assigning authorised users to the steps in a workflow specification, subject to some constraints arising from business rules and practices. The WSP has applications in information access control (e.g. see [1–3]), and it is extensively studied in the security research community [2,3,8,14]. In the WSP, we are given a set U of *users*, a set S of *steps*, a set $\mathcal{A} = \{A(u) : u \in U\}$ of *authorisation lists*, where $A(u) \subseteq S$ denotes the set of steps for which user u is authorised, and a set C of *(workflow) constraints*. In general, a *constraint* $c \in C$ can be described as a pair $c = (T, \Theta)$, where $T \subseteq S$ is the *scope* of the constraint and Θ is a set of functions from T to U which specifies those assignments of steps in T to users in U that satisfy the constraint (authorisations disregarded). Authorisations and constraints described in WSP

© Springer International Publishing Switzerland 2015
J. Wang and C. Yap (Eds.): FAW 2015, LNCS 9130, pp. 138–149, 2015.
DOI: 10.1007/978-3-319-19647-3_13

literature are relatively simple such that we may assume that all authorisations and constraints can be checked in polynomial time (in $|U|$, $|S|$ and $|C|$).

Given a *workflow* $W = (S, U, \mathcal{A}, C)$, W is *satisfiable* if there exists a function $\pi : S \to U$ such that

- for all $s \in S$, $s \in A(\pi(s))$ (each step is allocated to an authorised user);
- for all $(T, \Theta) \in C$, $\pi|_T \in \Theta$ (every constraint is satisfied).

A function $\pi : S \to U$ is an *authorised* (*eligible*, *valid*, respectively) *complete plan* if it satisfies the first condition above (the second condition, both conditions, respectively).

For example, consider the following instance of WSP. The step and user sets are $S = \{s_1, s_2, s_3, s_4\}$ and $U = \{u_1, u_2, \ldots, u_5\}$. The authorisation lists are $A(u_1) = \{s_1, s_2, s_3, s_4\}$, $A(u_2) = \{s_1\}$, $A(u_3) = \{s_2\}$, $A(u_4) = A(u_5) = \{s_3, s_4\}$. The constraints are $(s_1, s_2, =)$ (the same user must be assigned to s_1 and s_2), $(s_2, s_3, \neq)$ (s_2 and s_3 must be assigned to different users), $(s_3, s_4, \neq)$, and $(s_4, s_1, \neq)$. Since the function π assigning u_1 to s_1 and s_2, u_4 to s_3, and u_5 to s_4 is a valid complete plan, the workflow is satisfiable.

Clearly, not every workflow is satisfiable, and hence it is important to be able to determine whether a workflow is satisfiable or not and, if it is satisfiable, to find a valid complete plan. Unfortunately, the WSP is NP-hard [14] and, since the number k of steps is usually relatively small in practice (usually $k \ll n = |U|$ and we assume, in what follows, that $k < n$), Wang and Li [14] introduced its parameterisation[1] by k. Algorithms for this parameterised problem were also studied in [4–6,9]. While in general the WSP is W[1]-hard [14], the WSP restricted[2] to some practically important families of constraints is fixed-parameter tractable (FPT) [5,9,14]. (Recall that a problem parameterised by k is FPT if it can be solved by an FPT algorithm, i.e. an algorithm of running time $O^*(f(k))$, where f is an arbitrary function depending on k only, and O^* suppresses not only constants, but also polynomial factors in k and other parameters of the problem formulation.)

Many business rules are not concerned with the identities of the users that perform a set of steps. Accordingly, we say a constraint $c = (T, \Theta)$ is *user-independent* if, whenever $\theta \in \Theta$ and $\phi : U \to U$ is a permutation, then $\phi \circ \theta \in \Theta$. In other words, given a complete plan π that satisfies c and any permutation $\phi : U \to U$, the plan $\pi' : S \to U$, where $\pi'(s) = \phi(\pi(s))$, also satisfies c. The class of user-independent constraints is general enough in many practical cases; for example, all the constraints defined in the ANSI RBAC standard [1] are user-independent. Most of the constraints studied in [4,6,9,14] and other papers are also user-independent. Classical examples of user-independent constraints are the requirements that two steps are performed by either two different users (*separation-of-duty*), or the same user (*binding-of-duty*). More complex

---

[1] We use terminology of the recent monograph [10] on parameterised algorithms and complexity.

[2] While we consider special families of constraints, we do not restrict authorisation lists.

constraints state that at least/at most/exactly r users are required to complete some sensitive set of steps (these constraints belong to the family of *counting* constraints), where r is usually small. A simple reduction from GRAPH COLOURING shows that the WSP restricted to the separation-of-duty constraints is already NP-hard [14].

The WSP is an important applied problem and is thoroughly studied in the literature. However, as was shown by Cohen et al. [4], the methods developed so far were capable of solving user-independent WSP instances only for relatively small values of k. In this paper we propose a new approach that, compared to the existing solution methods, significantly extends the number of steps in practically solvable instances now covering the values of k expected in the majority of real-world instances. Importantly, the proposed method is relatively simple to implement or extend with new constraints, such that its accessibility is similar to that of SAT-solvers used by practitioners [14].

The proposed solution method is a deterministic algorithm that uses backtracking to browse the space of all the equivalence classes of partial solutions. We show that it is possible to test efficiently if there exists an authorised complete plan in a given equivalence class. This makes our algorithm FPT as the number of equivalence classes is exponential in k only.

2   Patterns and the User-Iterative Algorithm

A *plan* is a function $\pi : T \to U$, where $T \subseteq S$ (note that if $T = S$ then π is a complete plan). We define an *equivalence relation* on the set of all plans, which is a special case of an equivalence relation defined in [5]. For user-independent constraints, two plans $\pi : T \to U$ and $\pi' : T' \to U$ are *equivalent*, denoted by $\pi \approx \pi'$, if and only if $T = T'$, and $\pi(s) = \pi(t)$ if and only if $\pi'(s) = \pi'(t)$ for every $s, t \in T$. Assuming an ordering $s_1, s_2, \ldots, s_k$ of steps S, every plan $\pi : T \to U$ can be encoded into a *pattern* $P = P(\pi) = (x_1, \ldots, x_k)$ defined by:

$$
x_i = \begin{cases}
0 & \text{if } s_i \notin T, \\
1 & \text{if } i = 1 \text{ and } s_1 \in T, \\
x_j & \text{if } \pi(s_i) = \pi(s_j) \text{ and } j < i, \\
\max\{x_1, x_2, \ldots, x_{i-1}\} + 1 & \text{otherwise.}
\end{cases} \tag{1}
$$

The pattern $P(\pi)$ uniquely encodes the equivalence class of π, and $P(\pi) = P(\pi')$ for every π' in that equivalence class [5]. The pattern P represents an assignment of steps in T to some users in any plan of the equivalence class of π. We say that a pattern is *complete* if $x_i \neq 0$ for $i = 1, 2, \ldots, k$.

The state-of-the-art FPT algorithm for the WSP with counting constraints proposed in [4] and called here User-Iterative (UI), iterates over the set of users and gradually computes all encoded equivalence classes of valid plans until it finds a complete solution to the problem, or all the users have been considered. Effectively, it uses the breadth-first search in the space of plans. In the breadth-first search tree, equivalent plans can be generated but they are detected efficiently using patterns and the corresponding search branches are then merged together. Since the UI algorithm generates a polynomial number of plans per equivalence class and the number of equivalence classes is exponential in k only, the UI algorithm is FPT. The results of [4,6] show that the generic user-iterative FPT algorithm of [5] has a practical value, and its implementations are able to outperform the well-known pseudo-Boolean SAT solver SAT4J [13].

In this paper we propose a new FPT solution method for the WSP which also exploits equivalence classes and patterns but in a more efficient manner. Among other advantages, our algorithm never generates multiple plans within the same equivalence class. For further comparison of our algorithm with the UI algorithm, see Sect. 4.

3   The Pattern-Backtracking Algorithm

We call our new method *Pattern-Backtracking* (PB) as it uses the backtracking approach to browse the search space of patterns. To describe it, we introduce several additional notations. We will say that a plan $\pi : T \to U$ is *authorised* if $s \in A(\pi(s))$ for every $s \in T$, *eligible* if it does not violate any constraint in C, and *valid* if it is both authorised and eligible. Similarly, a pattern P is *authorised, eligible* or *valid* if there exists a plan π such that $P(\pi) = P$ and π is authorised, eligible or valid, respectively. By $P(s_i)$ we denote the value x_i in $P = (x_1, x_2, \ldots, x_k)$. We also use notations $A^{-1}(s) = \{u \in U : s \in A(u)\}$ for the set of users authorised for step $s \in S$ and $P^{-1}(x_i) = \{s \in S : P(s) = x_i\}$ for all the steps assigned to the same user encoded by the value of x_i in P. Note that $P^{-1}(x_i) \neq \emptyset$ for $i = 1, 2, \ldots, k$ for any complete pattern.

In Sect. 3.1 we show how to find a valid plan for an eligible pattern, which is an essential part of our algorithm, and in Sect. 3.2 we describe the algorithm itself.

3.1   Pattern Validity Test

The PB algorithm searches the space of patterns; once an eligible complete pattern P is found, we need to check if it is valid and, if it is, then to find a plan π such that $P = P(\pi)$. The following theorem allows us to address these two questions efficiently.

For a complete pattern $P = (x_1, x_2, \ldots, x_k)$, let $X = \{x_i : i = 1, 2, \ldots, k\}$ (note that the cardinality of the set X may be smaller than k). Let $G = (X \cup U, E)$ be a bipartite graph, where $(x_i, u) \in E$ if and only if $u \in A^{-1}(s)$ for each $s \in P^{-1}(x_i)$, $x_i \in X$.

Theorem 1. *A pattern P is authorised if and only if G has a matching of size $|X|$.*

Proof. Suppose M is a matching of size $|X|$ in G. Construct a plan π as follows: for each edge $(x_i, u) \in M$ and $s \in P^{-1}(x_i)$, set $\pi(s) = u$. Since M covers x_i for every $i = 1, 2, \ldots, k$ and P is a complete pattern, the above procedure defines $\pi(s)$ for every step $s \in S$. Hence, π is a complete plan. Now observe that, for each $x_i \in X$, all the steps $P^{-1}(x_i)$ are assigned to exactly one user, and if $x_i \neq x_j$ for some $i, j \in \{1, 2, \ldots k\}$, then $\pi(s_i) \neq \pi(s_j)$ by definition of the matching. Therefore $P(\pi) = P$. Observe also that π respects the authorisation lists; for each edge $(x_i, u) \in M \subseteq E$ and each step $s \in P^{-1}(x_i)$, we guarantee that $u \in A(s)$. Thus, plan π is authorised and, hence, pattern $P = P(\pi)$ is also authorised.

On the other hand, assume there exists an authorised plan π such that $P(\pi) = P$ for a given pattern P. Let $X = \{x_i : i = 1, 2, \ldots, k\}$. Construct a set M as follows: $M = \{(x_i, u) : x_i \in X \text{ and } \exists s \in P^{-1}(x_i) \text{ s.t. } u = \pi(s)\}$. Consider a pair $(x_i, u) \in M$, and find some $s \in P^{-1}(x_i)$. Note that, as $P = P(\pi)$ and by definition of pattern, $\pi(s') = u$ for every $s' \in P^{-1}(x_i)$. Since π is authorised, $u \in A(s')$ for every $s' \in P^{-1}(x_i)$, i.e. $(x_i, u) \in E$ and $M \subseteq E$. In other words, M is a subset of edges of G.

Now notice that, for each $x_i \in X$, there exists at most one edge $(x_i, u) \in M$ as $\pi(s') = u$ for every $s' \in P^{-1}(x_i)$. Moreover, for each $u \in U$, there exists at most one edge $(x_i, u) \in M$ as otherwise there would exist some $i, j \in \{1, 2, \ldots, k\}$ such that $\pi(s_i) = \pi(s_j)$ and $x_i \neq x_j$, which violates $P = P(\pi)$. Hence, the edge set M is disjoint. Finally, $|M| = |X|$ because $P^{-1}(x_i)$ is non-empty for every $x_i \in X$. We conclude that M is a matching in G of size $|X|$. $\qquad\square$

Theorem 1 implies that, to determine whether an eligible pattern P is valid, it is enough to construct the bipartite graph $G = (X \cup U, E)$ and to find a maximum size matching in G. It also provides an algorithm for converting a maximum matching M of size $|X|$ in G into a valid plan π such that $P(\pi) = P$.

The matching problem arising in Theorem 1 has some interesting properties:

– The bipartite graph $G = (X \cup U, E)$ is highly unbalanced as $|X| \leq k$, and we assume that $|U| \gg k$. It is easy to see that the maximum length of an augmenting path in G is $|X| \leq k$ and, hence, the time complexity of the Hungarian and Hopcroft-Karp methods are $O(k^3)$ and $O(k^{2.5})$, respectively.
– We are interested only in matchings of size $|X|$. If the maximum matching is of a smaller size, we do not need to retrieve it.
– Once a matching of size $|X|$ is found, the PB algorithm terminates since a valid plan is found. However, the algorithm might test an exponential number (in k) of graphs with the maximum matching of size smaller than $|X|$. Hence, we are mainly interested in time of checking whether the maximum matching is smaller than $|X|$.

To exploit the above features, we use the Hungarian method with a speed-up heuristic provided by the following proposition.

Proposition 1. *If $M = \{(x_{\chi(1)}, u_{\psi(1)}), \ldots, (x_{\chi(t)}, u_{\psi(t)})\}$, $t < |X|$, is a matching in the graph $G = (X \cup U, E)$ such that there exists no M-augmenting path in G starting at a vertex $x_{\chi(t+1)} \in X$, then there is no matching covering all the vertices of X in G.*

Proof. W.l.o.g, assume that $M = \{(x_1, u_1), \ldots, (x_t, u_t)\}$ and $x_{\chi(t+1)} = x_{t+1}$. Now suppose that G has a matching

$$M' = \{(x_1, y_1), (x_2, y_2), \ldots, (x_k, y_k)\}, \; y_i \in U, \; i = 1, 2, \ldots, k$$

covering all of X. Consider the symmetric difference of two matchings $H = (M \cup M') \setminus (M \cap M')$. Since every vertex of H has degree at most 2, the graph induced by H consists of some disjoint paths and even cycles having edges alternating between M and M'. Since x_{t+1} is not in M but covered by M', it is an end point of one of the alternating paths in H, say P_{t+1}. Now, it is possible to see that P_{t+1} is an augmenting path in G with respect to the matching M: since, starting at x_{t+1}, every time we use an edge of M' to go from a vertex in X to a vertex in U, P_{t+1} must have an end point in a vertex of U not covered by M (M' covers all the vertices in X). This contradicts the fact that there is no augmenting path in G starting at x_{t+1} with respect to M.   □

The result of Proposition 1 allows us to terminate the Hungarian algorithm as soon as a vertex $x_i \in X$ is found such that no augmenting path starting from x_i can be obtained. Construction of the graph takes $O(kn)$ time, and solving the maximum matching problem in G with the Hungarian method takes $O(k^3)$ time.

3.2   The Backtracking Algorithm

The PB algorithm uses a backtracking technique to search the space of patterns, and for each eligible pattern, it verifies whether such a pattern is valid. If a valid pattern P is found, the algorithm returns a complete plan π such that $P(\pi) = P$, see Sect. 3.1 for details. If no valid pattern is found during the search, the instance is unsatisfiable.

The calling procedure for the PB algorithm is shown in Algorithm 1, which in turn calls the recursive search function in Algorithm 2. The recursive function tries all possible extensions P' of the current pattern P by adding one new step s to it (line 12). The step s is selected heuristically (line 9), where function $\rho(s)$ is an empirically tuned function indicating the importance of step s in narrowing down the search space. The implementation of $\rho(s)$ depends on the specific types of constraints involved in the instance and should reflect the intuition regarding the structure of the problem. See (2) in Sect. 5 for a particular implementation of $\rho(s)$ for the types of constraints we used in our computational study. Note that our branching heuristic dynamically changes the steps ordering used in the pattern definition in Sect. 2. Nevertheless, this does not affect any theoretical properties of the pattern.

We use a heuristic (necessary but not sufficient) test (lines 13–15 of Algorithm 2) to check whether the pattern P' can be authorised; that allows us to prune branches which are easily provable to include no authorised patterns.

In line 16, the algorithm checks whether the new pattern P' violates any constraints and, if not, then executes the recursive call.

Algorithm 1. Backtracking search initialisation (entry procedure of PB)

---

input : WSP instance $W = (S, U, \mathcal{A}, C)$
output: Valid plan π or UNSAT
1 Initialise $P(s) \leftarrow 0$ for each $s \in S$;
2 $\pi \leftarrow$ Recursion(P);
3 **return** π (π may be UNSAT here);

---

4   Comparison of the PB and UI Algorithms

In this section we analyse the time and memory complexity of the PB algorithm and compare it to the UI algorithm.

Observe that each internal node (corresponding to an incomplete plan) in the search tree of the PB algorithm has at least two children, and each leaf in this tree corresponds to a complete pattern. Thus, the total number of patterns considered by the PB algorithm is less than twice the number of complete patterns. Observe that the number of complete patterns equals the number of partitions of a set of size k, i.e. the kth Bell number B_k. Finally, observe that the PB algorithm spends time polynomial in n on each node of the search tree.[3] Thus, the time complexity of the PB algorithm is $O^*(B_k)$. The PB algorithm follows the depth-first search order and, hence, stores only one pattern at a time. At each leaf node, it also solves the matching problem generating a graph with $O(kn)$ edges. Hence, the memory complexity of the algorithm is $O(kn)$.

It is interesting to compare the PB algorithm to the UI algorithm (briefly described in Sect. 2). Despite both algorithms using the idea of equivalence classes and being FPT, they have very different working principles and properties.

1. Observe that, in the worst case, the UI algorithm may store all patterns, and the number of patterns is B_{k+1}. Indeed, consider a pattern $P = (x_1, \ldots, x_k)$ and a set $\{s_1, \ldots, s_k, s_{k+1}\}$. Then each partition of the set corresponds to a pattern of P, where $x_i = 0$ if and only if s_i and s_{k+1} are in the same subset of the partition. Therefore, the UI algorithm takes $O(kB_{k+1})$ memory, which is in sharp contrast to the PB algorithm that requires very little memory. Considering that, e.g. $B_{20} = 51\,724\,158\,235\,372$, memory consumption poses a serious bottleneck for the UI algorithm as the RAM capacity of any mainstream machine is well below the value of B_{20}. Moreover, the UI algorithm accesses a large volume of data in a non-sequential order, which might have a dramatic effect on the algorithm's performance when implemented on a real machine as shown in [12].
2. From the practical point of view, the PB algorithm considers less patterns than the UI algorithm ($O(B_k)$ vs. $O(B_{k+1})$) as the PB algorithm assigns the steps in a strict order, avoiding generation of duplicate patterns. Moreover,

---

[3] Assuming that the WSP instance does not include any exotic constraints.

Algorithm 2. Recursion(P) (recursive function for backtracking search)

---

    input : Pattern P
    output: Eligible plan or UNSAT if no eligible plan exists in this branch of the
              search tree

1   Initialise the set $S' \subseteq S$ of assigned steps $S' \leftarrow \{s \in S : P(s) \neq 0\}$;
2   **if** $S' = S$ **then**
3   Verify if pattern P is valid (using the matching algorithm of Theorem 1);
4   **if** *pattern P is valid* **then**
5   **return** *plan π realising P*;

6   **else**
7   **return** *UNSAT*;

8   **else**
9   Select unassigned step $s \in S \setminus S'$ that maximises $\rho(s)$;
10   Calculate $k' \leftarrow 1 + \max_{s \in S} P(s)$;
11   **for** $x = 1, 2, \ldots, k'$ **do**
12   Set $P(s) \leftarrow x$ to obtain a new pattern P';
13   Compute the set of steps Q assigned to x: $Q = \{t \in S : P'(t) = x\}$;
14   **if** $|\bigcap_{t \in Q} A^{-1}(t)| = 0$ **then**
15   Proceed to the next value of x (reject P');

16   **if** P' *is an eligible pattern* **then**
17             $\pi \leftarrow$ Recursion(P');
18   **if** $\pi \neq UNSAT$ **then**
19   **return** π;

20 **return** *UNSAT (for a particular branch of recursion; does not mean that the whole instance is unsat)*;

---

    the PB algorithm generates each pattern at most once, while the UI algorithm is likely to generate a pattern several times rejecting the duplicates afterwards.
3. Both algorithms use heuristics to determine the order in which the search tree is explored. However, while the UI algorithm has to use a certain fixed order of users for all the search branches, the PB algorithm has the flexibility of changing the order of steps in each branch of the search. Note that the order of assignments is crucial to the algorithm's performance as it can help to prune branches early.

5   Computational Experiments

In this section we empirically verify the efficiency of the PB algorithm. We compare the following WSP solvers:

PB. The algorithm proposed in this paper;
UI. Another FPT algorithm proposed in [5] and evaluated in [4,6];
SAT4J. A pseudo-Boolean SAT formulation [4,6] of the problem solved with SAT4J.

Due to the difficulty of acquiring real-world WSP instances [4,14], we use the random instance generator described in [4]. Three families of user-independent constraints are used: *not-equals* (also called *separation-of-duty*) constraints $(s, t, \neq)$, *at-most-r* constraints $(r, Q, \leqslant)$ and *at-least-r* constraints $(r, Q, \geqslant)$. A not-equals constraint $(s, t, \neq)$ is satisfied by a complete plan π if and only if $\pi(s) \neq \pi(t)$. An at-most-r constraint $(r, Q, \leqslant)$ is satisfied if and only if $|\pi(Q)| \leq r$, where Q is the scope of the constraint. Similarly, an at-least-r constraint $(r, Q, \geqslant)$ is satisfied if and only if $|\pi(Q)| \geq r$. We do not explicitly consider the widely used binding-of-duty constraints, that require two steps to be assigned to one user, as those can be trivially eliminated during preprocessing. While the binding-of-duty and separation-of-duty constraints provide the basic modelling capabilities, the at-most-r and at-least-r constraints impose more general "confidentiality" and "diversity" requirements on the workflow, which can be important in some business environments.

The instance generator (available for downloading [11]) takes four parameters: the number of steps k, the number of not-equals constraints e, the number of at-most and at-least constraints c and the random generator seed value. Each instance has $n = 10k$ users. For each user $u \in U$, it generates a uniformly random authorisation list $A(u)$ such that $|A(u)|$ is selected uniformly from $\{1, 2, \ldots, \lceil 0.5k \rceil\}$ at random. It also generates e distinct not-equals, c at-most and c at-least constraints uniformly at random. All at-most and at-least constraints are of the form $(3, Q, \sigma)$, where $|Q| = 5$ and $\sigma \in \{\leqslant, \geqslant\}$.

Our test machine is based on two Intel Xeon CPU E5-2630 v2 (2.6 GHz) and has 32 GB RAM installed. Hyper-threading is enabled, but we never run more than one experiment per physical CPU core concurrently. The PB algorithm is implemented in C#, and the UI algorithm is implemented in C++. Concurrency is not exploited in any of the tested solution methods. The PB algorithm is also available for downloading [11].

The branching heuristic implemented in line 9 of Algorithm 2 selects a step $s \in S$ that maximises a ranking function $\rho(s)$:

$$\rho(s) = c_{\neq}(P) + \alpha c_{\leq}^0(P) + \beta c_{\leq}^1(P) + \gamma c_{\leq}^2(P), \tag{2}$$

where $c_{\neq}(P)$ is the number of not-equals constraints involving step s, $c_{\leq}^i(P)$ is the number of at-most-r constraints involving s such that $r - i$ distinct users are already assigned to it, and α, β and γ are parameters. The intuition is that the steps s that maximise $\rho(s)$ are tightening the search space quickly. The parameters α, β and γ were selected empirically. We found out that the algorithm is not very sensitive to the values of these parameters, and settled down at $\alpha = 100$, $\beta = 2$ and $\gamma = 1$. Note that the function does not account for at-least constraints. This reflects our empirical observation that the at-least constraints are usually relatively weak in our instances and rarely help in pruning branches of search.

We started from establishing what parameter values make the instances hard. However, due to the lack of space, we provide only the conclusions drawn from this series of experiments. As it could be expected, greatly under- and over-subscribed instances are easier to solve than the instances in the region between

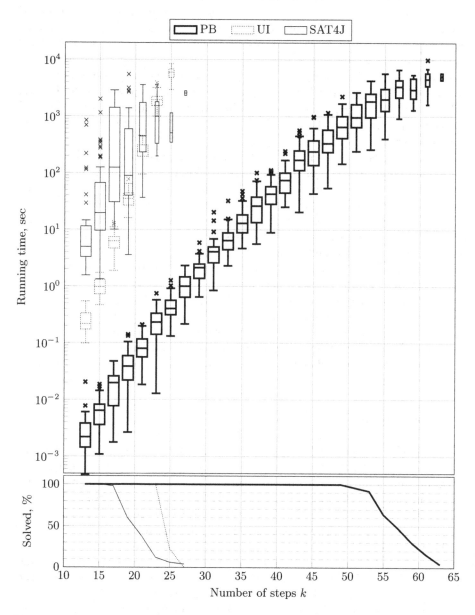

Fig. 1. Running time vs. number of steps k. For each k, we generate 50 instance sets with different seed values. The distributions are presented in the boxplot form, where the width of a box is proportional to the number of instance sets on which the solver succeeded. The plot at the bottom of the figure also shows the success rate of each solver.

those two extremes. The behaviour of the analysed solvers is consistent in this regard. The particular values of the number of not-equals constraints e and the number c of at-most and at-least constraints that make the instances most challenging depend on k. Thus, in our final experiment, which is to establish the maximum size k of instances practically solvable by each of the methods, we considered several instances with a range of parameters to ensure that at least one of them is hard. In particular, we fixed the *density* of not-equals constraints, calculated as $d = \frac{2e}{k(k-1)} \cdot 100\,\%$, at $d = 10\,\%$ and the number c of at-most and at-least constraints at each of $c = 1.0k$, $c = 1.2k$ and $c = 1.4k$, producing three instances for each k and seed value.

Each solver is given one hour limitation for each instance from the set. If a solver fails on at least one of the instances (could not terminate within 1 hour), we say that it fails on the whole set. The intention is to make sure that the solver can tackle hard satisfiable and unsatisfiable instances within a reasonable time. The results are presented in Fig. 1 in the form of boxplots. The percentage of runs in which the solver succeeded is shown as the width of the box. This information is also provided at the bottom of Fig. 1.

The PB algorithm, being faster than the two other methods by several orders of magnitude, reliably solves all the instances of size up to $k = 49$. Compare it to the UI and SAT4J solvers that succeed only for $k \leq 23$ and $k \leq 15$, respectively. Moreover, its running time grows slower than that of the UI and SAT4J solvers, which indicates that it has higher potential if more computational power is allocated. In other words, thanks to our new solution method, the previously unapproachable problem instances of practical sizes can now be routinely tackled.

6   Conclusion

We proposed a new FPT algorithm for the WSP with user-independent constraints. Our experimental analysis have shown that the new algorithm outperforms all the methods in the literature by several orders of magnitude and significantly extends the domain of practically solvable instances. Another advantage of the new FPT algorithm is that it is relatively easy to implement and extend; for example, it is straightforward to parallelise it.

Future research is needed to establish further potential to improve the algorithm's performance. Particular attention has to be paid to the branching heuristic. Thorough empirical analysis has to be conducted to investigate the performance of the algorithms on easy and hard instances.

Another relevant subject was recently studied in [7]; the paper introduces an optimisation version of WSP and proposes an FPT branch and bound algorithm inspired by Pattern Backtracking.

Acknowledgment. This research was partially supported by EPSRC grants EP/H000968/1 (for DK) and EP/K005162/1 (for AG and GG). The source codes of the Pattern Backtracking algorithm and the instance generator are publicly available [11].

References

1. American National Standards Institute. ANSI INCITS 359-2004 for Role Based Access Control (2004)
2. Basin, D.A., Burri, S.J., Karjoth, G.: Obstruction-free authorisation enforcement: aligning security and business objectives. J. Comput. Secur. **22**(5), 661–698 (2014)
3. Bertino, E., Ferrari, E., Atluri, V.: The specification and enforcement of authorisation constraints in workflow management systems. ACM Trans. Inf. Syst. Secur. **2**(1), 65–104 (1999)
4. Cohen, D., Crampton, J., Gagarin, A., Gutin, G., Jones, M.: Engineering algorithms for workflow satisfiability problem with user-independent constraints. In: Chen, J., Hopcroft, J.E., Wang, J. (eds.) FAW 2014. LNCS, vol. 8497, pp. 48–59. Springer, Heidelberg (2014)
5. Cohen, D., Crampton, J., Gagarin, A., Gutin, G., Jones, M.: Iterative plan construction for the workflow satisfiability problem. J. Artif. Intell. Res. **51**, 555–577 (2014)
6. Cohen, D., Crampton, J., Gagarin, A., Gutin, G., Jones, M.: Algorithms for the workflow satisfiability problem engineered for counting constraints. J. Combin. Optim. 22 (2015, to apppear). doi:10.1007/s10878-015-9877-7
7. Crampton, J., Gutin, G., Karapetyan, D.: Valued workflow satisfiability problem. In: Proceedings of ACM Symposium on Access Control Models and Technologies (SACMAT), Vienna, Austria, 1–3 June. ACM (2015, to appear)
8. Crampton, J.: A reference monitor for workflow systems with constrained task execution. In: Ferrari, E., Ahn, G.J. (eds.) SACMAT, pp. 38–47. ACM (2005)
9. Crampton, J., Gutin, G., Yeo, A.: On the parameterized complexity and kernelization of the workflow satisfiability problem. ACM Trans. Inf. Syst. Secur. **16**(1), 4 (2013)
10. Downey, R.G., Fellows, M.R.: Fundamentals of Parameterized Complexity. Springer, London (2013)
11. Karapetyan, D., Gutin, G., Gagarin, A.: Source codes of the Pattern Backtracking algorithm and the instance generator. doi:10.6084/m9.figshare.1360237. Accessed 31 March 2015
12. Karapetyan, D., Gutin, G., Goldengorin, B.: Empirical evaluation of construction heuristics for the multidimensional assignment problem. In: Proceedings of London Algorithmics 2008: Theory and Practice, Texts in Algorithmics 11, pp. 107–122. College Publications (2009)
13. Le Berre, D., Parrain, A.: The SAT4J library release 2.2. J. Satisf. Bool. Model. Comput. **7**, 59–64 (2010)
14. Wang, Q., Li, N.: Satisfiability and resiliency in workflow authorisation systems. ACM Trans. Inf. Syst. Secur. **13**(4), 40 (2010)

On the Sound Covering Cycle Problem
in Paired de Bruijn Graphs

Christian Komusiewicz[1]([⊠]) and Andreea Radulescu[2]

[1] Institut für Softwaretechnik und Theoretische Informatik,
TU Berlin, Berlin, Germany
`christian.komusiewicz@tu-berlin.de`
[2] LINA - UMR CNRS 6241, Université de Nantes, Saint-Nazaire, France
`andreea.radulescu@etu.univ-nantes.fr`

Abstract. Paired de Bruijn graphs are a variant of classic de Bruijn graphs used in genome assembly. In these graphs, each vertex v is associated with two labels $\mathcal{L}(v)$ and $\mathcal{R}(v)$. We study the NP-hard SOUND COVERING CYCLE problem which has as input a paired de Bruijn graph G and two integers d and ℓ, and the task is to find a length-ℓ cycle C containing all arcs of G such that for every vertex v in C and the vertex u which occurs exactly d positions after v in C, we have $\mathcal{R}(v) = \mathcal{L}(u)$. We present the first exact algorithms for this problem and several variants.

1 Introduction

DNA sequencing is the task of deciphering the sequence of a given DNA fragment. Most technologies approach this task by obtaining a collection of possibly overlapping small subfragments, called *reads*, of the given fragment. Genome assembly aims at recovering the original DNA fragment from the set of reads. When no reference genome is used, this is called *de novo* assembly.

Recent sequencing technologies, known as next-generation sequencing (NGS), create billions of very short erroneous reads. For this new type of data, *de novo* assembly is a challenging task [3,5,11,13]. The use of short reads makes it particularly difficult to correctly assemble repeated regions since the repeat may be longer than the reads. Therefore, many NGS methods generate pairs of reads separated by a known distance, called insert length, that is much longer than the read length. This new type of reads, called *paired-end reads*, is used to span regions that contain long repeats.

One classic approach in de novo genome assembly is the *de Bruijn* method [6,9] which computes a *de Bruijn graph* from the read set. This is done as follows. First, generate the set of k-mers of the reads, that is, the set of all length-k strings that occur as substrings of at least one read. Each k-mer in this set corresponds to exactly one vertex of the de Bruijn graph. Now draw an arc from a vertex u to a vertex v if and only if there is a $k + 1$-mer s in one of the input reads such

C. Komusiewicz – Partially supported by the DAAD Procope program (project 55934856).

J. Wang and C. Yap (Eds.): FAW 2015, LNCS 9130, pp. 150–161, 2015.
DOI: 10.1007/978-3-319-19647-3_14

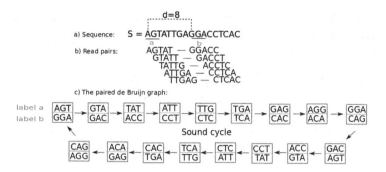

Fig. 1. A paired de Bruijn graph with a sound cycle: (a) The DNA fragment to be sequenced. (b) The set of paired reads with insert distance $d = 8$. (c) The paired de Bruijn graph constructed from the paired reads in (b) for a $k = 3$. The upper part of each vertex v is the label $\mathcal{L}(v)$, the lower part is the label $\mathcal{R}(v)$.

that the k-mer corresponding to u is a prefix of s and the k-mer corresponding to v is a suffix of s. A walk in this graph then corresponds to a DNA sequence. Classic de Bruijn assembling software uses paired-end information only in a post-processing step.

The *paired de Bruijn graph* model incorporates the paired-end information directly into the graph [8]. It is based on the classic de Bruijn graph, though now the overlaps are computed between pairs of k-mers separated by the insert length of the read pairs. This improves assembly quality and facilitates repeat detection but the computational problems involved in computing an assembly in these graphs become more challenging. In particular, the SOUND COVERING CYCLE which we define below is NP-hard [7]. In this work, we present the first exact algorithms for SOUND COVERING CYCLE and several variants.

Preliminaries. For a string s, let $s[i]$ denote the letter at position i of s and $s[i, j]$ the substring of s that starts at position i and ends at position j. We consider directed graphs $G = (V, A)$ with vertex set V and arc set A. A *walk* $(v_1, \ldots, v_p)$ is a tuple of vertices such that $(v_i, v_{i+1}) \in A$, $1 \le i < p$. The *length* $|W|$ of a walk $W := (v_1, \ldots, v_p)$ is the number p of tuple elements. A walk is *simple* if $i \ne j$ implies $v_i \ne v_j$. Given two walks $W_1 = (v_1, \ldots, v_p)$ and $W_2 = (u_1, \ldots, u_q)$ such that $(v_p, u_1) \in A$, let $W_1 \cdot W_2 := (v_1, \ldots, v_p, u_1, \ldots, u_q)$ denote the *concatenation* of W_1 and W_2. For a walk $W = (v_1, \ldots, v_p)$, let $W[i] := v_i$ denote the i-th vertex of W. Finally, let $A(W)$ denote the set of arcs contained in a walk W.

Paired de Bruijn Graphs. Before defining the problem, we describe how the paired de Bruijn graph is constructed from the input data; see Fig. 1 for an example. The input is a set $R := \{(r_1^{\mathcal{L}}, r_1^{\mathcal{R}}), \ldots, (r_m^{\mathcal{L}}, r_m^{\mathcal{R}})\}$ of paired-end reads and two integers d and k. Each $(r_i^{\mathcal{L}}, r_i^{\mathcal{R}})$ is a pair of strings of the same length over an alphabet Σ. The integer d is the *insert size* or *shift*. It specifies that the paired-end read corresponds to two substrings of the complete genome whose first letters have distance exactly d in the genome. The integer k is a user-defined parameter.

In a paired de Bruijn graph $G(R)$ constructed from a read set R, each vertex v is associated with a pair of k-mers $(\mathcal{L}(v), \mathcal{R}(v))$. This pair is called *bilabel*. Each node has a unique bilabel. For a read set R, the vertex set of $G(R)$ is defined as

$$V(R) := \{(s,t) \in \Sigma^k \times \Sigma^k \mid \exists (r_i^{\mathcal{L}}, r_i^{\mathcal{R}}) \in R, p \in \mathbb{N} :$$
$$s = r_i^{\mathcal{L}}[p, p+k-1] \wedge t = r_i^{\mathcal{R}}[p, p+k-1]\}. \quad (1)$$

An arc is drawn from a vertex u to a vertex v if some read in R contains the bilabels of u and v in consecutive positions. More precisely, the arc set of $G(R)$ is

$$A(R) := \{(u,v) \mid \exists (r_i^{\mathcal{L}}, r_i^{\mathcal{R}}) \in R, p \in \mathbb{N} :$$
$$\mathcal{L}(u) = r_i^{\mathcal{L}}[p, p+k-1] \wedge \mathcal{L}(v) = r_i^{\mathcal{L}}[p+1, p+k] \wedge$$
$$\mathcal{R}(u) = r_i^{\mathcal{R}}[p, p+k-1] \wedge \mathcal{R}(v) = r_i^{\mathcal{R}}[p+1, p+k]\}.$$

A walk in a paired de Bruijn graph directly corresponds to a string over Σ and thus to a DNA sequence. More precisely, a walk $W := (v_1, \ldots, v_p)$ in a paired de Bruijn graph spells the two strings $\mathcal{L}(v_1) \cdot \mathcal{L}(v_2)[k] \cdot \ldots \cdot \mathcal{L}(v_p)[k]$ and $\mathcal{R}(v_1) \cdot \mathcal{R}(v_2)[k] \cdot \ldots \cdot \mathcal{R}(v_p)[k]$. There is one walk in $G(R)$ that corresponds to the original DNA sequence. This walk fulfills several properties that we describe below. The computational task that we consider here is deciding whether a walk fulfilling these properties exists in $G(R)$. In the remainder of this work, the read set R is irrelevant in the computational problems, thus, we denote the paired de Bruijn graph $G = (V, A)$ instead of $G(R)$.

The Sound Covering Cycle Problem. In the case of organisms with a single circular chromosome, the walk that corresponds to the genome is a cycle. Slightly abusing notation, we define a *cycle* in a graph $G = (V, A)$ as a walk $(v_1, \ldots, v_p)$, $v_i \in V$, such that $(v_p, v_1) \in A$. This cycle should have a length ℓ which is the estimated genome length. As described above, the cycle spells two cyclic strings. The cycle that spells the genome should also spell every pair of observed $k + 1$-mers. By construction of G, a pair of $k + 1$-mers in the read set corresponds to an arc between two vertices. Thus, we demand the cycle to contain every arc of the graph. Accordingly, a cycle C is called *covering* if for each arc $(u, v) \in A$ there is a v_i such that $(u, v) = (v_i, v_{i+1})$ or $(u, v) = (v_p, v_1)$.

The above properties are also relevant in classic de Bruijn graphs. In a paired de Bruijn graph the walks should also fulfill the insert size constraint. Recall that the distance between the $\mathcal{L}$- and $\mathcal{R}$-label of a vertex is d. The soundness constraint will ensure that the strings spelled by these two labels are consistent with the insert size. More precisely, in a paired de Bruijn graph, a cycle is called *sound* if the pair of strings it spells matches with shift d. Let s and t be the two strings spelled by the bilabels of the cycle C. If $d \le \ell$, then we call C *sound* if

- $s[i+d] = t[i]$ for $1 \le i \le \ell - d$, and
- $s[i] = t[i+\ell-d]$ for $1 \le i \le d$.

Accordingly, we call a vertex v_i in a walk $(v_1, \ldots, v_q)$ *sound* if $\mathcal{R}(v_i) = \mathcal{L}(v_{i+d})$. The soundness definition corresponds to the one of Kapun and Tsarev [7]. This leads to our main problem definition.

SOUND COVERING CYCLE
Input: A paired de Bruijn graph $G = (V, A)$ and nonnegative integers d and ℓ.
Question: Does G contain a sound covering cycle of length ℓ?

Related Work. Kapun and Tsarev [7] show that SOUND COVERING CYCLE is NP-hard even if the values of d or k are small constants. They also claim that if $|\Sigma|$ and k are constants, which implies that the graph has constant size, then SOUND COVERING CYCLE cannot be NP-hard as the language defined by it is sparse since d is encoded in unary. We do not make this assumption, that is, in our case d and ℓ are encoded in binary. Thus, the complexity of SOUND COVERING CYCLE for fixed graph size is open in our encoding. A related graph-based approach of modeling the information of paired-end reads are rectangle graphs [1,10,12]. Computing a covering cycle of length *at most* ℓ in a directed graph is known as DIRECTED CHINESE POSTMAN and can be solved in polynomial time [4].

Contribution and Organization of the Paper. In Sect. 2, we describe a decomposition of cycles in directed graphs that we use throughout this work. Moreover, we describe an algorithm for computing a fixed-length covering cycle. This algorithm is used as a subroutine in Sect. 5 and may also be of independent interest. In Sect. 3, we present an algorithm for SOUND COVERING CYCLE that runs in $f(n, d) \cdot \text{poly}(\log \ell)$ time. In Sect. 4, we present similar algorithms for variants of SOUND COVERING CYCLE such as searching for a shortest covering sound cycle and dealing with relaxed models of soundness that model noisy input data. In Sect. 5, we present a special case of SOUND COVERING CYCLE that is solvable in $f(n) \cdot \text{poly}(\log \ell + \log d)$ time. Since paired de Bruijn graphs are very sparse, we use the maximum outdegree Δ in our running time bounds.

Due to space constraints, most proofs are deferred to an appendix. We use the following observations in our algorithms.

Lemma 1. *Let $G = (V, A)$ be a directed graph with maximum outdegree Δ and let ℓ be an integer. There are at most $n \cdot \Delta^{\ell-1}$ walks and cycles of length ℓ in G and they can be enumerated in $O(n \cdot \Delta^{\ell-1} \cdot (\ell + \Delta))$ time.*

This statement implies the following bound on the number of simple walks.

Lemma 2. *A directed graph $G = (V, A)$ with n vertices and maximum outdegree Δ has at most $2n \cdot \Delta^{n-1}$ different simple walks.*

2   Cycle-Walk Decompositions

Before presenting our algorithms, we describe a structured representation of cycles and walks.

First, we show that we can decompose any walk or cycle into maximal simple walks (denoted by Ω_i) and possibly empty simple walks between them (denoted by W_i). Herein, the term maximal refers to the property that in C each $\Omega_i = (u_1, \ldots, u_t)$ is followed by its first vertex u_1. This implies in particular that Ω_i is a cycle.

Lemma 3. *Let C be a walk in a graph G. Then C can be written as a concatenation of simple walks $\Omega_1 \cdot W_1 \cdot \ldots \cdot \Omega_q \cdot W_q$ such that*

1. $|\Omega_i| > 0$ *for each $i \in \{1, \ldots, q\}$, and*
2. *for each $\Omega_i := (u_1, u_2, \ldots, u_s)$, $1 \leq i \leq q$ it holds that $u_1 = v_1$ where v_1 is the first vertex of $W_i \cdot \Omega_{i+1}$.*

A representation adhering to Lemma 3 is called *cycle-walk decomposition* of C. Our next aim is to show the existence of cycle-walk decompositions with a compact description. The proof exploits the fact that if there are too many different cycles in the decomposition, then some of them can be replaced by repetitions of other cycles.

Before proving Lemma 5, we show the correctness of the following exchange operation.

Lemma 4. *Let C be a covering cycle of a graph G with cycle-walk decomposition $\Omega_1 \cdot W_1 \cdot \ldots \cdot \Omega_q \cdot W_q$. If C contains a cycle Ω_j such that*

- *there is a walk Ω_i, $i < j$, that has the same length as Ω_j, and*
- *for each arc $a \in A(\Omega_j)$, there is a walk Ω_p, $p \neq j$, such that $a \in A(\Omega_p)$,*

then $C' := \Omega_1 \cdot \ldots \cdot W_{i-1} \cdot \Omega_i^2 \cdot W_i \cdot \ldots \cdot W_{j-1} \cdot W_j \cdot \ldots \cdot W_q$ is a covering cycle of the same length in G.

Proof. Let $\Omega_i := (u_1, u_2, \ldots, u_s)$ and $W_i \cdot \Omega_{i+1} := (w_1, w_2, \ldots, w_t)$. Since Ω_i is a cycle, $\Omega_i \cdot \Omega_i \cdot W_i$ is a walk. Now consider $\Omega_{j-1} \cdot W_{j-1} := (x_1, \ldots, x_s)$, $\Omega_j := (y_1, \ldots, y_t)$, and $W_j \cdot \Omega_{j+1} := (z_1, \ldots, z_r)$. Since C is a walk we have $(x_s, y_1) \in A$ and by the properties of cycle-walk decompositions also $y_1 = z_1$. Therefore, $(x_s, z_1) \in A$ and thus $\Omega_{j-1} \cdot W_{j-1} \cdot W_j \cdot \Omega_{j+1}$ is a walk. Consequently, C' is a walk. Since Ω_i and Ω_j have the same length, C' has the same length as C. Moreover, C' is covering as every arc of Ω_j is contained in some Ω_p, $p \neq j$. □

Using the exchange operation described by Lemma 4, we now show that there are compact cycle-walk decompositions.

Lemma 5. *If a directed graph G has a covering cycle C of length ℓ, then it has a covering cycle C' of length ℓ such that C' has a cycle-walk decomposition $(\Omega_1)^{r_1} \cdot W_1 \cdot \ldots \cdot (\Omega_q)^{r_q} \cdot W_q$ where $q \leq n + m$.*

Proof. Assume that G has a covering cycle C of length ℓ. According to Lemma 3, C has a cycle-walk decomposition $(\Omega_1)^{r_1} \cdot W_1 \cdot \ldots \cdot (\Omega_q)^{r_q} \cdot W_q$ (Lemma 3 shows the existence of the special case $r_1 = \ldots = r_q = 1$). Now consider of all covering cycles of G one with a decomposition in which $q + \sum_{i=1}^{q} |W_i|$ is minimum.

Now assume towards a contradiction, that in this decomposition there are indices i and j, $i \neq j$, such that $|\Omega_i| = |\Omega_j|$ and each arc of Ω_j is contained in some Ω_p, $p \neq j$. Without loss of generality assume $i < j$. We transform C into a new cycle C' in which $q + \sum_{i=1}^{q} |W_i|$ is smaller. This contradicts our choice of C.

By the assumption on i and j and by Lemma 4, $C' := \Omega_1 \cdot \ldots \cdot W_{i-1} \cdot (\Omega_i)^{r_i + r_j} \cdot W_i \cdot \ldots \cdot \Omega_{j-1} \cdot W_{j-1} \cdot W_j \cdot \Omega_{j+1} \cdot \ldots \cdot W_q$ is also a covering cycle of G. Clearly, $|C| = |C'| = \ell$ and C' is also a covering cycle. Now consider two cases.

Case 1: $W_{j-1} \cdot W_j$ contains a simple cycle Ω^*. Let $W_{j-1} \cdot W_j = W_1^* \cdot \Omega^* \cdot W_2^*$ where W_1^* and W_2^* are not simple cycles. Then, $C' := \Omega_1 \cdot \ldots \cdot W_{i-1} \cdot (\Omega_i)^{r_i + r_j} \cdot W_i \cdot \ldots \cdot \Omega_{j-1} \cdot W_1^* \cdot \Omega^* \cdot W_2^* \cdot \Omega_{j+1} \cdot \ldots \cdot W_q$. In this decomposition, the overall number of Ω's has not changed but, since $|W_1^*| + |W_2^*| < |W_{j-1}| + |W_j|$, the sum of the lengths of the W_i's has decreased. This contradicts our choice of C.

Case 2: Otherwise. In this case, $W_{j-1} \cdot W_j$ is a simple walk. Thus, the number of Ω's has decreased by one while $\sum_{i=1}^{q} |W_i|$ remains the same. This contradicts our choice of C.

Since both cases lead to a contradiction to the choice of C we can assume that for each Ω_j in C there is either one arc a_j that is not contained in any other Ω_p, $p \neq j$, or there is no other Ω_i of the same length as Ω_j. By pigeonhole principle, there can be at most $|A| = m$ cycles Ω_j for which the first condition is true. For all further cycles, the first condition is false. Now since each Ω_j has length at most n there can be, again by pigeonhole principle, at most n further cycles for which the second condition is true. This implies that $q \leq m + n$.   □

The following lemma shows that the cycle lengths in a decomposition suffice to determine the possible overall cycle length.

Lemma 6. *A graph G has a covering cycle C of length ℓ if and only if it has a covering cycle C' with cycle-walk decomposition $\Omega_1 \cdot W_1 \cdot \ldots \cdot \Omega_q \cdot W_q$ such that*

1. *C' has length $x \leq 2n(m + n)$, and*
2. *there are nonnegative integers p_i, $1 \leq i \leq q$, such that $x + \sum_{1 \leq i \leq q} p_i \cdot |\Omega_i| = \ell$.*

We now bound the running time for determining the existence of such a cycle.

Theorem 1. *Let $G = (V, A)$ be a directed graph with n vertices and m arcs and let ℓ be an integer. Then, in $O(8^m \cdot 2^n) \cdot poly(n + \log \ell)$ time we can determine whether G contains a covering cycle of length exactly ℓ.*

3   An Algorithm for the Parameters n and d

We now describe our first algorithm for SOUND COVERING CYCLE. The running time of this algorithm is exponential in n and d. Thus, we avoid a combinatorial explosion in the number ℓ which is at least as large as d and usually much larger.

The algorithm exploits that in a sound walk, parts with distance more than d are "independent" with respect to the soundness property. To make the argument more precise, consider a yes-instance (G, d, ℓ) of SOUND COVERING CYCLE with a solution cycle $C = W_1 \cdot W_2 \cdot \ldots \cdot W_q \cdot W^*$, where $|W_i| = d$ for $1 \leq i \leq q$ and $|W^*| = \ell \mod d$. For each vertex v_j in W_i the vertex that is relevant to determine whether v_j is sound is contained in W_{i+1}. Thus, consider a graph $\mathcal{G} = (\mathcal{V}, \mathcal{A})$ which contains each length-d walk as a vertex. In particular, $\mathcal{G}$ contains each W_i. Moreover, assume that $\mathcal{G}$ contains an arc (W, W') if $W \cdot W'$ is a walk in G which is sound for all positions in W. Then $(W_i, W_{i+1}) \in \mathcal{A}$ for each $i < q$. Consequently, the walk $W_1 \cdot W_2 \cdot \ldots \cdot W_q$ in G corresponds to a walk $\mathcal{W}$ in $\mathcal{G}$ and $\mathcal{W} \cdot W^*$ is a sound covering cycle of length ℓ in G.

The algorithm outline hence is as follows: First construct the graph $\mathcal{G}$, called *walk graph* from now on. Second, compute "candidate" walks in $\mathcal{G}$. Finally, check for each candidate walk, whether there is some short walk W^* such that concatenating W^* at its end gives a sound covering cycle of the correct length.

Theorem 2. SOUND COVERING CYCLE *can be solved in* $O(8^{n \cdot \Delta} \cdot 2^{n \cdot \Delta^d}) \cdot \text{poly}(n \cdot \Delta^d + \log \ell)$ *time where* Δ *is the maximum outdegree of* G.

Proof. We describe each of the three main steps of the algorithm in detail and then bound its running time.

Constructing the Walk Graph $\mathcal{G}$. First, enumerate all walks of length d in G. Let $\mathcal{V}$ denote the set of these walks and make $\mathcal{V}$ the vertex set of $\mathcal{G}$. Now construct the arc set $\mathcal{A}$ of $\mathcal{G}$ as follows. For each pair of vertices W and W' in $\mathcal{V}$, check whether $W \cdot W' = (v_1, \ldots, v_{2d})$ is a walk in G and whether it is sound for each v_i, $1 \leq i \leq d$. That is, check whether $(v_d, v_{d+1}) \in A$ and whether $\mathcal{R}(v_i) = \mathcal{L}(v_{i+d})$ for each v_i, $1 \leq i \leq d$. If this is the case, then add the arc (W, W') to $\mathcal{G}$; otherwise, do not add this arc. This completes the construction of $\mathcal{G}$. Now "almost" sound walks in G correspond to walks in $\mathcal{G}$.

> *Observation 1:* A walk $W_1 \cdot \ldots \cdot W_i$ of length $d \cdot i$ in G with $|W_j| = d$, $1 \leq j \leq i$, is sound for all of its first $d \cdot (i - 1)$ positions $\Leftrightarrow (W_1, \ldots, W_i)$ is a walk in $\mathcal{G}$.

Dynamic Programming. Now, for a walk $(W_1, \ldots, W_i)$ in $\mathcal{G}$, let $A(W_1, \ldots, W_i)$ denote the arcs of $W_1 \cdot \ldots \cdot W_i$ in G. Moreover, for an arc (W, W') in $\mathcal{G}$ with $W = (v_1, \ldots, v_d)$ and $W' = (v_{d+1}, \ldots, v_{2d})$ let $\text{arc}(W, W')$ denote the arc (v_d, v_{d+1}) in G (by the construction of the walk graph, this arc is present in G).

Following the discussion above, we now solve SOUND COVERING CYCLE by determining whether there is a walk $(W_1, \ldots, W_q)$ of length $q := \lfloor \ell/d \rfloor$ in $\mathcal{G}$ and a walk W^* of length $\ell \mod d$ in G such that (1) $W_q \cdot W^* \cdot W_1$ is a walk of length $2d + (\ell \mod d)$ in G which is sound for its first $d + (\ell \mod d)$ positions, and (2) every arc of G is contained in $A(W_1, \ldots, W_q)$ or in $W_q \cdot W^* \cdot W_1$. This is done by a dynamic programming algorithm that fills a table $\mathcal{T}$ with entries of the type $\mathcal{T}[W, W', A', \Lambda, y]$ where

- W and W' are vertices of $\mathcal{G}$,
- A' is a subset of A (note that A is the arc set of G not of $\mathcal{G}$),

- Λ is a subset of $\{1, \ldots, |\mathcal{V}|\}$, and
- y is a nonnegative integer of value at most $|\mathcal{V}| \cdot (|\mathcal{V}| + |\mathcal{A}|)$.

Each entry in $\mathcal{T}$ is either true or false. The aim of the algorithm is to fill the table $\mathcal{T}$ such that $\mathcal{T}[W, W', A', \Lambda, y]$ is true if and only if $\mathcal{G}$ contains a walk $(W, \ldots, W')$ with cycle-walk decomposition $\Omega_1 \cdot \Psi_1 \cdot \ldots \cdot \Omega_i \cdot \Psi_i$ such that

- $A(W, \ldots, W') = A'$, that is, the walk $W \cdot \ldots \cdot W'$ in G contains exactly the arcs of A',
- $\Lambda = \{|\Omega_j| \mid 1 \leq j \leq i\}$, and
- $(W, \ldots, W')$ has length y.

The idea behind $\mathcal{T}$ is that, by Lemma 6, it suffices to consider walks of length at most $|\mathcal{V}| \cdot (|\mathcal{V}| + |\mathcal{A}|)$ and then to extend them by using the cycle lengths in Λ. In a preprocessing, we compute a table $\mathcal{D}$. For the correctly filled table $\mathcal{D}$, an entry $\mathcal{D}[W, W', A', y]$ is true if and only if $\mathcal{G}$ contains a walk $(W, \ldots, W')$ of length y such that $A(W, \ldots, W') = A'$. The table is filled for all $A' \subseteq A$ and for increasing $y < |\mathcal{V}|$. Initially, set $\mathcal{D}[W, W, A(W), 1]$ to true for each $W \in \mathcal{V}$. Then the recurrence for $\mathcal{D}$ is

$$\mathcal{D}[W, W', A', y] := \begin{cases} \text{true} & \begin{aligned} &\text{if } \exists (\tilde{W}, W') \in \mathcal{A}, \tilde{A} \subseteq A' : \\ &\tilde{A} \cup \{\text{arc}(\tilde{W}, W')\} \cup A(W') = A' \wedge \\ &\mathcal{D}[W, \tilde{W}, \tilde{A}, y-1], \end{aligned} \\ \text{false} & \text{otherwise.} \end{cases}$$

After $\mathcal{D}$ is completely filled, compute the table $\mathcal{T}$. The recurrence is

$$\mathcal{T}[W, W', A', \Lambda, y] := \begin{cases} \text{true} & \text{if } \Lambda = \{y\} \wedge \mathcal{D}[W, W', A', y] \wedge (W', W) \in \mathcal{A}, \\ \text{true} & \begin{aligned} &\text{if } \exists (\tilde{W}, W') \in \mathcal{A}, \tilde{A} \subseteq A' : \\ &\tilde{A} \cup \{\text{arc}(\tilde{W}, W')\} \cup A(W') = A' \wedge \\ &\mathcal{T}[W, \tilde{W}, \tilde{A}, \Lambda, y-1], \end{aligned} \\ \text{true} & \begin{aligned} &\text{if } \exists \mathcal{T}[W, \tilde{W}, \tilde{A}, \tilde{\Lambda}, y-z], \mathcal{D}[\hat{W}, W', A^*, z] : \\ &\mathcal{T}[W, \tilde{W}, \tilde{A}, \tilde{\Lambda}, y-z] \wedge \mathcal{D}[\hat{W}, W', A^*, z] \wedge \\ &(\tilde{W}, \hat{W}) \in \mathcal{A} \wedge \tilde{A} \cup \{\text{arc}(\tilde{W}, \hat{W})\} \cup A^* = A' \wedge \\ &\Lambda \setminus \{z\} \subseteq \tilde{\Lambda} \subseteq \Lambda, \end{aligned} \\ \text{false} & \text{otherwise.} \end{cases}$$

Determining the Possible Lengths for Candidate Entries. The next step is to compute for each entry whether it can be extended, by repeating cycles of the cycle-walk decomposition, to obtain a walk of length $\ell - (\ell \mod d)$.

This is done by reducing to MONEY CHANGING [2]. More precisely, for each true entry $\mathcal{T}[W, W', A', \Lambda, y]$, we check whether there is a walk of length $\ell - (\ell \mod d)$ that can be obtained from repeating the simple cycles in a length-y walk $\mathcal{W}$ whose existence is implied by the table entry. The set of different lengths of simple cycles in $\mathcal{W}$ is $\Lambda = \{\lambda_1, \ldots, \lambda_{|\Lambda|}\}$. Thus, determining the existence of a cycle of length $\ell - (\ell \mod d)$ is equivalent to checking whether the equation

$$p_1\lambda_1 + p_2\lambda_2 + \ldots + p_{|\Lambda|}\lambda_{|\Lambda|} = \ell - (\ell \mod d) - y$$

has a solution in which each p_i is a nonnegative integer. Each table entry for which the above equation has such a solution is labeled as *candidate entry*. By Lemma 6, the existence of a walk of length $\ell - (\ell \mod d)$ implies that there is a corresponding candidate entry.

Closing the Cycle. The final step of the algorithm is to check whether any of the candidate entries can be completed to a sound cycle of length ℓ by adding a "short" walk of length $\ell \mod d$. To this end, first enumerate all walks W^* of length $\ell \mod d$ in G. Now, for each candidate entry $\mathcal{T}[W, W', A', \Lambda, y]$ and each W^* and for each enumerated short walk W^* do the following. Check whether $W' \cdot W^* \cdot W$ is a walk in G. If yes, then $W \cdot \ldots \cdot W' \cdot W^*$ corresponds to a cycle C in G. This cycle has length $y + (\ell \mod d)$ but since we consider only candidate entries in $\mathcal{T}$, the walk $W \cdot \ldots \cdot W'$ which has length y can be extended to one of length $\ell - (\ell \mod d)$. Thus, the existence of C implies the existence of a cycle C' of length ℓ in G. It remains to check whether C' is sound and covering. To check whether C' is sound it is sufficient to check whether the walk $W' \cdot W^* \cdot W$ is sound for its first $d + (\ell \mod d)$ positions (as every walk in $\mathcal{G}$ corresponds to a walk in G whose positions are sound except for the last d). Finally, it remains to check whether C' is covering. This is done by checking whether $A = A' \cup \{(w'_d, w^*_1), (w^*_{\ell \mod d}, w_1)\} \cup A(W^*)$ where w'_d is the last vertex of W' and w_1 is the first vertex of W in G. If yes, C' is a solution. If none of the combinations of candidate entry and W^* yields a solution, then the instance is a no-instance.

Running Time Analysis. By Lemma 1 the number of different walks of length d in G is at most $n \cdot \Delta^{d-1}$ and thus $|\mathcal{V}| \leq n \cdot \Delta^{d-1}$. Consequently, the construction of $\mathcal{G}$ can be performed in $\text{poly}(n \cdot \Delta^{d-1})$ time.

The running time in the dynamic programming part is dominated by the time for filling the table $\mathcal{T}$. This is dominated by the time needed to check whether the third case of the recursion applies. To do this, one needs to consider, for each table entry, $O(|\mathcal{V}|^2 \cdot 4^{n \cdot \Delta})$ possibilities (recall that $n \cdot \Delta \geq |A|$). For each possibility, the check can be performed in $\text{poly}(|\mathcal{V}|)$ time and there are $O(|\mathcal{V}|^2 \cdot 2^{n \cdot \Delta} \cdot 2^{|\mathcal{V}|} \cdot |\mathcal{V}| \cdot (|\mathcal{V}| + |\mathcal{E}|)) = 2^{n \cdot \Delta} \cdot 2^{|\mathcal{V}|} \cdot \text{poly}(|\mathcal{V}|)$ table entries to compute. Thus, the overall running time in the dynamic programming part is $8^{n \cdot \Delta} \cdot 2^{|\mathcal{V}|} \cdot \text{poly}(|\mathcal{V}|)$.

Next, we solve $O(2^{n \cdot \Delta} \cdot 2^{|\mathcal{V}|})$ MONEY CHANGING instances, each in $\text{poly}(|\mathcal{V}| + \log \ell)$ time [2]. The checks in the final stage require $\text{poly}(|\mathcal{V}|)$ time for each candidate entry since less than $|\mathcal{V}|$ different short walks are considered and each check needs $\text{poly}(|\mathcal{V}|)$ time. The overall running time follows. $\square$

4   Shortest Sound Cycles and Approximately Sound Cycles

The idea described above can be used in several problem variants. One possibility is to find a shortest sound covering cycle instead of one with fixed length.

SHORTEST SOUND COVERING CYCLE
Input: A paired de Bruijn graph $G = (V, A)$ and a nonnegative integer d.
Task: Find a sound covering cycle of minimum length in G.

We first bound the length of a shortest sound covering cycle G.

Lemma 7. *Let $G = (V, A)$ be a directed graph with maximum outdegree Δ. If G has a sound covering cycle, then it has a sound covering cycle of length at most $2d(n \cdot \Delta + 1) \cdot (n \cdot \Delta^{d-1}) + d$.*

Proof. Consider the walk graph $\mathcal{G}$ of G. This graph has $n \cdot \Delta^{d-1}$ vertices. Now consider a walk $\mathcal{W}$ in $\mathcal{G}$ such that $\mathcal{W} \cdot W^*$ is a sound covering cycle in G, where $|W^*| < d$ and assume that $\mathcal{W}$ has minimum length with this property.

Let $\Omega_1 \cdot W_1 \cdot \ldots \cdot \Omega_q \cdot W_q$ be the cycle-walk decomposition of $\mathcal{W}$ which exists due to Lemma 3. Then, for each Ω_i, $i < q$, there is some $a \in A(\Omega_i)$ which is not contained in any Ω_j, $j \neq i$, otherwise removing Ω_i from $\mathcal{W}$ yields a shorter walk in $\mathcal{G}$ whose corresponding walk in G also covers all arcs. By pigeonhole principle this implies $q \leq |A| + 1$. Thus, the length of $\mathcal{W}$ in $\mathcal{G}$ is at most $2(|A| + 1) \cdot (n \cdot \Delta^{d-1})$ as each Ω_i and W_i are simple walks in $\mathcal{G}$. The corresponding walk in G has length at most $2d(|A| + 1) \cdot (n \cdot \Delta^{d-1})$ since each vertex of $\mathcal{G}$ corresponds to a length-d walk in G. The overall bound now follows from the fact that $|A| \leq n \cdot \Delta$ and $|W^*| < d$. $\qquad\square$

Using this bound, we now derive a dynamic programming algorithm that computes walks of increasing length in G. In this computation, we store the last d vertices of the walk, since these have influence on the soundness condition. Moreover, we store which arcs of G are already covered by the walk.

Theorem 3. SHORTEST SOUND COVERING CYCLE *can be solved in $O(n^4 \cdot d \cdot \Delta^{3d})$ time.*

The second variant relaxes the soundness constraint. This is motivated by the fact that the insert size between the paired end reads is not always exactly d. This relaxed notion of soundness is defined as follows. Recall that a cycle in a paired de Bruijn graph spells two strings s and t. Now, a length-ℓ cycle C is called *x-approximately sound* if

$$\forall i \in \{1 \leq i \leq\} : t[i \mod \ell] \in \{s[i + d - x \mod \ell], \ldots, s[i + d \mod \ell]\}.$$

Informally, this means that the right label has distance at least $d - x$ and distance at most d to the left label. This definition leads to the following problem.

APPROXIMATELY SOUND COVERING CYCLE
Input: A paired de Bruijn graph $G = (V, A)$ and nonnegative integers d, x, and ℓ.
Question: Does G contain an x-approximately sound covering cycle of length ℓ?

We slightly modify the algorithm for SOUND COVERING CYCLE to obtain the following.

Theorem 4. APPROXIMATELY SOUND COVERING CYCLE *can be solved in* O $(8^{n \cdot \Delta} \cdot 2^{n \cdot \Delta^d}) \cdot \text{poly}(n \cdot \Delta^d + \log \ell)$ *time.*

Finally, we consider the combination of finding a short *and* approximately sound cycle. In this variant, we can even allow a number y of mismatches, that is, there can be y positions that are not approximately sound. More formally, a length-ℓ walk W is called x-*approximately sound with cost at most y* if there is a set $M \subseteq \{1, \ldots, \ell\}$ of size at most y such that

$$\forall i \in \{1, \ldots, \ell\} \setminus M : t[i \mod \ell] \in \{s[i + d - x \mod \ell], \ldots, s[i + d \mod \ell]\}.$$

The COST-BOUNDED SHORTEST APPROXIMATELY SOUND COVERING CYCLE problem now is to find a shortest covering cycle that is x-approximately sound with cost at most y (if such a cycle exists).

To obtain an algorithm for this variant, first note that the bound of Lemma 7 also holds for shortest approximately sound cycles of bounded cost: the replacement argument in the proof only removes cycles in the walk graph which does not increase the cost of the solution.

Theorem 5. COST-BOUNDED SHORTEST APPROXIMATELY SOUND COVERING CYCLE *can be solved in* $O(n^4 \cdot d^2 \cdot \Delta^{3d})$ *time.*

5   A Tractable Special Case for the Parameter n

Finally, we present an $f(n) \cdot \text{poly}(\log \ell)$-time algorithm for a special case of SOUND COVERING CYCLE. To describe the structure of this special case, we introduce the following notion: the *compatibility graph* of a paired de Bruijn graph $G = (V, A)$ is a graph $H = (V, B)$ such that $(a, b) \in B$ if $\mathcal{L}(a) = \mathcal{R}(b)$.

We now exploit this structure by presenting an algorithm for the case that H is a union of loops. Then, each pair of vertices with distance d within a sound cycle is identical. Due to this periodic behavior, we obtain the following relationship between sound cycles and shorter covering cycles.

Lemma 8. *Let G be a paired de Bruijn graph such that its compatibility graph H is a union of loops. Then, G has a sound covering cycle of length ℓ with shift d if and only if G has a covering cycle of length $\gcd(d, \ell)$.*

Here, $\gcd(d, \ell)$ denotes the greatest common divisor of d and ℓ.

Theorem 6. SOUND COVERING CYCLE *can be solved in* $O(8^{n \cdot \Delta} \cdot 2^n) \cdot \text{poly}(n + \log \ell)$ *time if the compatibility graph H of G is a union of loops.*

Proof. By Lemma 8 the problem reduces to one of finding a covering cycle of the correct length $\ell' \leq \ell$. Since G has n vertices and at most $n \cdot \Delta$ arcs, this can be done in $O(8^{n \cdot \Delta} \cdot 2^n) \cdot \text{poly}(n + \log \ell)$ time by Theorem 1.   □

6   Outlook

It would be clearly desirable to improve the presented algorithms. Any substantial improvement would need to avoid the enumeration of all length-d walks in G. Also it would be interesting to extend the algorithm for the case that H is a disjoint union of loops, for example to the case that every vertex in H has outdegree one. Finally, in a subroutine we consider the problem of computing a covering cycle of fixed length. It is open whether this problem is NP-hard or solvable in polynomial time.

References

1. Bankevich, A., Nurk, S., Antipov, D., Gurevich, A.A., Dvorkin, M., Kulikov, A.S., Lesin, V.M., Nikolenko, S.I., Pham, S., Prjibelski, A.D., et al.: SPAdes: a new genome assembly algorithm and its applications to single-cell sequencing. J. Comput. Biol. **19**(5), 455–477 (2012)
2. Böcker, S., Lipták, Z.: A fast and simple algorithm for the money changing problem. Algorithmica **48**(4), 413–432 (2007)
3. Earl, D., Bradnam, K., John, J.S., Darling, A., Lin, D., Fass, J., Yu, H.O.K., Buffalo, V., Zerbino, D.R., Diekhans, M., et al.: Assemblathon 1: a competitive assessment of de novo short read assembly methods. Genome Res. **21**(12), 2224–2241 (2011)
4. Edmonds, J., Johnson, E.L.: Matching, Euler tours and the Chinese postman. Math. Program. **5**(1), 88–124 (1973)
5. Haiminen, N., Kuhn, D.N., Parida, L., Rigoutsos, I.: Evaluation of methods for de novo genome assembly from high-throughput sequencing reads reveals dependencies that affect the quality of the results. PLoS One **6**(9), e24182 (2011)
6. Idury, R.M., Waterman, M.S.: A new algorithm for DNA sequence assembly. J. Comput. Biol. **2**(2), 291–306 (1995)
7. Kapun, E., Tsarev, F.: On NP-hardness of the paired de Bruijn sound cycle problem. In: Darling, A., Stoye, J. (eds.) WABI 2013. LNCS, vol. 8126, pp. 59–69. Springer, Heidelberg (2013)
8. Medvedev, P., Pham, S., Chaisson, M., Tesler, G., Pevzner, P.: Paired de Bruijn graphs: a novel approach for incorporating mate pair information into genome assemblers. J. Comput. Biol. **18**(11), 1625–1634 (2011)
9. Pevzner, P.A., Tang, H., Waterman, M.S.: An Eulerian path approach to DNA fragment assembly. Proc. Nat. Acad. Sci. **98**(17), 9748–9753 (2001)
10. Prjibelski, A.D., Vasilinetc, I., Bankevich, A., Gurevich, A., Krivosheeva, T., Nurk, S., Pham, S., Korobeynikov, A., Lapidus, A., Pevzner, P.A.: ExSPAnder: a universal repeat resolver for DNA fragment assembly. Bioinformatics **30**(12), i293–i301 (2014)
11. Salzberg, S.L., Phillippy, A.M., Zimin, A., Puiu, D., Magoc, T., Koren, S., Treangen, T.J., Schatz, M.C., Delcher, A.L., Roberts, M., et al.: GAGE: a critical evaluation of genome assemblies and assembly algorithms. Genome Res. **22**(3), 557–567 (2012)
12. Vyahhi, N., Pyshkin, A., Pham, S., Pevzner, P.A.: From de Bruijn graphs to rectangle graphs for genome assembly. In: Raphael, B., Tang, J. (eds.) WABI 2012. LNCS, vol. 7534, pp. 249–261. Springer, Heidelberg (2012)
13. Zhang, W., Chen, J., Yang, Y., Tang, Y., Shang, J., Shen, B.: A practical comparison of de novo genome assembly software tools for next-generation sequencing technologies. PLoS One **6**(3), e17915 (2011)

Approximation Algorithms for the Multilevel Facility Location Problem with Linear/Submodular Penalties

Gaidi Li[1], Dachuan Xu[1(✉)], Donglei Du[2], and Chenchen Wu[3]

[1] Department of Information and Operations Research,
Beijing University of Technology, 100 Pingleyuan, Chaoyang District,
Beijing 100124, People's Republic of China
xudc@bjut.edu.cn
[2] Faculty of Business Administration, University of New Brunswick,
Fredericton, NB E3B 5A3, Canada
[3] College of Science, Tianjin University of Technology,
Tianjin 300384, People's Republic of China

Abstract. We consider two multilevel facility location problems with linear and submodular penalties respectively, and propose two approximation algorithms with performance guarantee 3 and $1 + \frac{2}{1-e^{-2}} (\approx 3.314)$ for these two problems.

Keywords: Multilevel facility location problem · Submodular function · Approximation algorithm

1 Introduction

Facility location is an important area in combinatorial optimization with vast applications in operation research, computer science and management science.

The k-level facility location problem (k-LFLP) is a classical and extensively investigated problem, which has many applications. For example, typical products are shipped through manufacturers, warehouses and retailers before they can reach customers. One of the problems facing the business is to minimize the cost across these multi-level facilities. Formally, we are given a set $\mathcal{D}$ of clients, and k pairwise disjoint facility sets $\mathcal{F}^\ell$ ($\ell = 1, 2, \ldots, k$). Assume that the set $\mathcal{D} \cup \left(\cup_{\ell=1}^{k} \mathcal{F}^\ell \right)$ constitutes a metric space; that is, if $i, j \in \mathcal{D} \cup \left(\cup_{\ell=1}^{k} \mathcal{F}^\ell \right)$ are connected, then we pay a connected cost c_{ij}, which is symmetric, nonnegative and satisfies triangle inequalities. Each facility has an open cost $f_{i_\ell}, i_\ell \in \mathcal{F}^\ell$. Define a facility path as a sequence of k facilities $(i_1, i_2, \ldots, i_k)$ such that $i_\ell \in \mathcal{F}^\ell$, ($\ell = 1, 2, \ldots, k$). A path is *open* if all its facilities are open. If client j is connected to a path $p = (i_1, i_2, \ldots, i_k)$, the corresponding connection cost is defined as $c_{jp} = c_{ji_1} + \sum_{\ell=1}^{k-1} c_{i_\ell i_{\ell+1}}$. The goal of the k-LFLP is to serve all the clients in the set $\mathcal{D}$ by connecting each of them to an open path so as to minimize the total cost, including both connection and open cost. When k is equal to 1,

© Springer International Publishing Switzerland 2015
J. Wang and C. Yap (Eds.): FAW 2015, LNCS 9130, pp. 162–169, 2015.
DOI: 10.1007/978-3-319-19647-3_15

the problem is reduced to the classic uncapacitated facility location problem, which is proved to be NP-hard. The FLP has been widely studied. Shmoys et al. [14] gave the first constant 3.16-approximation algorithm using LP-rounding technique, and Li [12] proposed the current best approximation factor 1.488, close to the lower bound 1.463 of this problem given by Guha and Khuller [7]. For the k-LFLP, Aardal et al. [1] gave the current best approximation ratio 3 using the stochastic LP-rounding technique, while Ageev et al. [3] presented the current best combinatorial algorithm with performance guarantee 3.27. Based on the stochastic LP-rounding, Wu and Xu [15] proposed a bifactor $\left(\frac{\ln(1/\beta)}{1-\beta}, 1 + \frac{2}{1-\beta}\right)$-approximation algorithm[1], where $\beta \in (0,1)$ is a constant. In addition, Gabor et al. [6] gave a 3-approximation algorithm by adopting a new integer programming formulation with a polynomial number of variables and constraints. Li et al. [11] gave a cross-monotonic cost sharing method for the multi-level economic lot-sizing game. On the negative side, Krishnaswamy and Sviridenko [10] proved that the lower bound for the k-LFLP is 1.61. In light of the 1.488-approximation algorithm of Li [12] for the FLP, the k-LFLP obviously is harder to approximate.

The focus of this work is on the k-LFLP with penalties, where each client is either connected to an open facility path by paying connection cost or rejected for service with a penalty cost. The goal of the problem is to open a subset of facilities from $\mathcal{F}^\ell$, $\ell = 1, 2, \ldots, k$, such that each client $j \in \mathcal{D}$ is either connected to an open facility path, or rejected with a penalty cost so as to minimize the total facility cost, connection cost and penalty cost. If each client j has a fixed penalty cost q_j, the corresponding problem is called the k-LFLP with linear penalties. On the other hand, the k-LFLP with submodular penalties treats the penalty cost as a monotone increasing submodular function $h(\cdot)$ defined on the client set $\mathcal{D}$; that is, $h(A) \leq h(B)$ and $h(A \cup \{j\}) - h(A) \geq h(B \cup \{j\}) - h(B)$ for all $A \subseteq B$, $j \notin B$. Li et al. [8] extended the primal-dual technique of Jain and Vazirani [9] to the k-LFLP with submodular penalties and obtained a combinatorial 6-approximation algorithm. Based on LP-rounding, Asadi et al. [2] gave a 4-approximation algorithm for the k-LFLP with linear penalties. For any given k, Byrka et al. [4] considered the k-LFLP with linear penalties and gave an approximation algorithm whose approximation ratio converges monotonically to three when k tends to infinity.

Our main contribution is to give two approximation algorithms for the k-LFLP with submodular/linear penalties, which are the current best approximation ratio respectively. The rest of the paper is organized as follows. We present an $\left(1 + \frac{2}{1-e^{-2}}\right)$-approximation algorithm for the k-FLP with submodular penalties in Sect. 2, and a 3-approximation algorithm for the k-FLP with linear penalties in Sect. 3.

---

[1] Let us denote the facility cost by F^* and the connection cost by C^* in the optimal solution. If an algorithm can get an integer feasible solution to the k-LFLP with the total cost no more than $aF^* + bC^*$ in polynomial time, then this algorithm is called a bifactor (a, b)-approximate algorithm for k-LFLP.

2 Multilevel Facility Location Problem with Submodular Penalties

In this section, we consider the k-LFLP with submodular penalties. We introduce several binary decision variables as follows: x_{jp} indicates whether client j is connected to path p or not; z_S indicates whether the client set S is penalized or not; and y_{i_ℓ} indicates whether facility i_ℓ is open or not. The k-LFLP with submodular penalties can be formulated as an integer linear program:

$$\min \sum_{\ell=1}^{k} \sum_{i_\ell \in \mathcal{F}^l} f_{i_\ell} y_{i_\ell} + \sum_{p \in \mathcal{P}} \sum_{j \in \mathcal{D}} c_{jp} x_{jp} + \sum_{S \subseteq \mathcal{D}} h(S) z_S$$

$$\text{s. t. } \sum_{p \in \mathcal{P}} x_{jp} + \sum_{S \subseteq \mathcal{D}: j \in S} z_S \geq 1, \ \forall j \in \mathcal{D},$$

$$\sum_{p: i_\ell \in p} x_{jp} - y_{i_\ell} \leq 0, \ \forall j \in \mathcal{D}, i_\ell \in \mathcal{F}^l, \ell = 1, 2, \ldots, k, \quad (1)$$

$$x_{jp} \in \{0, 1\}, \ \forall p \in \mathcal{P}, j \in \mathcal{D},$$

$$y_{i_\ell} \in \{0, 1\}, \ \forall i_\ell \in \mathcal{F}^\ell, \ell = 1, 2, \ldots, k,$$

$$z_S \in \{0, 1\}, \ \forall S \subseteq \mathcal{D},$$

where the first constraints say that client j is either connected to a facility path or penalized, and the second constraints imply that client j must be connected to an open facility path.

Now we present the LP-based approximation algorithm for k-LFLP with submodular penalties. We consider the convex relaxation of problem (1):

$$\min \sum_{\ell=1}^{k} \sum_{i_\ell \in \mathcal{F}^l} f_{i_\ell} y_{i_\ell} + \sum_{p \in \mathcal{P}} \sum_{j \in \mathcal{D}} c_{jp} x_{jp} + h'(z)$$

$$\text{s. t. } \sum_{p \in \mathcal{P}} x_{jp} + z_j \geq 1, \ \forall j \in \mathcal{D},$$

$$\sum_{p: i_\ell \in p} x_{jp} - y_{i_\ell} \leq 0, \ \forall j \in \mathcal{D}, i_\ell \in \mathcal{F}^l, \ell = 1, 2, \ldots, k, \quad (2)$$

$$x_{jp} \geq 0, \ \forall p \in \mathcal{P}, j \in \mathcal{D},$$

$$y_{i_\ell} \geq 0, \ \forall i_\ell \in \mathcal{F}^\ell, \ell = 1, 2, \ldots, k,$$

$$z_j \geq 0, \ \forall j \in \mathcal{D},$$

where $h'(z)$ is the Lovász extension function of any given submodular function $h(\cdot)$(cf. [13]),

$$h'(z) = \max \sum_{j \in \mathcal{D}} \theta_j z_j$$

$$\text{s. t. } \sum_{j \in S} \theta_j \leq h(S), \ \forall S \subseteq \mathcal{D},$$

$$\theta_j \geq 0, \ j \in \mathcal{D}.$$

It follows from Li et al. [13] that the function $h'(\cdot)$ has the following properties.

Property 1. $h'(I(S)) = h(S)$, and $h'(0) = h(\emptyset)$.

Property 2. $h'(aw) = ah'(w)$, $a \geq 0$ is a number; $w \geq 0$ is a vector.

Property 3. $h'(\cdot)$ is a monotonically increasing convex function.

Property 4. Given the optimal solution (x^*, y^*, z^*) to problem (2), and let $\theta^* := \arg\max h'(z^*)$. If $j \in \mathcal{D}$, $x_{jp}^* > 0$ and $z_j^* > 0$, then we have $\theta_j^* \geq c_{jp}$.

Now we illustrate the high level idea of our algorithm. Our algorithm is motivated by Li et al. [8] for the one-level FLP (1-LFLP) with submodular penalties. By considering the Lovász extension of submodular function, we overcome the difficulties brought by submodular penalties and separate the client set into the penalized set and the serviced set according to the optimal solution to the linear programming relaxation. First, we solve the convex relaxation problem (2) to get an optimal solution (x^*, y^*, z^*). Then based on the penalized/connected proportion of each client, we partition the client set $\mathcal{D}$ into two sets: the connected set $\mathcal{D}_\gamma$ and the penalized set $\bar{\mathcal{D}}_\gamma$ (where γ is a pre-specified parameter). Next, we consider an instance of the k-LFLP with the facility sets $\mathcal{F}_1, \ldots, \mathcal{F}_k$, and the client set $\mathcal{D}_\gamma$. Then, based on the optimal solution to problem (2), we construct a feasible solution $(\hat{x}, \hat{y})$ to the linear program relaxation for the k-LFLP with client set $\mathcal{D}_\gamma$. Finally, we call the bifactor $\left(\frac{\ln(1/\beta)}{1-\beta}, 1 + \frac{2}{1-\beta}\right)$-approximation algorithm for the k-LFLP by Wu and Xu [15] as a subroutine (cf. Steps 4–6 of Algorithm 1) to get a feasible integer solution $(\bar{x}, \bar{y})$ to the k-LFLP with the client set $\mathcal{D}_\gamma$. Combining with the penalized client set $\bar{\mathcal{D}}_\gamma$, we get a feasible integer solution $(\bar{x}, \bar{y}, \bar{z})$ to our problem.

The algorithm is given below.

Algorithm 1

Step 1. *Solve the LP relaxation (2) to obtain the fractional optimal solution* (x^*, y^*, z^*).

Step 2. *According to the penalized proportion of client j, we partition the clients set into two sets* $\mathcal{D}_\gamma = \left\{ j \in \mathcal{D} : \sum_{p \in \mathcal{P}} x_{pj}^* \geq \frac{1}{\gamma} \right\}$, *and* $\bar{\mathcal{D}}_\gamma = \mathcal{D} \setminus \mathcal{D}_\gamma$. *Penalize the clients in $\bar{\mathcal{D}}_\gamma$ and set $\bar{x}_{jp} := 0$, $\forall j \in \bar{\mathcal{D}}_\gamma, p \in \mathcal{P}$,*
$$\bar{z}_S := \begin{cases} 1, & \text{if } S = \bar{\mathcal{D}}_\gamma \\ 0, & \text{otherwise} \end{cases}$$

Step 3. *For the served client set $\mathcal{D}_\gamma$, set*
$$\hat{x}_{jp} := \frac{x_{jp}^*}{1 - z_j^*}, \forall j \in \mathcal{D}_\gamma, p \in \mathcal{P}, \quad \hat{y}_i := \min\{\gamma y_i^*, 1\}, \forall i \in \cup_{\ell=1}^k \mathcal{F}^\ell.$$
Then $(\hat{x}, \hat{y})$ is a feasible solution for the relaxed k-LFLP with the client set $\mathcal{D}_\gamma$ and facility sets $\mathcal{F}_1, \ldots, \mathcal{F}_k$.

Step 4. *Given a parameter $\beta \in (0, 1)$, choose randomly and uniformly $\alpha \in (\beta, 1)$.*

Step 5. *$\forall j \in \mathcal{D}_\gamma$, let $\mathcal{P}_j := \{p \in \mathcal{P} : \hat{x}_{jp} > 0\}$. Sort the paths in $\mathcal{P}_j$ in an nondecreasing distance to client j; that is, $c_{jp_1^j} \leq c_{jp_2^j} \leq \cdots \leq c_{jp_{|\mathcal{P}_j|}^j}$. Let s_j^* be the number satisfying $\sum_{s \leq s_j^*} \hat{x}_{jp_s^j} \geq \alpha > \sum_{s \leq s_j^* - 1} \hat{x}_{jp_s^j}$ and define $c_j(\alpha) = c_{jp_{s_j^*}^j}$. Calculate the average connection cost of client j,*

$$D_{av}(j) := \frac{\sum_{s=1}^{s_j^*} c_{jp_s^j} \hat{x}_{jp_s^j}}{\sum_{s=1}^{s_j^*} \hat{x}_{jp_s^j}}.$$

Step 6. *For the served client set $\mathcal{D}_\gamma$, iteratively generate greedily cluster centers as follows.*

Step 6.1 *Set $t = 1, \mathcal{C} := \emptyset, \mathcal{C}' = \mathcal{D}_\gamma$.*

Step 6.2 *Choose cluster center $j_t := \arg\min_{j \in \mathcal{C}'} c_j(\alpha) + D_{av}(j)$.*

Step 6.3 *Set $P_t := \{p_1^{j_t}, p_2^{j_t}, \ldots, p_{s_{j_t}^*}^{j_t}\}$,*

$F_t := \cup_{\ell=1}^k \{i_\ell \in \mathcal{F}^\ell : \exists p \in P_t, s.t. i_\ell \in p\}$,

$\mathcal{C}_t := \{j \in \mathcal{C}' : \exists s \leq s_j^* \ s.t. \ p_s^j \cap F_t \neq \emptyset\}$.

Step 6.4 *Set $\mathcal{C}' := \mathcal{C}' \setminus \mathcal{C}_t$, $\mathcal{C} := \mathcal{C} \cup \{j_t\}$.*

Step 6.5 *We randomly open a facility path from the path set P_t; namely choose a path $p \in P_t$ with probability $\hat{x}_{j_t p}/\sum_{s=1}^{s_{j_t}^*} \hat{x}_{j_t p_s^{j_t}}$; round all variables $\{\bar{y}_{i_\ell}\}_{i_\ell \in p}$ and $\{\bar{y}_{i_\ell}\}_{i_\ell \in F_t \setminus p}$ to 1 and 0 respectivly; connect the cluster center j_t to this open path; and set $\bar{x}_{j_t p} = 1$ and $\bar{x}_{j_t p'} = 0$, $\forall p' \in \mathcal{P} \setminus p$.*

Step 6.6 *If $\mathcal{C}' \neq \emptyset$, set $t = t + 1$ and go to Step 6.2.*

Step 6.7 *Connect all the clients $j \in \mathcal{D}_\gamma \setminus \mathcal{C}$ to the closest open path.*

Since Steps 4–6 of Algorithm 1 is Wu and Xu [15]'s bifactor $\left(\frac{\ln(1/\beta)}{1-\beta}, 1 + \frac{2}{1-\beta}\right)$-approximation algorithm for the k-LFLP with feasible LP solution $(\hat{x}, \hat{y})$, the cost of integer feasible solution $(\bar{x}, \bar{y})$ generated is no more than $\frac{\ln(1/\beta)}{1-\beta}\hat{F} + 1 + \frac{2}{1-\beta}\hat{C}$, where $\hat{F}, \hat{C}$ are the facility cost and connection cost respectively of solution $(\hat{x}, \hat{y})$.

The performance of Algorithm 1 is summarized in the following theorem.

Theorem 1. *When $\beta = e^{-2}, \gamma = \frac{3-e^{-2}}{2}$, Algorithm 1 is an $\left(1 + \frac{2}{1-e^{-2}}\right)$-approximation algorithm for the k-LFLP with submodular penalties.*

3   Multilevel Facility Location Problem with Linear Penalties

Analogous to the k-LFLP with submodular penalties, the k-LFLP with linear penalties can also be formulated as an integer linear program:

$$\min \sum_{\ell=1}^k \sum_{i_\ell \in \mathcal{F}^l} f_{i_\ell} y_{i_\ell} + \sum_{p \in \mathcal{P}} \sum_{j \in \mathcal{D}} c_{jp} x_{jp} + \sum_{j \in \mathcal{D}} q_j z_j$$

$$\text{s. t.} \sum_{p \in \mathcal{P}} x_{jp} + z_j \geq 1, \ \forall j \in \mathcal{D},$$

$$\sum_{p:i_\ell \in p} x_{jp} - y_{i_\ell} \leq 0, \ \forall j \in \mathcal{D}, i_\ell \in \mathcal{F}^l, \ell = 1, 2, \ldots, k, \quad (3)$$

$$x_{jp} \in \{0, 1\}, \ \forall p \in \mathcal{P}, j \in \mathcal{D},$$

$$y_{i_\ell} \in \{0, 1\}, \ \forall i_\ell \in \mathcal{F}^\ell, \ell = 1, 2, \ldots, k,$$

$$z_j \in \{0, 1\}, \ \forall j \in \mathcal{D}.$$

The linear programming relaxation of the above program is as follows:

$$\min \sum_{\ell=1}^{k} \sum_{i_\ell \in \mathcal{F}^l} f_{i_\ell} y_{i_\ell} + \sum_{p \in \mathcal{P}} \sum_{j \in \mathcal{D}} c_{jp} x_{jp} + \sum_{j \in \mathcal{D}} q_j z_j$$

$$\text{s. t.} \sum_{p \in \mathcal{P}} x_{jp} + z_j \geq 1, \ \forall j \in \mathcal{D},$$

$$\sum_{p:i_\ell \in p} x_{jp} - y_{i_\ell} \leq 0, \ \forall j \in \mathcal{P}, i_\ell \in \mathcal{F}^l, \ell = 1, 2, \ldots, k, \qquad (4)$$

$$x_{jp} \geq 0, \ \forall p \in \mathcal{P}, j \in \mathcal{D},$$

$$y_{i_\ell} \geq 0, \ \forall i_\ell \in \mathcal{F}^\ell, \ell = 1, 2, \ldots, k,$$

$$z_j \geq 0, \ \forall j \in \mathcal{D}.$$

The dual of the problem (4) is

$$\max \sum_{j \in \mathcal{D}} \alpha_j$$

$$\text{s. t.} \ \alpha_j - \sum_{i_\ell \in p} \beta_{i_\ell j} \leq c_{jp}, \ \forall p \in \mathcal{P}, j \in \mathcal{D},$$

$$\sum_{j \in \mathcal{D}} \beta_{i_\ell j} \leq f_{i_\ell}, \ \forall i_\ell \in \mathcal{F}^l, \ell = 1, 2 \ldots, k, \qquad (5)$$

$$\alpha_j \leq q_j, \ \forall j \in \mathcal{D},$$

$$\alpha_j \geq 0, \beta_{i_\ell j} \geq 0, \ \forall j \in \mathcal{D}, i_\ell \in \mathcal{F}^\ell, \ell = 1, 2 \ldots, k.$$

Lemma 2. *Let (x^*, y^*, z^*) and (α^*, β^*), respectively, be the optimal solutions to the primal and dual LP. Then we have*

(1) $x_{jp}^* > 0 \Rightarrow \alpha_j^* - \sum_{i_\ell \in p} \beta_{i_\ell j}^* = c_{jp}$, $c_{jp} \leq \alpha_j^*$, $\forall p \in \mathcal{P}, j \in \mathcal{D}$;

(2) $x_{jp}^* > 0$, and $z_j^* > 0 \Rightarrow c_{jp} \leq q_j = \alpha_j^*$, $\forall p \in \mathcal{P}, j \in \mathcal{D}$.

Proof. If $x_{jp}^* > 0$, it follows from the complementary slackness condition that

$$\alpha_j^* - \sum_{i_\ell \in P} \beta_{i_\ell j}^* = c_{jp}.$$

Combining with $\beta_{i_\ell j}^* \geq 0, i_\ell \in p$, we get $c_{pj} \leq \alpha_j^*$.

If $x_{pj}^* > 0$ and $z_j^* > 0$, we have $c_{jp} \leq \alpha_j^*$ from Case 1. On the other hand, we have the complementary slackness condition $z_j^*(\alpha_j^* - q_j) = 0$, implying that $q_j = \alpha_j^*$.

Our algorithm is motivated by Aardal et al. [1]. First, we solve the linear program (4) and its relaxation (5) to obtain the optimal solutions (x^*, y^*, z^*) and (α^*, β^*). Two clients j and j' are neighbors if there exist two paths p and p' such that $x_{jp}^* > 0$, $x_{j'p'}^* > 0$ and $p \cap p' \neq \emptyset$. Set $(\bar{x}, \bar{y}, \bar{z}) := (x^*, y^*, z^*)$. Iteratively revise $(\bar{x}, \bar{y}, \bar{z})$ untill we obtain a feasible integer solution to problem (3). In each iterative step, we maintain the feasibility of the solution $(\bar{x}, \bar{y}, \bar{z})$. Based on the optimal solution (x^*, y^*, z^*), we partition the set $\mathcal{D}$ of clients into the penalized set $\bar{\mathcal{D}}_\gamma$ of clients and the served set $\mathcal{D}_\gamma$ of clients.

Next for the client set $\mathcal{D}_\gamma$, we construct pairwise disjoint cluster sets, each containing a client called cluster center, all its neighbors and facilities from the path of fractionally serving that cluster center in the solution (x^*, y^*, z^*).

For each cluster set, we randomly choose an open path from the path set serving cluster center fractionally, and connect cluster center to this path. Because a non-cluster-center client j is possible to be the neighbor of two or more cluster centers, we connect these clients to the closest open path finally.

Introduce the following notation:

$$D_{av}(j) = \left(\sum_{p \in P} c_{pj} x^*_{pj} \right) / (1 - z^*_j).$$

Algorithm 2

Step 1. *Solve the LP relaxation (4) and its dual (5) to obtain the optimal solutions of them* (x^*, y^*, z^*) *and* (α^*, β^*). *Set* $(\bar{x}, \bar{y}, \bar{z}) := (x^*, y^*, z^*)$.

Step 2. *According to the penalized proportion of client j, we partition the clients set into two sets* $\mathcal{D}_\gamma = \{j \in \mathcal{D} : \sum_{p \in P} x^*_{jp} \geq \frac{1}{\gamma}\}$, *and* $\bar{\mathcal{D}}_\gamma = \mathcal{D} \setminus \mathcal{D}_\gamma$. *We penalized the clients in the set* $\bar{\mathcal{D}}_\gamma$ *and set* $\bar{z}_j := 1, \bar{x}_{jp} := 0, \forall j \in \bar{\mathcal{D}}_\gamma, p \in \mathcal{P}$.

Step 3. *For the served clients set* $\mathcal{D}_\gamma$, *generate greedily cluster center as follows.*

    Step 3.1 *Set* $t := 1, \mathcal{C} := \emptyset, \mathcal{C}' := \mathcal{D}_\gamma$.

    Step 3.2 *Choose cluster center* $j_t := \arg \min_{j \in \mathcal{C}'} \alpha^*_j + D_{av}(j)$.

    Step 3.3 *Set* $P_t := \{p \in \mathcal{P} : x^*_{j_t p} > 0\}$,
      $F_t := \cup^k_{\ell=1} \{i_\ell \in \mathcal{F}^\ell : \exists p \in P_t, \text{s. t. } i_\ell \in p\}$,
      $\mathcal{C}_t := \{j \in \mathcal{D} : \exists p \in \mathcal{P} \text{s. t. } x^*_{jp} > 0, p \cap F_t \neq \emptyset\}$.

    Step 3.4 *Set* $\mathcal{C}' := \mathcal{C}' \setminus \mathcal{C}_t, \mathcal{C} := \mathcal{C} \cup \{j_t\}$.

    Step 3.5 *We randomly open a path with probability 1 from the path set* P_t. *Formally, choose a path* $p \in P_t$ *with probability* $\frac{x^*_{j_t p}}{1 - z^*_{j_t}}$ *and perform the following operations: round all the variables* $\{\bar{y}_{i_\ell}\}_{i_\ell \in p}$ *to 1 and all other variables* $\{\bar{y}_{i_\ell}\}_{i_\ell \in F_t \setminus p}$ *to 0; connect the cluster center* j_t *to this open path; and set* $\bar{x}_{j_t p} := 1$ *and* $\bar{x}_{j_t p'} := 0, \forall p' \in \mathcal{P} \setminus p$.

    Step 3.6 *If* $\mathcal{C}' \neq \emptyset$, *set* $t := t + 1$, *go to Step 3.2.*

    Step 3.7 *Connect all the clients* $j \in \mathcal{D}_\gamma \setminus \mathcal{C}$ *to the closest open path.*

The performance of Algorithm 2 is given below.

Theorem 3. *If* $\gamma = 1$, *then the expected total cost of* $(\bar{x}, \bar{y}, \bar{z})$ *is no more than 3 times the optimum cost, which implies that Algorithm 2 is a 3-approximation algorithm for the k-LFLP with linear penalties.*

Acknowledgments. The first author's research is supported by the NSF of China (No. 11201013). The research of the second author is supported by the NSF of China (No. 11371001). The third authors research is supported in part by the Natural Sciences and Engineering Research Council of Canada (NSERC) grant 283106.

References

1. Aardal, K., Chudak, F., Shmoys, D.: A 3-approximation algorithm for the k-level uncapacitated facility location problem. Inf. Process. Lett. **72**, 161–167 (1999)
2. Asadi, M., Niknafs, A., Ghodsi, M.: An approximation algorithm for the k-level uncapacitated facility location problem with penalties. In: Sarbazi-Azad, H., Parhami, B., Miremadi, S.-G., Hessabi, S. (eds.) Proceeding of CSICC. Communications in Computer and Information Science, vol. 6, pp. 41–49. Springer, Heidelberg (2008)
3. Ageev, A., Ye, Y., Zhang, J.: Improved combinatorial approximation algorithms for the k-FLP. SIAM J. Discrete Math. **18**, 207–217 (2007)
4. Byrka, J., Li, S., Rybicki, B.: Improved approximation algorithm for k-Level UFL with penalties, a simplistic view on randomizing the scaling parameter. In: Kaklamanis, C., Pruhs, K. (eds.) WAOA 2013. LNCS, vol. 8447, pp. 85–96. Springer, Heidelberg (2014)
5. Fujishige, S.: Submodular Functions and Optimization, 2nd edn. Elsevier, Amsterdam (2005)
6. Gabor, F., Van Ommeren, J.: A new approximation algorithm for the multilevel facility location problem. Discrete Appl. Math. **158**, 453–460 (2010)
7. Guha, S., Khuller, S.: Greedy strikes back: improved facility location algorithms. In: Proceeding of SODA, pp. 649–657 (1998)
8. Li, G., Wang, Z., Xu, D.: An approximation algorithm for the k-FLP with submodular penalties. J. Ind. Manage. Optim. **18**, 521–529 (2012)
9. Jain, K., Vazirani, V.: Approximation algorithms for metric facility location and k-median problems using the primal-dual schema and Lagrangian relaxation. J. ACM **48**, 274–276 (2001)
10. Krishnaswamy, R., Sviridenko, M.: Inapproximability of the multi-level uncapacitated facility location problem. In: Proceeding of SODA, pp. 718–734 (2012)
11. Li, G., Du, D., Xu, D., Zhang, R.: A cost-sharing method for the multi-level economic lot-sizing game. Sci. China Inf. Sci. **57**, 1–9 (2014)
12. Li, S.: A 1.488 Approximation Algorithm for the Uncapacitated Facility Location Problem. In: Aceto, L., Henzinger, M., Sgall, J. (eds.) ICALP 2011, Part II. LNCS, vol. 6756, pp. 77–88. Springer, Heidelberg (2011)
13. Li, Y., Du, D., Xiu, N., Xu, D.: Improved approximation algorithms for the facility locaiton problem with linear/submodular penalties, Algorithmica. doi:10.1007/s00453-014-9911-7
14. Shmoys, D., Tardos, E., Aardal, K.: Approximation algorithms for facility location problem (extended abstract). In: Proceeding of STOC, pp. 265–274 (1997)
15. Wu, C., Xu, D.: An improved approximation algorithm for the k-FLP with soft capacities, accepted by Acta Mathematicae Applicatae Sinica

Smaller Kernels for Several FPT Problems Based on Simple Observations

Wenjun Li$^{(\boxtimes)}$ and Shuai Hu

School of Information Science and Engineering, Central South University,
Changsha 410083, People's Republic of China
liwenjun@csu.edu.cn

Abstract. In the field of parameterized computation and theory, as a pre-processing technique of algorithms, kernelization has received considerable attention. In this paper, we study the kernelization algorithms for several fixed parameter tractable problems, including Co-Path Set, Path-Contractibility and Connected Dominating Set on G_7 Graphs. For these three problems, based on simple observations, we give simple kernelization algorithms with kernel size of $4k, 3k + 4$ and $O(k^2)$ respectively, which are smaller than the previous corresponding smallest kernels $6k, 5k + 3$, and $O(k^3)$.

1 Introduction

With the development of modern industry, more and more computation problems has been emerged. Due to massive data input, many of them are hard to be solved. For this reason, how to decrease the size of instances of problems effectively has become a research focus. Be accompanied with the Parameterized Computation and Complexity Theory coming to the world [5], as a pre-processing technique of algorithms for decreasing the size of instances of problems effectively, kernelization has drawn considerable attention, and become a significant research direction.

In this paper, we consider kernelization algorithms for several FPT problems, including Co-path Set, Path-Contractibility and Connected Dominating Set on G_7 graphs (graphs with girth at least 7).

The Minimum Co-Path Set problem is to delete a minimum set of edges in a given graph G such that each connected component in the remaining graph is a path. This problem has a lot of practical applications [2,3,22,23]. Garey and Johnson [10] proved that Co-path Set problem is NP-complete. The parameterized version of this problem is to ask whether deleting at most k edges in a given graph G such that each connected component in the remaining graph is a path. As to the kernelization algorithms for this problem, Jiang et al. [16] gave a $5k$-weak kernel. Roughly speaking, Weak kernel is the potential search space, which is different from the traditional kernel. Usually, the size of the formal

This work is supported by the National Natural Science Foundation of China under Grants (61232001, 61472449, 61420106009, 61402054).

J. Wang and C. Yap (Eds.): FAW 2015, LNCS 9130, pp. 170–179, 2015.
DOI: 10.1007/978-3-319-19647-3_16

is smaller than the later. Zhang et al. [24] presented a kernelization algorithm with kernel size of 22k. Recently, Feng et al. [6] improved the kernel to 9k. It is worthy to be remarked that, earlier in [7], Fernau presented a 6k kernel for the parameterized problem named LINEAR ARRANGEMENT BY DELETING EDGES. We find that this problem is actually equivalent to the parameterized CO-PATH SET problem.

The parameterized PATH-CONTRACTIBILITY problem is to decide if the given graph could be contracted into a path through at most k edge contractions. Edge contraction is one type of graph modification operation, and it is also one of the three minor operations (vertex deletion, edge deletion and edge contraction). A lot of problems with respect to vertex deletion and edge deletion are proved to be NP-hard and well studied. However, the problems with respect to edge contraction do not have received extensive attention. P. Heggernes et al. [13] proved that BIPARTITE GRAPH-Contractibility is FPT. Recently, P. Heggernes et al. [14] presented a kernelization algorithm for PATH-CONTRACTIBILITY problem with linear kernel (5k + 3), and shown that TREE-CONTRACTIBILITY does not have a polynomial kernel, unless coNP $\subseteq$ NP/poly.

In a graph $G = (V, E)$, a *dominating set* is a set of vertices $S \subseteq V$ such that every vertex in $G \backslash S$ is adjacent to some vertex in S. A dominating set S of G is said to be a *connected dominating set* if the subgraph $G[S]$ induced by S is connected. The parameterized (CONNECTED) DOMINATING SET problem is to decide if G has a (connected) dominating set with at most k vertices? The girth of a graph G is the length of a shortest cycle in G. The notation G_r denotes the set of all graphs with girth at least $r \in \mathbb{N}$. On general graphs, the CONNECTED DOMINATING SET problem is NP-hard [10], and the parameterized version of the problem is known to be W[2]-complete [19]. This problem remains W[2]-complete even in many restricted classes of graphs, such as graphs of bounded average degree, bipartite graphs, triangle free graphs, split graphs, chordal graphs and G_4 graphs [11,20,21]. But, the parameterized CONNECTED DOMINATING SET problem has FPT algorithms in certain restricted families of graphs, such as planar graphs [8,12,17,18], graphs of bounded genus [1], apex-minor-free graphs [9], nowhere-dense classes of graphs [4], graphs of bounded degeneracy [11] and graphs of bounded treewidth [15]. Misra et al. [20] studied the kernelization complexity of the CONNECTED DOMINATING SET problem. They showed that this problem has no polynomial kernel on graphs containing cycles of length at most 6, unless the Polynomial Hierarchy (PH) collapses to the third level, and gave an $O(k^3)$-kernelization algorithm for the problem on G_7 graphs.

Based on simple observations, we give some improved kernelization algorithms for the problems mentioned above. More precisely, our contributions in this paper consist of the following four items:

- We find that the LINEAR ARRANGEMENT BY DELETING EDGES problem is equivalent to the CO-PATH SET problem.
- We show that parameterized CO-PATH SET problem has a 4k-kernel.

- We give a $(3k + 4)$-kernelization algorithm for the parameterized PATH-CONTRACTIBILITY problem.
- We present an $O(k^2)$-kernelization algorithm for the parameterized CONNECTED DOMINATING SET problem on G_7 Graphs, which can also lead to an $O(k^2)$-kernel for the parameterized DOMINATING SET problem on G_7 Graphs.

The paper is organized as follows. Section 2 gives some related terminologies. Section 3 proves that the LINEAR ARRANGEMENT BY DELETING EDGES problem is equivalent to the CO-PATH SET problem, and shows a kernelization algorithm for the parameterized CO-PATH SET problem. Section 4 gives a kernelization algorithm for the parameterized PATH-CONTRACTIBILITY problem. Section 5 presents a kernelization algorithm for the CONNECTED DOMINATING SET problem on G_7 Graphs.

2 Preliminaries

Let $G = (V, E)$ be a finite, undirected and simple graph. For two vertices $u, v \in V$, $[u, v]$ denotes the edge between u and v. The neighborhood set of a vertex v in G is $N(v) = \{w \in V | [v, w] \in E\}$. For a vertex subset $S \subseteq V$, $G[S]$ denotes the subgraph of G induced by S. For a subgraph G' of G, $E(G')$ consists of all the edges in G'. If the degree of vertex v is r, we say v is a degree-r vertex. Furthermore, if the degree of vertex v is not less than r, then v is a degree$_{\geq}r$ vertex, and if the degree of vertex v is not larger than r, then v is a degree$_{\leq}r$ vertex. For the convenience, we use $G - v$ to denote $G[V \backslash \{v\}]$ for a vertex $v \in V$, and $G - S$ to denote $G[V \backslash S]$ for a vertex subset $S \subseteq V$. Similarly, for an edge $e \in E$ and edge subset $E' \subseteq E$, $G - e$ and $G - E'$ denote the subgraph $G' = (V, E \backslash e)$ and $G'' = (V, E \backslash E')$ respectively. We use $P = v_1 v_2 v_3 ... v_l$ to denote a path with l vertices.

3 CO-PATH SET and LINEAR ARRANGEMENT BY DELETING EDGES

In this section, we prove that the CO-PATH SET problem is equivalent to the LINEAR ARRANGEMENT BY DELETING EDGES problem first. Later, we show an improved kernelization algorithm for these two problems, which consists of 6 reduction rules and builds a $4k$ kernel for them. The formal definition of parameterized LINEAR ARRANGEMENT BY DELETING EDGES is as follows.

> Parameterized LINEAR ARRANGEMENT BY DELETING EDGES: Given an undirected graph $G = (V, E)$ and an integer k, is there an edge subset $E' \subseteq E$ with $|E'| \leq k$ and a one-to-one mapping $\sigma : V \longrightarrow \{1, \ldots, |V|\}$ such that $\sum_{[u,v] \in E \backslash E'} |\sigma(u) - \sigma(v)| = |E \backslash E'|$?

Theorem 1. *The* CO-PATH SET *problem is equivalent to the* LINEAR ARRANGEMENT BY DELETING EDGES *problem.*

Proof. Let $(G = (V, E), k)$ be a *yes*-instance of the parameterized CO-PATH SET problem, and E' be a subset of E with $|E'| \leq k$ such that each connected component of $G - E'$ is a path. Assume that $G - E' = \{P_1, \ldots, P_i, \ldots, P_l\}$ is a set of paths and path $P_i = v_{i,1} \ldots v_{i,j} \ldots v_{i,r_i}$, where $1 \leq i \leq l$. Then we can construct a one-to-one mapping σ as follows: $\sigma(v_{1,j}) = j$, and $\sigma(v_{i,j}) = \sum_{m=1}^{i-1} r_m + j$, where $2 \leq i \leq l$. It is easy to see that $\sum_{[u,v] \in E \setminus E'} |\sigma(u) - \sigma(v)| = |E \setminus E'|$.

For the other direction, let $(G = (V, E), k)$ be a *yes*-instance of the parameterized LINEAR ARRANGEMENT BY DELETING EDGES problem, E' be a subset of E with $|E'| \leq k$ and $\sigma : V \longrightarrow \{1, \ldots, |V|\}$ be a one-to-one mapping such that $\sum_{[u,v] \in E \setminus E'} |\sigma(u) - \sigma(v)| = |E \setminus E'|$. Assume one of the component C in $G - E'$ is not a path, then there is a degree≥ 3 vertex in C or C itself is a cycle. From the definition of σ, we know that, for any edge $[u, v] \in E \setminus E'$, $|\sigma(u) - \sigma(v)| \geq 1$. Since $\sum_{[u,v] \in E \setminus E'} |\sigma(u) - \sigma(v)| = |E \setminus E'|$, we can say that for each edge $[u, v] \in E \setminus E'$, $|\sigma(u) - \sigma(v)| = 1$. If there is a degree$\geq 3$ vertex w in C, from the definition of σ, we know that there is at least one vertex w' such that, for the edge $[w, w'] \in E \setminus E'$, $|\sigma(w) - \sigma(w')| > 1$, a contradiction. If C itself is a cycle, then $|C| \geq 3$ for G is a simple graph. In the following, we will prove that $\sum_{[u,v] \in E(C)} |\sigma(u) - \sigma(v)| > |C|$. Assume that $\sum_{[u,v] \in E(C)} |\sigma(u) - \sigma(v)| = |C|$. Let $C = \{v_1, v_2, \ldots, v_{|C|}, v_1\}$ be a cycle. Obviously, $C - [v_{|C|}, v_1]$ is a path. From the definition of σ, we know that if $\sum_{[u,v] \in E(C) \setminus [v_{|C|}, v_1]} |\sigma(u) - \sigma(v)| = |C| - 1$, then $|\sigma(v_{|C|}) - \sigma(v_1)| > 1$. Thus, $\sum_{[u,v] \in E(C)} |\sigma(u) - \sigma(v)| > |C|$, which means $\sum_{[u,v] \in E \setminus E'} |\sigma(u) - \sigma(v)| = |E \setminus E'|$, also a contradiction, which finishes the proof of the theorem. □

3.1   Reduction Rules

In this part, we give a kernelization algorithm for the parameterized CO-PATH SET problem, which consists of 6 reduction rules. It should be mentioned that the former 5 rules have been given in [6,7]. Thus, we just prove the correctness of the last rule.

Rule 3.1. Delete vertices of degree-0.

Rule 3.2. If a degree-1 vertex v is adjacent to a degree≤ 2 vertex, then delete v.

Rule 3.3. If a degree-2 vertex v is adjacent to a degree-2 vertex w such that $N(v) \cap N(w) = \emptyset$, then contract the edge $[v, w]$.

Rule 3.4. If a degree-2 vertex v is adjacent to a degree-2 vertex w such that $N(v) \cap N(w) \neq \emptyset$ (assume $u = N(v) \cap N(w)$), then delete the edge $[u, v]$ and decrease the parameter k by one.

Rule 3.5. If v is a degree≥ 3 vertex such that $N(v)$ includes two degree-1 vertices x, y, then delete all the edges adjacent to v but $[v, x]$ and $[v, y]$, decrease the parameter k by $|N(v)| - 2$.

Rule 3.6. If v is a degree≥ 3 vertex such that $N(v)$ includes a degree-1 vertex x and a degree-2 vertex y, then delete all the edges adjacent to v except $[v, x]$ and $[v, y]$, decrease the parameter k by $|N(v)| - 2$.

Lemma 1. *Rule 3.6 is safe.*

Proof. For a *yes*-instance $(G = (V, E), k)$ of parameterized CO-PATH SET problem, suppose $G - E'$ is a set of paths with $|E'| \leq k$. Let v be a degree$_{\geq}3$ vertex in G, $x, y \in N(v)$, where x is a degree-1 vertex, y is a degree-2 vertex. Now, We are going to prove that there is an edge set E'' with $|E''| \leq |E'|$ such that (1) $[v, x] \notin E''$ and $[v, y] \notin E''$, and (2) each connected component of $G - E''$ is a path.

Assume $[v, x] \in E'$. If v is a degree-1 vertex in $G - E'$, then each connected component of $G - (E' \backslash [v, x])$ is also a path, which means that deleting all the edges in $E'' = E' \backslash [v, x]$ makes each connected component of $G - E''$ is a path. Furthermore, $|E''| = |E'| - 1$. If v is a degree-2 vertex in $G - E'$, let v_1, v_2 be the two neighbors of v in $G - E'$, then deleting all the edges in $E'' = E' \backslash [v, x] \cup [v, v_1]$ makes each connected component of $G - E''$ is also a path, and $|E''| = |E'|$.

Suppose $[v, y] \in E'$. From the above proof, we can assume that $[v, x] \notin E'$. Since y is a degree-2 vertex in G, y is a degree$_{\leq}1$ vertex in $G - E'$. If v is a degree-1 vertex in $G - E'$, then each connected component of $G - (E' \backslash [v, y])$ is also a path, i.e., deleting all the edges in $E'' = E' \backslash [v, y]$ makes each connected component of $G - E''$ is a path. Furthermore, $|E''| = |E'| - 1$. If v is a degree-2 vertex in $G - E'$, let w be a neighbor of v in $G - E'$ (x is the other neighbor of v in $G - E'$), then deleting all the edges in $E'' = E' \backslash [v, y] \cup [v, w]$ makes each connected component of $G - E''$ is also a path, and $|E''| = |E'|$.

Above all, there is an edge set E'' with $|E''| \leq |E'|$ such that (1) $[v, x] \notin E''$ and $[v, y] \notin E''$, and (2) each connected component of $G - E''$ is a path. Since the degree of each vertex on a path is no larger than 2, deleting all the edges adjacent to v but $[v, x]$ and $[v, y]$ is safe.   □

3.2   Kernel Analysis

Let (G, k) be an instance of parameterized Co-Path Set problem. Starting from (G, k), we repeatedly apply the Rules 3.1–3.6 if possible. Obviously, the total running time of reduction process is polynomial.

If none of Rules 3.1–3.6 can be applied on graph G, then we say G is irreducible. For any irreducible graph G, it holds the following two propositions, which are the base of the kernel analysis.

Proposition 1. *If v is degree-2 vertex, then the two neighbors of v are degree$_{\geq}3$ vertices.*

Proof. If one neighbor of v is a degree-1 vertex, then Rule 3.2 could be applied. And if one neighbor of v is a degree-2 vertex, then Rule 3.3 or 3.4 could be applied. Therefore, the two neighbors of v are degree$_{\geq}3$ vertices.

Proposition 2. *If v is a degree-1 vertex, then the only neighbor w of v is a degree$_{\geq}3$ vertex and each vertex in $N(w) \backslash v$ is a degree$_{\geq}3$ vertex.*

Proof. By Rule 3.2, the only neighbor w of v must be a degree$_{\geq}3$ vertex. If there is a degree-1 vertex in $N(w)\backslash v$, then Rule 3.5 could be applied. And if there is a degree-2 vertex in $N(w)\backslash v$, then Rule 3.6 could be applied. Therefore, each vertex in $N(w)\backslash v$ is a degree$_{\geq}3$.

Theorem 2. *The parameterized Co-Path Set problem admits a kernel with size of $4k$.*

Proof. For a *yes*-instance $(G = (V, E), k)$ of parameterized CO-PATH SET problem, let $G - E'$ be a set of paths with $|E'| \leq k$. If vertex v is an endpoint of an edge in E', then we say v is an influenced vertex; otherwise, it is an un-influenced vertex. Obviously, each degree$_{\geq 3}$ vertex must be an influenced vertex. Furthermore, there is no degree-0 vertex in G, otherwise, Rule 3.1 could be applied. Thus, we can divide the vertices in V into three disjoint parts: (1) A: influenced vertices; (2) B: un-influenced degree-1 vertices; and (3) C: un-influenced degree-2 vertices. Now, we are going to bound the size of V:

- $|A| \leq 2k$: For the reason that $|E'| \leq k$;
- $|B| + |C| \leq 2k$: Let $A_B = \{v|v \in A, N(v) \cap B \neq \emptyset\}$ and $A_C = \{v|v \in A, N(v) \cap C \neq \emptyset\}$. Obviously, $A_B \cup A_C \subseteq A$. By Proposition 2, $A_B \cap A_C = \emptyset$. Therefore, $|A_B| + |A_C| \leq |A|$. From Proposition 2, we know that $|B| = |A_B|$. By Proposition 1, each vertex of C is adjacent to exactly two vertices of A_C in G. Since each vertex of A_C is a degree$_{<2}$ vertex in $G - E'$, it is adjacent to at most two vertices of C in G. Therefore, we have $|C| \leq |A_C|$. Thus, the inequality $|B| + |C| \leq |A_B| + |A_C| \leq |A| \leq 2k$ holds.

Summarizing above, $|V| = |A| + |B| + |C| \leq 2k + 2k = 4k$. □

4   Path-Contractibility

In this section, we consider kernelization algorithm for the parameterized PATH-CONTRACTIBILITY Problem. In the following, some necessary definitions and notations are showed, which are identical with that in [14].

A graph is connected if there exists a path between every pair of vertices, and disconnected otherwise. The connected component of graph G is its maximal connected subgraph. We say that a vertex subset S of graph G is connected if $G[S]$ is connected. A bridge in a connected graph is an edge e such that $G - e$ is disconnected. A graph is 2-edge connected if it has no bridge. A 2-edge connected component of graph G is a maximal 2-edge connected subgraph of G.

The contraction of edge $[u, v]$ in G removes both u and v from G, and adds a new vertex w, which is made adjacent to vertices in $N_G(u) \cup N_G(v)$ but u and v. A graph G is contractible to a graph H, if H can be obtained from G by a sequence of edge contractions. Equivalently, G is H-contractible if there is a surjection $\phi : V(G) \rightarrow V(H)$, with $W(h) = \{v \in V(G) \mid \phi(v) = h\}$ for every $h \in V(H)$, that satisfies the following two conditions: (1) for every $h \in V(H)$, $W(h)$ is (non-empty) a connected set in G; (2) for every pair $h_i, h_j \in V(H)$, there is an edge in G between a vertex of $W(h_i)$ and a vertex of $W(h_j)$ if and only if $[h_i, h_j] \in E(H)$. $\mathcal{W} = \{W(h) \mid h \in V(H)\}$ is an H-witness structure of G, and the sets $W(h)$, for $h \in V(H)$, are called witness sets of $\mathcal{W}$. If a vertex in G is contained in some witness set which contains more than one vertex of G, then we call it a contracted vertex; otherwise, it is an un-contracted vertex.

4.1   Reduction Rule and Kernel Analysis

The main idea of our kernelization algorithm for the parameterized PATH-CONTRACTIBILITY Problem is derived from the reduction rule in [14]. But it includes the ideas of the following two observations. For any instance ($G = (V, E), k$) of parameterized PATH-CONTRACTIBILITY Problem, we assume that G is connected. Otherwise, (G, k) is a trivial *no*-instance.

> **Observation 1:** For an arbitrary 2-edge connected graph $G = (V, E)$, if G is contractible to a path P by q edge contractions, then $q \geq (|V|-1)/3$.

For a given connected graph G, it is easy to see that after an edge contraction, the newly constructed graph is also connected and the number of vertices is decreased by 1. We first claim that for any edge $[u, v]$ of G, at least one of u and v is a contracted vertex. Assume that u and v are both un-contracted vertices. By the definition of 2-edge connected graph, $G \backslash [u, v]$ is connected. Then $G \backslash [u, v]$ remains connected after q edge contractions. As edge $[u, v]$ is not contracted, the new graph created by q edge contractions from G contains a cycle, which contradicts to the fact that it is a path. Assume P be the path contracted from G by q edge contractions. Let $V_1(P) = \{v | v \in V(P), |W(v)| \geq 2\}$ and $V_2(P) = V \backslash V_1(P)$. Then $|V_1(P)| \geq \lceil (|V(P)|-1)/2 \rceil$. Since the induced subgraph $G(W(x))$ contracted to x needs a sequence of $|W(x)| - 1$ edge contractions, then $|V(P)| \leq (2|V| + 1)/3$, which means $q = |V| - |V(P)| \geq |V| - (2|V| + 1)/3 = (|V| - 1)/3$.

> **Observation 2:** In a given connected graph $G = (V, E)$, C_0 is a 2-edge connected component or a single vertex such that each edge adjacent to it is a bridge, $G - C_0 = \{C_1, ..., C_h\}$, where $h \geq 2$. If G is contractible to a path P, then there are at least $h - 2$ components of $G - C_0$ such that all the vertices in each component are contracted.

Suppose there are no more than $h - 3$ components in $G - C_0$ such that all the vertices in each component are contracted, which means that there are at least three components in $G - C_0$ such that there is at least one un-contracted vertex in each component. We know that $G(W(x))$ is connected in G for every vertex $x \in V(P)$. Obviously, in this case, there is a degree$_{\geq 3}$ vertex on P, a contradiction.

Based on the above two observations, we can refine the reduction rule for the parameterized PATH-CONTRACTIBILITY Problem in [14] easily, and obtain the following rule. The proof of its correctness is omitted here.

Rule 4.1. For an induced subgraph C_0 of G, where C_0 is a 2-edge connected component or a single vertex such that each edge adjacent to it is a bridge, let $G - C_0 = \{C_1, ..., C_h\}$, where the number of vertices in $C_1, ..., C_h$ is in decreasing order and $h \geq 1$, e be a bridge between C_0 and C_1. If (i) $|V\backslash V(C_1)| \geq k+2$, and (ii) $|V(C_1)| + (|V(C_0)| - 1)/3 \geq k + 2$, when $1 \leq h \leq 3$; $|V(C_1)| + (|V(C_0)| - 1)/3 + \sum_{i=4}^{h} |V(C_i)| \geq k + 2$, when $h \geq 4$, then return (G', k), where G' is the graph resulting from the contraction of edge e.

Theorem 3. *Parameterized* PATH-CONTRACTIBILITY *problem has a kernel with at most $3k + 4$ vertices.*

Proof. Let G be a graph such that Rule 4.1 cannot be applied on it. For any induced subgraph C_0 of G, where C_0 is a 2-edge connected component or a single vertex such that each edge adjacent to it is a bridge, $G - C_0 = \{C_1, ..., C_h\}$ is a set of components, where the number of vertices in $C_1, ..., C_h$ is in decreasing order and $h \geq 1$, $e = [u, v]$ is a bridge between C_0 and C_1. Obviously, for any graph G, $|V(C_1)|$ is decided by C_0. Assume C_0^* be a 2-edge connected component or a single vertex such that each edge adjacent to it is a bridge such that $|V(C_1^*)|$ is minimum.

If $|V\backslash V(C_1^*)| \leq k+1$ and $|V(C_1^*)| \leq k+1$, then $|V| = |V\backslash V(C_1^*)| + |V(C_1^*)| \leq 2k + 2$. If $|V\backslash V(C_1^*)| \leq k + 1$ and $|V(C_1^*)| \geq k + 2$, then we can find another induced subgraph C_0' satisfying that $(1) C_0'$ is a 2-edge connected component or a single vertex such that each edge adjacent to it is a bridge; and $(2) |V(C_1')| < |V(C_1^*)|$. But this contradicts to the fact that $|V(C_1^*)|$ is minimum. Thus, in this case, we have $|V| \leq 2k + 2$.

If $|V\backslash V(C_1^*)| \geq k+2$ and $1 \leq h \leq 3$, then we must have $|V(C_1^*) + (|V(C_0^*)| - 1)/3 \leq k + 1$. By definition, $|V(C_1^*)| \geq |V(C_2^*)| \geq |V(C_3^*)|$, here $V(C_2^*)$ and $V(C_3^*)$ may be empty. Thus, $|V| = |V(C_0^*)| + |V(C_1^*)| + |V(C_2^*)| + |V(C_3^*)| \leq 3|V(C_1^*)| + |V(C_0^*)| = 3(|V(C_1^*) + (|V(C_0^*)| - 1)/3) + 1 \leq 3k + 4$.

If $|V\backslash V(C_1^*)| \geq k + 2$ and $h \geq 4$, then $|V(C_1^*) + (|V(C_0^*)| - 1)/3 + \sum_{i=4}^{h} |V(C_i^*)| \leq k + 1$. Thus, $|V| = |V(C_0^*)| + |V(C_1^*)| + |V(C_2^*)| + |V(C_3^*)| + \sum_{i=4}^{h} |V(C_i^*)| \leq 3|V(C_1^*)| + |V(C_0^*)| + 3\sum_{i=3}^{h} |V(C_i^*)| \leq 3k + 4$. $\square$

5   CONNECTED DOMINATING SET on G_7 graphs

In this section, we give a kernelization algorithm for the parameterized CONNECTED DOMINATING SET problem on G_7 Graphs. This algorithm consists of a simple reduction rule, and leads to an $O(k^2)$-kernel. Furthermore, such a kernelization algorithm can also lead to an $O(k^2)$-kernel for the parameterized DOMINATING SET problem on G_7 Graphs.

5.1   Reduction Rule and Kernel Analysis

In our kernelization algorithm for parameterized CONNECTED DOMINATING SET problem, there is only one reduction rule. Furthermore, this rule is simple and trivial, and it is as follows.

Rule 5.1 If there are more than one degree-1 vertex adjacent to a vertex u, then delete all these degree-1 vertices except one.

Let $G = (V, E)$ be a graph such that Rule 5.1 is not applicable on G, D be a dominating set of G, where $|D| \leq k$. We divide the vertices in $V \backslash D$ into three disjoint parts: (1) V_1: degree-1 vertices; (2) V_2: each vertex has no neighbor in $V \backslash D$ and has at least two neighbors in D; and (3) V_3: other vertices in $V \backslash D$. Then, we have

- V_1: By Rule 5.1, there are no two degree-1 vertices adjacent to a vertex in G. Thus, there are at most k degree-1 vertices in $V \backslash D$;
- V_2: For each two vertices a and b in D, if there are two vertices u and v in $V \backslash D$ such that vertices u and v are neighbors of a and b, then there is a cycle $\{a, u, b, v, a\}$ with size 4, which contradicts to the fact that G is a G_7 graph. Therefore, we can conclude that there are no two vertices in $V \backslash D$ having the same two neighbors in D. Since there are at most $k(k-1)/2$ different vertex pairs in D, there are at most $k(k-1)/2$ vertices in V_2.
- V_3: Let u be a vertex in V_3. Then u has at least one neighbor v in V_3, i.e., there is an edge $[u, v]$ in $G[V_3]$. Otherwise, v is a vertex in V_1 or V_2. Assume a and b are the neighbors of u and v in D respectively. Then, $a \neq b$, otherwise, $\{u, v, a, u\}$ is a triangle, a contradiction. Furthermore, there is no edge $[u', v']$ in $G[V_3]$ such that a and b are the neighbors of u' and v' respectively. Otherwise, if $\{u, v\} \cap \{u', v'\} = \emptyset$, then there is a cycle $\{u, v, b, v', u', a, u\}$ with girth of 6, a contradiction. Similarly, If $\{u, v\} \cap \{u', v'\} \neq \emptyset$, then there is a cycle with girth of 4, also a contradiction. Since there are at most $k(k-1)/2$ vertex pairs in D, there are at most $k(k-1)/2$ edges in $G[V_3]$, i.e., there are at most $k(k-1)$ vertices in V_3.

Above all, there are at most $|V_1| + |V_2| + |V_3| + |D| = k + k(k-1)/2 + k(k-1) + k = 3/2k^2 + k/2$ vertices in G.

References

1. Bodlaender, H.L., Fomin, F.V., Lokshtanov, D., Penninkx, E., Saurabh, S., Thilikos, D.M.: (Meta) kernelization. In: FOCS 2009, pp. 629–638 (2009)
2. Cheng, Y., Cai, Z., Goebel, R., Lin, G., Zhu, B.: The radiation hybrid map construction problem: recognition, hardness, and approximation algorithms (2008). (unpublished manuscript)
3. Cox, D.R., Burmeister, M., Price, E.R., Kim, S., Myers, R.M.: Radiation hybrid mapping: a somatic cell genetic method for constructing high resolution maps of mammalian chromosomes. Science **250**, 245–250 (1990)
4. Dawar, A., Kreutzer, S.: Domination problems in nowhere-dense classes. In: FSTTCS2009, pp. 157–168 (2009)
5. Downey, R.G., Fellows, M.R.: Parameterized Complexity. Monographs in Computer Science. Springer, New York (1999)
6. Feng, Q., Zhou, Q., Li, S.: Randomized parameterized algorithms for co-path set problem. In: Chen, J., Hopcroft, J.E., Wang, J. (eds.) FAW 2014. LNCS, vol. 8497, pp. 82–93. Springer, Heidelberg (2014)

7. Fernau, H.: Parameterized algorithmics for linear arrangement problems. Discrete Appl. Math. **156**(17), 3166–3177 (2008)
8. Fernau, H.: Parameterized algorithmics: a graph-theoretic approach. Habilitationsschrift, Universität Tübingen, Germany (2005)
9. Fomin, F.V., Lokshtanov, D., Saurabh, S., Thilikos, D.M.: Bidimensionality and kernels. In: SODA2010, pp. 503–510. SIAM, Philadelphia (2010)
10. Garey, M.R., Johnson, D.S.: Computers and Intractability: A Guide to the Theory of NP Completeness. W.H. Freeman, New York (1979)
11. Golovach, P.A., Villanger, Y.: Parameterized complexity for domination problems on degenerate graphs. In: Broersma, H., Erlebach, T., Friedetzky, T., Paulusma, D. (eds.) WG 2008. LNCS, vol. 5344, pp. 195–205. Springer, Heidelberg (2008)
12. Gu, Q., Imani, N.: Connectivity is not a limit for kernelization: planar connected dominating set. In: López-Ortiz, A. (ed.) LATIN 2010. LNCS, vol. 6034, pp. 26–37. Springer, Heidelberg (2010)
13. Heggernes, P., Hof, P.V.T., Lokshtanov, D., Paul, C.: Obtaining a bipartite graph by contracting few edges. SIAM J. Discrete Math. **27**(4), 2143–2156 (2013)
14. Heggernes, P., Hof, P.V., Lokshtanov, D., Paul, C.: Contracting graphs to paths and trees. Algorithmica **68**(1), 109–1320 (2014)
15. Hermelin, D., Mnich, M., van Leeuwen, E.J., Woeginger, G.J.: Domination when the stars are out. In: Aceto, L., Henzinger, M., Sgall, J. (eds.) ICALP 2011, Part I. LNCS, vol. 6755, pp. 462–473. Springer, Heidelberg (2011)
16. Jiang, H., Zhang, C., Zhu, B.: Weak kernels. ECCC Report, TR10-005 (2010)
17. Lokshtanov, D., Mnich, M., Saurabh, S.: A linear kernel for a planar connected dominating set. Theor. Comput. Sci. **412**(23), 2536–2543 (2011)
18. Luo, W., Wang, J., Feng, Q., Guo, J., Chen, J.: An improved kernel for planar connected dominating set. In: Ogihara, M., Tarui, J. (eds.) TAMC 2011. LNCS, vol. 6648, pp. 70–81. Springer, Heidelberg (2011)
19. Marx, D.: Chordal deletion is fixed-parameter tractable. In: Fomin, F.V. (ed.) WG 2006. LNCS, vol. 4271, pp. 37–48. Springer, Heidelberg (2006)
20. Misra, N., Philip, G., Raman, V., Saurabh, S.: The kernelization complexity of connected domination in graphs with (no) small cycles. Algorithmica **68**(2), 504–530 (2014)
21. Raman, V., Saurabh, S.: Short cycles make W-hard problems hard: FPT algorithms for W-hard problems in graphs with no short cycles. Algorithmica **52**(2), 203–225 (2008)
22. Richard, C.W., Withers, D.A., Meeker, T.C., Maurer, S., Evans, G.A., Myers, R.M., Cox, D.R.: A radiation hybrid map of the proximal long arm of human chromosome 11 containing the multiple endocrine neoplasia type 1 (MEN-1) and bcl-1 disease loci. Am. J. Hum. Genet. **49**(6), 1189 (1991)
23. Slonim, D., Kruglyak, L., Stein, L., Lander, E.: Building human genome maps with radiation hybrids. J. Comput. Biol. **4**, 487–504 (1997)
24. Zhang, C., Jiang, H., Zhu, B.: Radiation hybrid map construction problem parameterized. In: Lin, G. (ed.) COCOA 2012. LNCS, vol. 7402, pp. 127–137. Springer, Heidelberg (2012)

Parameterized Minimum Cost Partition
of a Tree with Supply and Demand

Mugang Lin[1,2(✉)], Wenjun Li[1], and Qilong Feng[1]

[1] School of Information Science and Engineering,
Central South University, Changsha, China
[2] Department of Computer Science, Hengyang Normal University,
Hengyang, China
linmu718@csu.edu.cn

Abstract. In this paper, we study the minimum cost partition problem of a tree with supply and demand. For the kernelizaton of the problem, several reduction rules are given, which result in a kernel of size $O(k^2)$ for the problem. Based on the branching technique, a parameterized algorithm of running time $O^*(2.828^k)$ is presented.

1 Introduction

In practical applications, many problems can be efficiently formulated as a graph partitioning problem and its variants, such as VLSI circuit layout [12], clustering [2], and distributing workloads for parallel computation [4]. In this paper, we are focused on a graph partitioning problem called *Minimum Cost Partition of a Tree with Supply and Demand*(MCPTSD), which has applications in the feeder reconfiguration of power delivery networks [3,14,18] and the self-adequacy of interconnected smart micro-grids [1,9]. The problem can be modeled by a directed tree $T = (V, E)$ with node set V and edge set E, where $V = V_d \cup V_s$, and $V_d \cap V_s = \emptyset$. Each node $v \in V_d$ is called a demand node, and is assigned a positive real number $dem(v)$, called a *demand* of v. Each node $u \in V_s$ is called a supply node, and is assigned a positive real number $sup(u)$, called a *supply* of u. Every edge $e \in E$ is also assigned a positive integer $c(e)$, called the *cost* of e, which represents the cost deleting e from T or reversing the direction of e in T. A subset $E' \subseteq E$ is called a *critical edge set*, if by deleting or reversing the edges in E', a forest $\{T_1, T_2, \cdots, T_l\}$ satisfying following properties can be obtained: (1) $l = |V_s|$, $\bigcup_{i=1}^l V(T_i) = V$, and $V(T_i) \cap V(T_j) = \emptyset$ ($1 \leq i < j \leq l$); (2) each T_i ($1 \leq i \leq l$) is the tree which contains only one supply node $u_i \in V_s$, and $\sum_{v \in V(T_i) \setminus \{u_i\}} dem(v) \leq sup(u_i)$; (3) each edge of T_i is directed away from the supply node u_i. The definition of the MCPTSD problem is given as follows.

This work is supported by the National Natural Science Foundation of China under Grants (61232001, 61472449, 61420106009, 61402054), and the Hengyang Foundation for Development of Science and Technology under Grant(2014KJ21).

J. Wang and C. Yap (Eds.): FAW 2015, LNCS 9130, pp. 180–189, 2015.
DOI: 10.1007/978-3-319-19647-3_17

Minimum Cost Partition of a Tree with Supply and Demand (MCPTSD):
Given a directed tree $T = (V, E)$, where $V = V_d \cup V_s$, and $V_d \cap V_s = \emptyset$,
find a critical edge set E' in T whose total cost $\sum_{e \in E'} c(e)$ is minimum
for T.

Ito et al. [5] studied the problem of partitioning a tree with supply and demand
such that the total obtained demand of all of subtrees is maximized, which allows
that some subtrees have no supply node. They showed that the problem is NP-
complete even for series-parallel graphs. However, if the problem is to decide
whether a given tree has at least one partition, it was proved in [5] that the
problem can be solved in linear time. A pseudo polynomial-time algorithm and
a FPTAS algorithm to solve the maximum partition problem for tree T are
given in [5]. In [6,7], the problem proposed in [5] is extended to general graphs,
which was proved to be MAXSNP-hard. Pseudo polynomial-time algorithms and
FPTAS algorithms for series-parallel graphs and partial k-trees were also studied in [6,7]. Ito et al. [8] proved that the MCPTSD problem is NP-hard, and
gave a pseudo polynomial-time algorithm and a FPTAS algorithm for the problem. Recently, lots of attention [9–11,13,15–17] have been paid on the partition
problem with supply and demand from heuristic and approximation algorithm
perspective.

   In this paper, we study the parameterized Minimum Cost Partition of a Tree
with Supply and Demand problem. The definition of the problem is given as
follows.

Parameterized Partition of a Tree with Supply and Demand(PPTSD):
Given a directed tree $T = (V, E)$, and a parameter k, $V = V_s \cup V_d$,
where $V_s \cap V_d = \emptyset$, find a critical edge set $E' \subseteq E$ with $\sum_{e \in E'} c(e) \leq k$,
or report that no such subset exists.

In the paper, we give several polynomial-time data reduction rules, which result
in a kernel of size $O(k^2)$ for the PPTSD problem. Based on the analysis of branching cases of the problem, a parameterized algorithm of running time $O^*(2.828^k)$
is presented.

2   Preliminaries

Let $T = (V, E)$ be a directed tree with $|V| = n$ nodes and $|E| = m$ edges. We
use (u, v) to denote a directed edge in T from u to v. For a node $v \in V(T)$, let
$N(v)^- = \{u : (u, v) \in E(T), u \in V(T)\}$ and $N(v)^+ = \{u : (v, u) \in E(T), u \in V(T)\}$ be the in-neighbors and out-neighborhs of v, respectively. The neighbors
of v is defined as $N(v) = N(v)^- \cup N(v)^+$. The in-degree and out-degree of
v are defined as $|N(v)^-|$ and $|N(v)^+|$, respectively. Thus, the degree of v is
$|N(v)^-| + |N(v)^+|$. If a node v has degree d, then v is called a *degree-d node*. For
a directed tree T, tree T' is called a *line tree* of T if T' is obtained by ignoring
all the directions of the edges in T. By designating one node r in T' as root, for
a node u in T', the parent of u is the node connected to it on the path to the

root r in T'. For two node u, v in T, let u', v' be two corresponding nodes in T', respectively. If u' is the parent of v' in T', for simplicity, we say that u and v are the parent-children relationship in T, i.e., u is the parent of v in T. For directed tree T, a node with degree-1 is called a leaf of T. For an edge e in T, the contraction of edge e is defined that e is moved and its two incident nodes u and v are merged into a new node w such that the edges incident to either u or v are incident to w. In the paper, many edge contraction operations are applied in the kernelization and branching algorithm. Assume that forest $F = \{T_1, T_2, \cdots, T_l\}$ can be obtained by deleting or reversing the edges in a critical edge set of T. The edges contracted in the kernelization and branching algorithm are still in some trees of F by reversing the process of contraction.

3   Kernelization for PPTSD

In this section, we study the kernelization for PPTSD problem. Assume that (T, k) is the given instance of the PPTSD problem. We start with three simple observations to decide whether (T, k) is a NO-instance.

Rule 1. (1) If there are more than $k + 1$ supply nodes in T, then return "NO". (2) If the sum of in-degrees of all supply nodes in T is larger than k, then return "NO". (3) If there are more than k demand nodes in T with in-degrees at least 2, then return "NO".

Lemma 1. *Rule 1 is correct and executable in $O(m + n)$ time.*

Proof. Assume that $F = \{T_1, T_2, \cdots, T_l\}$ is a forest obtained by deleting or reversing the edges in a critical edge set E' of T. If there are more than $k + 1$ supply nodes in T, we need to delete at least $k + 1$ edges of T to get F. Thus, the cost of deleting edges is larger than k, and (T, k) is a NO-instance. Since every edge is directed away from the supply node in each T_i $(1 \leq i \leq l)$ of F, in each T_i, the in-degree of supply node is zero and the in-degree of every demand node is one. Thus, for Rule 1 (2), if the sum of in-degrees of all supply nodes in T is larger than k, then (T, k) is a NO-instance. Similarly, for Rule 1(3), if there are more than k demand nodes in T with in-degrees at least 2, then (T, k) is a NO-instance. Obviously, Rule 1 can be done in $O(m + n)$ time. □

Rule 2. Let u be a leaf node of tree T and v be the parent node of u in T. If u and v are both demand nodes, then for the edge e constructed by u and v: (1) If $e = (v, u)$, then contract edge (v, u). Let v' be the new demand node, and $dem(v') = dem(u) + dem(v)$; (2) If $e = (u, v)$, then reverse edge e, and contract e. Let v' be a new demand node, $dem(v') = dem(u) + dem(v)$, and $k = k - c(e)$.

Rule 3. Let u be a leaf of tree T and node v be the parent node of u in T. If u is a demand node and v is a supply node, then:

(1) If $sup(v) < dem(u)$, then return "NO".
(2) If $sup(v) = dem(u)$ and node v is adjacent to at least one other demand leaf except u, then return "NO".

(3) If $sup(v) > dem(u)$, then check the direction of edge e constructed by u, v.

(3.1) If $e = (v, u)$, then contract e. Let v' be the new supply node, and $sup(v') = sup(v) - dem(u)$;

(3.2) If $e = (u, v)$, then reverse edge e, and contract e. Let v' be the new supply node, $sup(v') = sup(v) - dem(u)$, and $k = k - c(e)$.

Lemma 2. *Rules 2 and 3 are correct and can be done in $O(m + n)$ time.*

Proof. Assume that $F = \{T_1, T_2, \cdots, T_l\}$ is a forest obtained by deleting or reversing the edges in a critical edge set E' of T. We first prove that edge (v, u) must exist in a tree T_i of F if the input instance is a YES-instance. Suppose that edge e is deleted to get forest F. Then, there exists a tree in F without a supply node, contradicting the definition of the PPTSD problem. Therefore, edge (v, u) must exist in a tree T_i of F. In order to guarantee that the node u is away from the supply node in the forest F, the direction of the edge e should be from v to u in F.

From the above fact, the correctness of Rule 2 is obvious.

For Rule 3, if $sup(v) < dem(u)$, the demand node u cannot get enough demand from supply node v, whatever the direction of the edge e between u and v is in F. Thus, instance (T, k) is a NO-instance. If $sup(v) = dem(u)$ and node v is adjacent to at least one other demand leaf except u, then at least one demand node does not obtain its demand. Thus (T, k) is a NO-instance. If $sup(v) > dem(u)$, from the above fact, the case is correct. It is easy to see that the Rule 2–3 can be done in $O(m + n)$ time. □

Lemma 3. *After applying Rules 2 and 3, there are at most $k + 1$ leaves in the reduced instance.*

Proof. Assume that (T', k') is the reduced instance of the PPTSD problem after applying Rules 2–3. If there still exist demand leaves in (T', k'), then the parent node of every demand leaf is a supply node and each supply node has only one demand leaf by Rule 2 and Rule 3(2). Therefore, the number of leaves is no more than that of supply nodes in the instance (T', k'). By Rule 1, (T', k') contains at most $k + 1$ supply nodes. Thus, there are at most $k + 1$ leaves in (T', k'). □

Rule 4. Let u be a leaf of tree T, node v be the parent node of u, and node w be the parent node of v. If u is a supply node, then:

(1) If v is also a supply node, then delete node u and $k = k - c(e)$;

(2) If v is a demand node, then compare $sup(u)$ and $dem(v)$;

(2.1) If $sup(u) < dem(v)$, then delete node u and $k = k - c(e)$.

(2.2) If $sup(u) > dem(v)$, v is a degree-2 node, $e = (u, v)$, and $e' = (v, w)$, then contract edge e. Let u' be the new supply node, and $sup(u') = sup(u) - dem(v)$.

(2.3) If $sup(u) = dem(v)$, v is a degree-2 node, $e = (u, v)$, and $e' = (v, w)$, then delete node u and v, and $k = k - c(e')$.

Lemma 4. *Rule 4 is correct and can be done in* $O(m+n)$.

Proof. Assume that $F = \{T_1, T_2, \cdots, T_l\}$ is a forest obtained by deleting or reversing the edges in a critical edge set E' of T. If both u and v are supply nodes, then u and v are in different trees in F. Thus, edge e must be deleted in T, and $k = k - c(e)$. If v is a demand node and $sup(u) < dem(v)$, then the supply node u cannot provide enough demand for node v, and u and v are in different trees in F. Therefore, node u can be deleted, and $k = k - c(e)$. For the cases (2.2) and (2.3) of Rule 4, it is clear that the edge (u, v) cannot be reversed. Suppose that $F_{min} = \{T_1, \cdots, T_i, \cdots, T_j, \cdots, T_l\}$ is a forest obtained by deleting or reversing the edges in a critical edge set E_{min} with minimum cost for T, by deleting edge $(u, v) \in E_{min}$ and reversing edge $(v, w) \in E_{min}$. Assume that $T_i = \{u\}$, and v is in T_j. If edge (v, w) deleted and edge (u, v) is not deleted, then two new trees $T_i' = \{u, v\}$ and $T_j' = T_j \setminus \{v\}$ can be obtained. Thus, a new forest $F' = \{T_1, \cdots, T_i', \cdots, T_j', \cdots, T_l\}$ can be obtained by replacing T_i and T_j with T_i' and T_j', respectively. Assume that E'' is the critical edge set obtaining F'. Then, the total cost of E'' is less than that of E_{min}, contradicting the assumption that the cost of E_{min} is minimum for T. Therefore, case (2.2) and (2.3) of Rule 4 are correct. Obviously, Rule 4 can be done in $O(m+n)$ time.   □

Rule 5. Let l be the number of all supply nodes in T and $P = (u_1, u_2, \cdots, u_{h-1}, u_h)$ be a directed path in T, where u_i $(1 \leq i \leq h)$ is degree-2 demand node. If $h > k - l + 2$, then contract all the edges in $\{(u_i, u_{i+1}) | i = 1, \cdots, h - k + l - 2\}$. Let u the new demand node obtained, and $dem(u) = \sum_{j=1}^{h-k+l-2} dem(u_j)$.

Lemma 5. *Rule 5 is correct and can be done in* $O(m+n)$.

Proof. Assume that $F = \{T_1, T_2, \cdots, T_l\}$ is a forest obtained by deleting or reversing the edges in a critical edge set E' of T. Since each tree T_i $(1 \leq i \leq l)$ has exactly one supply node, $l - 1$ edges of T must be deleted. Thus the cost of deleting edges in T is at least $l - 1$ and the cost of reversing edges is at most $k - l + 1$. We now prove that no edge in $\{(u_i, u_{i+1}) | i = 1, \cdots, h - k + l - 3\}$ of P can be deleted or reversed. Assume that edge e from $\{(u_i, u_{i+1}) | i = 1, \cdots, h - k + l - 3\}$ is deleted. In order to satisfying all the demands of the nodes in $\{u_{h-k+l-1}, \cdots, u_h\}$, all the edges in $\{u_j, u_{j+1} | j = h - k + l - 2, \cdots, u_{h-1}\}$ should be reversed with cost at least $k - l + 2$. Similarly, we can prove that no edges $\{(u_i, u_{i+1}) | i = 1, \cdots, h - k + l - 3\}$ of P can be reversed. Therefore, all the edges in $\{(u_i, u_{i+1}) | i = 1, \cdots, h - k + l - 2\}$ can be contracted. It is easy to see that Rule 5 can be done in $O(m+n)$ time.   □

Theorem 1. *PPTSD problem has a kernel with* $O(k^2)$ *nodes.*

Proof. Let (T', k') be the reduced instance obtained by applying Rules 1–5. By Lemma 3, there are at most $k + 1$ supply nodes and at most $k + 1$ leaves in T'. Therefore, at most k internal nodes with degree at least three are contained in T'. Assume that S is the set of supply nodes in T', L is the set of supply leaves in T', and A is the set of demand leaves of T'. For the internal nodes, assume that B is the set of degree-2 internal demand nodes whose in-degree or out-degree is two, and C is set of degree-2 internal demand nodes whose in-degree and out-degree are both one. For the internal supply nodes, let D_1 be the set of degree-3 supply nodes, each of which has a leaf as neighbor, and D_2 be the set of degree-2 supply nodes. Assume that F is the set of internal nodes of degree at least three. Let $D = D_1 \cup D_2$ and $I = F \backslash D_1$. Clearly, $V(T') = L \cup A \cup I \cup B \cup C \cup D$, and $|D| \leq |S| - |L|$. Since every demand node must receive supply from exact one supply node and the total cost is at most k, for each demand node u in B, at least one edge incident to u must be deleted or reversed. Thus, $|B| \leq k$. By Rule 2 and Rule 3(2), the parent node of every demand leaf is a supply node and each supply node has only one demand leaf. Therefore, $|L| + |A| \leq |S|$ and $|I| \leq |S| - 1$. Let p be the number of the directed paths and t be the number of nodes in the longest directed path. By Lemma 5, $t \leq k - |S| + 2$. At most $|L| + |I| - 1$ directed paths are contained in T' if T' do not contain nodes in $A \cup B \cup D$, denoted by $\mathcal{P}$. For a directed path P in $\mathcal{P}$, if one node of $B \cup D$ is added to P, then the number of directed paths is increased by at most one. Thus, $p \leq |L| + |I| - 1 + |B| + |D|$.

$$\begin{aligned} |V(T)| &= |L| + |A| + |I| + |B| + |C| + |D| \\ &\leq |S| + |L| + k - 1 + (k - |S| + 2) \cdot (|S| + |L| + k - 2) + |S| \\ &\leq -2(|S| - \frac{k+9}{4})^2 + \frac{9k^2 + 26k + 41}{8} \\ &\leq \frac{9k^2 + 26k + 41}{8} \end{aligned}$$

Hence, there are at most $O(k^2)$ nodes in the reduced instance. □

4   A Parameterized Algorithm for PPTSD

In this section, we present a parameterized algorithm for PPTSD based on branching. Note that in our algorithm, the kernelization algorithm is applied whenever possible. Let (T, k) be a reduced instance of PPTSD and u be a leaf of tree T. Assume that the parent node of u is denoted by v. We solve the PPTSD problem by the following cases.

Case 1: Node u is a demand node.

Since (T, k) be a reduced instance of PPTSD, v is a supply node and $sup(v) = dem(u)$ by Rule 3(2). Let e be the edge between u and v, and E_v be the edge set adjacent to v. Then delete u and v, and $k = k - \sum_{e' \in E_v \backslash e} c(e')$.

Case 2: Node u is a supply node and node v is adjacent to at least two leaves.

Let w be another leaf of node v, e be the edge between u and v, and e' be the edge between v and w. We can have the following two branchings: (1) delete edge e; (2) delete edge e'. Thus, the branching vector is $(1, 1)$ and the branching number is $\alpha = 2$.

Case 3: Node u is a supply node and node v is a degree-2 node adjacent to only one leaf u.

Let w be the parent node of node v, e be the edge between u and v, and e' be the edge between v and w. Note that when $e = (u, v)$ and $e' = (v, w)$, Rule 4 can be applied. According to the direction of edges e and e', we have the following three subcases.

Case 3.1: $e = (v, u)$ and $e' = (w, v)$.

We can have the following two branchings: (1) delete (v, u); (2) reverse (v, u), and merge v, u into a new supply node v', $sup(v') = sup(u) - dem(v)$. Thus, the branching vector is $(1, 1)$ and the branching number is $\alpha = 2$.

Case 3.2: $e = (v, u)$ and $e' = (v, w)$.

We can have the following two branchings: (1) delete (v, u) and reverse (v, w); (2) reverse (v, u), and merge v, u into a new supply node v', $sup(v') = sup(u) - dem(v)$. Thus, the branching vector is $(2, 1)$ and the branching number is $\alpha = 1.618$.

Case 3.3: $e = (u, v)$ and $e' = (w, v)$.

Let x be the parent of node w. We deal with **Case 3.3** by the following subcases.

Case 3.3.1: Node w is a supply node.

We can have the following two branchings: (1) delete edge (u, v); (2) delete edge (w, v). Thus, the branching vector is $(1, 1)$ and the branching number is $\alpha = 2$.

Case 3.3.2: Node w is a degree-2 demand node, and the edge between w and x is (x, w).

We can have the following four branchings: (1) delete edge (u, v); (2) delete edge (w, v); (3) delete edge (x, w) and reverse edge (w, v); (4) reverse edges (w, v) and (x, w), merge u, v and w into a new supply node w', $sup(w') = sup(u) - dem(v) - dem(w)$. Thus, the branching vector is $(1, 1, 2, 2)$ and the branching number is $\alpha = 2.732$.

Case 3.3.3: Node w is a degree-2 demand node, and the edge between w and x is (w, x).

We can have the following three branchings: (1) delete edge (u, v) and reverse (w, x); (2) delete edge (w, x) and reverse (w, v); (3) reverse (w, v) and merge u, v and w into a new supply node w', $sup(w') = sup(u) - dem(v) - dem(w)$. Thus, the branching vector is $(2, 2, 1)$ and the branching number is $\alpha = 2$.

Case 3.3.4: Node w is a demand node of degree at least three.

We deal with **Case 3.3.4** by the following three subcases.

Case 3.3.4.1: Node w is adjacent to a supply leaf y, and the edge between w and y is (w, y).

We can have the following two branchings: (1) delete edge (w, y); (2) reverse (w, y), merge nodes w and y into a new supply node w', $sup(w') = sup(y) - dem(w)$. Thus, the branching vector is $(1, 1)$ and the branching number is $\alpha = 2$.

Case 3.3.4.2: Node w is adjacent to a supply leaf y, and edge (y, w) is an edge between nodes w and y.

We can have the following three branchings: (1) delete edge (u, v); (2) delete edge (w, v); (3) delete edge (y, w) and reverse edge (w, v), merge nodes u, v and w into a new supply node w', $sup(w') = sup(u) - dem(v) - dem(w)$. The branching vector is $(1, 1, 2)$ and the branching number is $\alpha = 2.414$.

Case 3.3.4.3: Node w is adjacent to a node y which is a degree-2 demand node adjacent to a leaf z.

Under this case, the edge between w and y is (w, y) and the edge between z and y is (z, y). Therefore, we can have the following eight branchings: (1) delete edges (w, v) and (w, y); (2) delete edge (w, y) and reverse edge (w, v), merge nodes u, v, w into a new supply node w', $sup(w') = sup(u) - dem(v) - dem(w)$; (3) delete edge (w, v) and reverse edge (w, y), merge nodes w, y, z into a new supply node w', $sup(w') = sup(z) - dem(y) - dem(w)$; (4) delete edges (u, v) and (z, y); (5) delete edges (w, v) and (z, y); (6) delete edges (w, y) and (u, v); (7) delete edge (u, v), reverse edge (w, y), and merge nodes v, w, y, z into a new supply node w', $sup(w') = sup(z) - dem(y) - dem(w) - dem(v)$; (8) delete edge (z, y), reverse edge (w, v), and merge nodes u, v, w, y into a new supply node w', $sup(w') = sup(u) - dem(y) - dem(v) - dem(w)$. Thus, the branching vector is $(2, 2, 2, 2, 2, 2, 2, 2)$ and the branching number is $\alpha = 2.828$.

Note that in the above branchings, when we merge some nodes into a new supply node u, we check its supply $sup(u)$. If $sup(u) < 0$, then return "NO". If $sup(u) = 0$, then delete node u and decrease k accordingly.

Based on the above branchings, we can obtain the following theorem.

Theorem 2. *PPTSD can be solved in $O^*(2.828^k)$ time.*

Proof. Given an instance (T, k) of the PPTSD problem, if (T, k) is a NO-instance, after doing Rules 1–5 and Cases 1–3.3.3, no critical edge set with cost at most k can be found, and "NO" will be returned.

We now assume that (T, k) is a Yes-instance. By applying the reduction rules in kernelization process until no rule is applicable, T only contains supply leaf and demand leaf. Particularly, for each demand leaf u in T, its parent is a supply node and has unique child u, which can be handled correctly by Case 1. Now we can get that all the leaves of T are supply nodes. For an internal node v, if $N(v)$ contains at least two leaves of T, then for any two leaves u, w in $N(v)$, only one of u and w is kept in $N(v)$, which corresponds to the two branchings in Case 2. After doing the branchings in Case 2 whenever possible, for each supply leaf node u in T, its parent is a demand node and has unique child u. For a supply leaf u, its parent v and the parent w of v, based on the degree of node w and whether w is a supply or demand node, we can get Case 3.1–3.3 and Case 3.3.1–3.3.3. Assume that $F' = \{T_1, T_2, \cdots, T_i\}$ $(1 \leq i \leq l)$ is a forest obtained in the branching process, where T_j $(1 \leq j \leq i)$ is a directed tree. If T_j contains only one supply node, then we can reduce it to a node by Rules 1–5. For a tree T_j in F', if the number of nodes in T_j is larger than one, we now prove that either some rules from Rules 1–5, or some cases from Cases 1–3.3.4.3 can be applied. Since T_j has at least two nodes, a leaf node u can be found in T_j. It is easy to see that if the conditions of Rules 1–5 are satisfied, then Rules 1–5 can be applied. Now suppose that no rule of Rules 1–5 is applicable. Let v be the parent of u. By the above discussion, if u and v satisfies the conditions of Case 1, or the conditions of Case 2, it can be handled correctly, respectively. For the parent w of v, if w is a supply node, no matter the degree of w, it can be handled by Case 3.1, 3.2, or 3.3.1. If w is a demand node, when the degree of w is two, it can be handled by Case 3.3.2 or 3.3.3. If w is a demand node with degree larger than two, it can be handled by Case 3.3.4. We now prove that when no Rules 1–5 and no case from Case 1 to Case 3.3.3 can be applied, if the number of nodes in T_j is larger than two, there must exist a demand node w that one of Cases 3.3.4.1–3.3.4.3 can be applied. For a internal demand node w, let v the child of w, and u be the unique child of v. If v is a degree-2 node and w is a supply node, then u is called the indirect-leaf of w. For the tree T_j, since no Rules 1–5 and no case from Case 1 to Case 3.3.3 can be applied, there must exist a internal demand node w that has at least one leaf and one indirect-leaf, or has at least two indirect-leaves, which can be handled by 3.3.4.1–3.3.4.2 and 3.3.4.3, respectively. Therefore, if (T, k) is Yes-instance, then by applying Rules 1–5 and Cases 1–3.3.4.3, a forest $F = \{T_1, T_2, \cdots, T_l\}$ can be obtained such that each tree in F contains only one supply node.

Now we analyze the complexity of the branching process. From the above branching, the worst case branching number is $\alpha = 2.828$. Therefore, PPTSD problem can be solved in $O^*(2.828^k)$ time.   □

5   Conclusion

The minimum cost partition problem is an important optimization problem in the reconfiguring power delivery networks and the self-adequacy smart microgrids. In this paper, we first study the partition of trees problem from parameterized perspective. We preprocess the input instance through some reduction rules and derive an $O(k^2)$ kernel of the problem, and then present a FPT algorithm to solve the problem in time $O^*(2.828^k)$.

References

1. Arefifar, S.A., Mohamed, Y.A.I., El-Fouly, T.H.: Supply-adequacy-based optimal construction of microgrids in smart distribution systems. IEEE Trans. Smart Grid **3**(3), 1491–1502 (2012)
2. Berkhin, P.: A survey of clustering data mining techniques. In: Kogan, J., Nicholas, C., Teboulle, M. (eds.) Grouping Multidimensional Data, pp. 25–71. Springer, Heidelberg (2006)
3. Boulaxis, N.G., Papadopoulos, M.P.: Optimal feeder routing in distribution system planning using dynamic programming technique and gis facilities. IEEE Trans. Power Delivery **17**(1), 242–247 (2002)
4. Hendrickson, B., Kolda, T.G.: Graph partitioning models for parallel computing. Parallel Comput. **26**(12), 1519–1534 (2000)
5. Ito, T., Zhou, X., Nishizeki, T.: Partitioning Trees of Supply and Demand. Int. J. Found. Comput. Sci. **16**(04), 803–827 (2005)

6. Ito, T., Demaine, E.D., Zhou, X., Nishizeki, T.: Approximability of partitioning graphs with supply and demand. J. Discrete Algorithms **6**(4), 627–650 (2008)
7. Ito, T., Zhou, X., Nishizeki, T.: Partitioning graphs of supply and demand. Discrete appl. Math. **157**(12), 2620–2633 (2009)
8. Ito, T., Hara, T., Zhou, X., Nishizeki, T.: Minimum cost partitions of trees with supply and demand. Algorithmica **64**(3), 400–415 (2012)
9. Jovanovic, R., Bousselham, A.: A greedy method for optimizing the self-adequacy of microgrids presented as partitioning of graphs with supply and demand. In: Proceedings of the 2nd International Renewable and Sustainable Energy Conference(IRSEC 2014), pp. 17–19, Ouarzazate, October 2014
10. Jovanovic, R., Bousselham, A., Voss, S.: A Heuristic Method for Solving the Problem of Partitioning Graphs with Supply and Demand. arXiv preprint arXiv:1411.1080 (2014)
11. Kawabata, M., Nishizeki, T.: Partitioning trees with supply, demand and edge-capacity. IEICE Trans. Fundam. Electron. Commun. Comput. Sci. **96**(6), 1036–1043 (2013)
12. Lienig, J., Markov, I.L., Hu, J.: VLSI Physical Design: From Graph Partitioning to Timing Closure. Springer, The Netherlands (2011)
13. Morishita, S., Nishizeki, T.: Parametric power supply networks. J. Comb. Optim. **29**(1), 1–15 (2015)
14. Morton, A.B., Mareels, I.M.: An efficient brute-force solution to the network reconfiguration problem. IEEE Trans. Power Delivery **15**(3), 996–1000 (2000)
15. Narayanaswamy, N.S., Ramakrishna, G.: Tree t-spanners in Outerplanar Graphs via Supply Demand Partition. arXiv preprint arXiv:1210.7919 (2012)
16. Popa, A.: Modelling the power supply network – hardness and approximation. In: Chan, T.-H.H., Lau, L.C., Trevisan, L. (eds.) TAMC 2013. LNCS, vol. 7876, pp. 62–71. Springer, Heidelberg (2013)
17. Taoka, S., Watanabe, K., Watanabe, T.: Experimental evaluation of maximum-supply partitioning algorithms for demand-supply graphs. IEICE Trans. Fundum. Electron. Commun. Comput. Sci. **89**(4), 1049–1057 (2006)
18. Teng, J.H., Lu, C.N.: Feeder-switch relocation for customer interruption cost minimization. IEEE Trans. Power Delivery **17**(1), 254–259 (2002)

The Online Storage Strategy for Automated Storage and Retrieval System with Single Open in One Dimension

Henan Liu[1,2](✉) and Yinfeng Xu[1,2]

[1] School of Management, Xi'an Jiaotong University, Xi'an 710049, China
boomword@stu.xjtu.edu.cn
[2] The State Key Lab for Manufacturing Systems Engineering, Xi'an 710049, China
yfxu@mail.xjtu.edu.cn

Abstract. In this paper, we study the online storage strategy problem for one-dimensional automated storage and retrieval system with m storage units. The items released online overlist need to be assigned to the storage units via a picker. The departure time of the item is known at the moment that it is released. Our goal is to minimize the total travel distance of the picker. We show that without transposition process, there is no online strategy with a competitive ratio less than m. We also present an online strategy that includes transposition process has competitive ratio $(\frac{\sqrt{5}-1}{2})m$. The ratio between the number of sorted units and m is around $\frac{3-\sqrt{5}}{2}$.

Keywords: Automated storage and retrieval system · Storage strategy · Online algorithm · Competitive analysis

1 Introduction

Automated Storage and Retrieval Systems are warehousing systems that are widely used for the storage and retrieval of products in both production and distribution environments. This article studies the online storage strategy for one-dimensional automated storage and retrieval system. We are asked to assign the current item to a unit without any information about subsequent items. Since it is an automated system, an item can always be moved into another idle unit before its departure via a picker. Our goal is to minimize the total travel distance of the picker.

There is a rich literature on the storage strategies for automated storage and retrieval system. Tompkins et al. show that the dominant component of the picking time is travel time (50 %)[1]. The travel time is closely related to the travel distance. Bartholdi and Hackman point out that since travel time costs hours but does not add value, it is wasted [2]. A storage strategy attempts to provide an effective way of locating products in order to reduce the travel distance of the picker. Several methods exist for assigning products to storage locations in the system. Hausman et al. summarize five storage strategies which are often used for automated storage and retrieval systems [3,4]. These strategies are:

© Springer International Publishing Switzerland 2015
J. Wang and C. Yap (Eds.): FAW 2015, LNCS 9130, pp. 190–197, 2015.
DOI: 10.1007/978-3-319-19647-3_18

- dedicated storage strategy
- random storage strategy
- closest open location storage strategy
- full-turnover-based storage strategy
- class-based storage strategy

For the *dedicated storage strategy* each product type is assigned to a fixed location. The *random storage strategy* is widely used in many automated storage and retrieval system because it is simple to use, often requires less space than other storage methods, and results in a better level utilization of all units(see Petersen and Schmenner [5]). If the *closest open location storage strategy* is applied, the first empty location that is encountered will be used to store the products. The *full-turnover-based storage strategy* determines storage locations for loads based on their demand frequency. Heskett presents the cube-per-order index (COI) rule, which is a form of full-turnover storage [6,7]. To reduce space requirements and periodic repositioning while maintaining most of the efficiency gains, *class-based storage strategy* can be used.

As a similar research, Wen et al. study the online traveling salesman problem that the server is moving in a line, their goal is to minimize the server's costs [8,9]. The server can be treated as the picker in our study which is also moving in a line.

Frazelle and Sharp prove that the offline storage assignment problem is NP-hard problems [10]. Usually, it is impossible to produce the best solution when computation occurs online. To gauge the performance of an online strategy A, the concept of competitive ratio (see Borodin and El-yaniv [11]) is often used, which is the worst case ratio of the online algorithm's performance versus that of the optimal offline algorithm. For this problem, denote by $C_I(OPT)$ the performance of an optimal offline algorithm on an instance I, by $C_I(A)$ the performance of an online approximation algorithm A, where the performance refers to the total travel distance of the picker. The competitive ratio of an online algorithm A is defined to be

$$r_A = sup_I \frac{C_I(A)}{C_I(OPT)}$$

2   Problem Statement and Notations

We consider an automated storage and retrieval system which is rectangular with a picker and m storage units. The input/output (I/O) point is located at the left end of the system, as illustrated in Fig. 1. The picker is the fully automated storage and retrieval machine that can autonomously move, pick up and drop off loads. An item is delivered by the picker between the I/O point and the assigned storage unit. The capacity of a picker is just to handle one item in a delivery tour. And one storage unit could be used to store only one item at one time.

In this problem, the input is a list of items which arrive in an online manner. Each item is associated with a departure time (when the item will be taken away). Once a new item arrives, it would get placed on the I/O point if there is

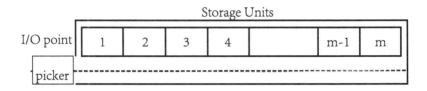

Fig. 1. An automated storage and retrieval system in 1-D

an idle unit in the system. The picker will transport the item to the assigned unit (storage process). Similarly, the picker would get the item out of the storage unit and get it back to the I/O point at its departure time (retrieval process). If necessary, the system will adjust the position of some of the items (transposition process). After the storage or retrieval process of this item is completion, a new item is released. The system decides whether to perform transposition process before the storage process of the new item. Since the items released individually, the first m items will fill the system. And after an item left at its departure time, a storage unit will be idle to store a new item. The cost of the operations is the travel distance of the picker. The total number of items is n. Our aim is to minimize the total travel distance of the picker in this online problem.

In particular, the cost of the operations -pick up and drop off-is ignored. The operation mode of the picker follows *restoration strategy* - it is waiting at the I/O point while idling, and returns back to the I/O point after a storage or a retrieval process completion. In the storage or retrieval process the picker will travel a distance which is twice the distance from the I/O point to the assigned storage unit. In transposition process, the picker will pick up the item to be transposed and assign it to another idle storage unit, and then return to the I/O point.

We denote by d_j the distance between the I/O point and storage unit j d_j equals to j, $j = \{1, 2, 3, \ldots, m\}$. Denoted by τ_i the departure time of item i, $\tau_i > 0, i = \{1, 2, 3, \ldots\}$. Denote by t^j the departure time of the item which is currently stored in the jth storage unit. As items arrive and leave the system t^j is dynamic. When the ith item is stored in the jth storage unit, $t^j = \tau_i$. When the jth storage unit is idle, $t^j = 0$. We denote by $C_{i \in [p,q]}(A)$ the total travel distance of the picker to transport from the pth to the qth item in the online strategy A and $C_I(A)$ the total travel distance to transport all the items in the list I.

3 $(\frac{3-\sqrt{5}}{2})m$-Sorted Strategy

As items are released overlist, at first the system is to be filled by the first m items. After the item with the smallest τ_i in the system is retrieved at its departure time, the storage unit is idle and a new item will be released.

We consider such a question: once we stored an item into such a system, should we change its location (i.e., transpose it) during the storage period or not?

Theorem 1. *Without transposition, there is no online strategy with a competitive ratio less than m.*

Proof. The theorem is proved by considering a worst case input.

The reasoning behind the proposed worst case is as follows: The picker keeps transporting items until all the m storage units filled with items. Assume that the item which is stored in the mth storage unit needs to leave the system. Then a large number of items released with the same departure time τ_i^* that is smaller than any of the others stored in the first to the $(m-1)$th storage units. Since all the strategies do not include transposition process, the picker should travel the whole system to retrieve the item which is stored in the mth storage unit and to store a new one in this storage unit. As all the information is known in advance by the optimal offline strategy, the offline strategy will let the first storage unit to store the item which is stored in the mth storage unit by the online strategy. Then the picker of the optimal offline strategy only needs to take the distance of 2 to retrieve or store a new item.

The worst case could be as follows. Firstly, m items with distinct departure times are released one by one. Filling m storage units costs $C_{i \in [1,m]}$ travel distance in any arbitrary strategy, $C_{i \in [1,m]}(A) = \Sigma_{j=1}^m 2j = m(m+1)$. Let $\mathcal{A}$ be the set of τ_i which is less than t^m, $\mathcal{A} = \{\tau_i | \tau_i < t^m, 1 \leq i \leq m\}, |\mathcal{A}| = a$, Let $\mathcal{B}$ be the set of τ_i which is greater than t^m, $\mathcal{B} = \{\tau_i | \tau_i > t^m, 1 \leq i \leq m\}, |\mathcal{B}| = b$. Note that $a+b+1 = m$. Denote by $t^{min}(\mathcal{B}) = min\{\tau_i | \tau_i \in \mathcal{B}\}$. Secondly, a items are released with the same departure time $\tau_i' > t^{min}(\mathcal{B})$, and they will take the place of the items with the departure times in set $\mathcal{A}$. $C_{i \in [m+1,m+a]}(A) = \Sigma\{2d_j | \tau_i \in \mathcal{A}\}$. Therefore, the current $t^m = min\{t^j\}$, and all the other items will stay longer than the one that is stored in the mth storage unit . Thirdly, n items with $\tau_i^* (i \geq m+1)$ are released, $\tau_i^* = t^m + i\Delta$, where i is increasing and the number Δ is small enough to guarantee $\tau_i^* < t^{min}(\mathcal{B})$. Because that t^m is always the smallest of all the $t^j (1 \leq j \leq m)$ in the system, the picker in an online strategy has to go through all the items to store or retrieve every new item with τ_i^*. $C_{i \in [m+a+1,m+a+n]}(A) = 2m * n = 2nm$. Since the offline player has the information of all the items in advance, it can decrease the total travel distance of the picker through setting the item in the first storage unit which is stored in the mth storage unit by the online algorithm A. The competitive ratio is

$$\frac{C_I(A)}{C_I(OPT)} = \frac{C_{i \in [1,m]}(A) + C_{i \in [m+1,m+a]}(A) + C_{i \in [m+a+1,m+a+n]}(A)}{C_{i \in [1,m]}(OPT) + C_{i \in [m+1,m+a]}(OPT) + C_{i \in [m+a+1,m+a+n]}(OPT)}$$

$$\geq \frac{m(m+1) + 2am + 2nm}{m(m+1) + 4a + 2n}$$

When $n \to \infty$,

$$\frac{C_I(A)}{C_I(OPT)} \geq m. \qquad \square$$

Therefore, the online storage strategies which could transpose the item to the appropriate storage unit are proposed. The transposition process is to avoid the item which stays for a long time occupying the storage unit near the I/O point as the worst case illustrated above. Actually, there is a trade-off between the strategy that includes transposition process and the one does not. If we take as much as transposition, the situation is avoidable that the item with a big departure time occupies the storage unit near the I/O point. The cost is the increasing travel distance brought by the transposition process. Thus, we try to find a strategy with appropriate transposition process.

Denote by $\mathcal{S}$ the set of τ_i, the departure times of the items currently in the system. $\mathcal{S}$ contains m numbers. Let the $\overline{\mathcal{S}}(\mathbf{x})$ denote the set that contains the top x larger τ_i in $\mathcal{S}$, $0 \leq x \leq m$.

Algorithm 1. The x–Sorted Strategy

The online strategy that includes transposition process keeps the items with $\tau_i \in \overline{\mathcal{S}}(x)$ into the right-part in the order of $t^j \leq t^{j+1} \leq t^m$.
1. Divide the system into two parts—the left-part that contains $(m - x)$ storage units and the right-part that contains x storage units.
2. A new item i' gets placed on the I/O point, If $\tau_{i'}$ is greater than any $\tau_i \in \overline{\mathcal{S}}(x)$, the system performs sorting for the right-part considering the new item and put the new item into the right-part(in order) . Otherwise, put the new item into the idle unit directly.

Theorem 2. *The competitive ratio is max $\{m - x, \frac{-x^2+2mx+3m+1}{m+1}\}$ for a x–Sorted Strategy.*

Proof. Two kinds of cases may happen when the x–Sorted Strategy divides the system into two parts with different storage rules.

Case 1: Since the x–Sorted Strategy divides the system into two parts and the left-part does not have to be kept in order, the situation could be as bad as the example illustrated in the proof of the Theorem 1. The special instance is that the departure time of the item which is stored in the $(m - x)$th storage unit is the smallest in the left-part, and n items are released with $\tau_i^* = t^{m-x}+i\Delta$ which is always less than any of the others' stored in the system. Thus, the picker has to go through all the items to store or retrieve an item. The total travel distance of the online stratgy is

$$C_I(A) \leq 2(\Sigma_{j=1}^m d_j + (m - x) * n).$$

The total travel distance of the offline stratgy is

$$C_I(OPT) \geq 2(\Sigma_{j=1}^m d_j + 1 * n).$$

The competitive ratio of case 1 is

$$\rho_1 = \frac{C_I(A)}{C_I(OPT)} \leq m - x.$$

To retrieve	To store		1	2	...	m-x	m-x+1	...	m-1	m
Item 1	Item m+1	State 1	τ_1	τ_2	...	τ_{m-x}	τ_{m-x+1}	...	τ_{m-1}	τ_m
Item 2	Item m+2	State 2	τ_{m-x+1}	τ_2	...	τ_{m-x}	τ_{m-x+2}	...	τ_m	τ_{m+1}
Item 3	Item m+3	State 3	τ_{m-x+1}	τ_{m-x+2}	...	τ_{m-x}	τ_{m-x+3}	...	τ_{m+1}	τ_{m+2}

Fig. 2. The x–sorted strategy

To retrieve	To store		1	2	...	m-x	m-x+1	...	m-1	m
Item 1	Item m+1	State 1	τ_1	τ_2	...	τ_{m-x}	τ_{m-x+1}	...	τ_{m-1}	τ_m
Item 2	Item m+2	State 2	τ_{m+1}	τ_2	...	τ_{m-x}	τ_{m-x+1}	...	τ_{m-1}	τ_m
Item 3	Item m+3	State 3	τ_{m+1}	τ_{m+2}	...	τ_{m-x}	τ_{m-x+1}	...	τ_{m-1}	τ_m

Fig. 3. The offline optimal strategy

Case 2: Since a x–Sorted Strategy divides the system into two parts and the right-part should have to be kept in order, the worst case could make it to transpose the items every time.

Such an instance $I^* = \{\tau_i \mid \tau_{i+1} > \tau_i\}$ has been found that the total travel distance of the optimal offline strategy is the minimal without transposition process. The x–Sorted Strategy should take the most amount of transpositions to make sure the order is kept in the right-part. In a sorting process, the system will perform transpositions no more than x times. The state of storage situation changing with the items releasing is showed in Figs. 2 and 3.

After item 1 left the system, item $(m + 1)$ is released. From the first row to the second row in Fig. 2, it means that the departure time of the $(m + 1)$th item is greater than the item m. So, in the online strategy, the $(m + 1)$th item will occupy the mth unit and all the items stored in the right part should be transposed. The item $(m - x + 1)$ is moved to the first unit. Such that the travel distance of the picker includes three parts:

(1) the travel distance of the retrieval process equals to $2d_j$,
(2) the travel distance of the transposition process equals to $2(\Sigma^m_{j=m-x+1}d_j)$,
(3) the travel distance of the storage process equals to $2m$.

After $(m - x)$ items released, the situation is as same as the beginning. Every $(m - x)$ items is a cycle for the online algorithm. And every m items is a cycle for the offline algorithm. With no loss of generality, we assume that n equals to $n^*(m - x)m$. The other instances which make the online player performing the transposition process would not take a longer travel distance than the worst case in $Case2$. Thus the total travel distance of the x–Sorted Strategy is

$$C_{I^*}(A) \leq 2[\Sigma^{m-x}_{j=1}d_j + (m - x)(\Sigma^m_{j=m-x+1}d_j) + (m - x)m] \cdot mn^*.$$

The total travel distance of the optimal offline strategy is

$$C_{I^*}(OPT) = 2(\Sigma^m_{j=1}d_j) \cdot (m - x)n^*.$$

The competitive ratio of case 2 is

$$\rho_2 = \frac{C_{I^*}(A)}{C_{I^*}(OPT)} \leq \frac{-x^2 + 2mx + 3m + 1}{m+1}$$

$\square$

Corollary 1. *The minimal competitive ratio of the x–Sorted Strategy is* $(\frac{\sqrt{5}-1}{2})m$ *where x is similar to* $(\frac{3-\sqrt{5}}{2})m$.

Notice that ρ_1 is decreasing with x and ρ_2 is increasing, we could get the minimal competitive ratio of the x–Sorted Strategy when $\rho_1 = \rho_2$.
  Let $F(x) = \rho_1 - \rho_2, 0 \leq x \leq m, x \in Z^+, m \in Z^+$,

$$F(x) = \frac{x^2 - (3m+1)x + m^2 - 2m - 1}{m+1},$$

according to the characters of this quadratic equation, the x^*–sorted strategy has the minimal competitive ratio where $x^* = 2m + 1 - \sqrt{2m^2 + 5m + 2}$.
  When m is increasing, $x^*/m \rightarrow \frac{3-\sqrt{5}}{2}$. Actually, x^* is a positive integer close to $(\frac{3-\sqrt{5}}{2})m$, and the competitive ratio is $(m - x^*)$ which is around $(\frac{\sqrt{5}-1}{2})m$.

4   Conclusion

In this paper, we present a strategy for the online storage strategy problem of the one-dimensional automated storage and retrieval system with single open that improves the competitive ratio from m to $(\frac{\sqrt{5}-1}{2})m$. We prove it is beneficial to do some appropriate adjustment where the departure times of the items have been known. Although the strategy allows the transposition, we also prove that it's not necessary to keep all the items in order. The ratio between the number of the sorted units and the total units is around to $(\frac{3-\sqrt{5}}{2})$.
  Next, we will extend this problem to the multidimensional automated storage and retrieval system. We will also explore how to extend this storage strategy problem with different information of the items.

Acknowledgments. The authors would like to acknowledge the financial support of Grants (No.61221063) from NSF of China and (No.IRT1173) from PCSIRT of China.

References

1. Tompkins, J.A., White, J.A., Bozer, Y.A., Frazelle, E.H., Tanchoco, J.M.A.: Facilities Planning. John Wiley, New York (2003)
2. Bartholdi, J.J., Hackman, S.T.: Warehouse & Distribution Science. Georgia Institute of Technology, Atlanta (2011)
3. Hausman, W.H., Schwarz, L.B., Graves, S.C.: Optimal storage assignment in automatic warehousing systems. Manage. Sci. **22**(6), 629–638 (1976)

4. Graves, S.C., Hausman, W.H., Schwarz, L.B.: Storage-retrieval interleaving in automatic warehousing systems. Manage. Sci. **23**(9), 935–945 (1977)
5. Petersen II, C.G., Schmenner, R.W.: An evaluation of routing policies in an order picking operation. Decis. Sci. **30**, 481–501 (1999)
6. Heskett, J.L.: Cube-per-order index-a key to warehouse stock location. Transp. Distrib. Manage. **3**, 27–31 (1963)
7. Heskett, J.L.: Putting the cube-per-order index to work in warehouse layout. Transp. Distrib. Manage. **4**, 23–30 (1964)
8. Wen, X., Xu, Y., Zhang, H.: Online traveling salesman problem with deadline and advanced information. Comput. Ind. Eng. **63**(4), 1048–1053 (2012)
9. Wen, X., Xu, Y., Zhang, H.: Online traveling salesman problem with deadlines and service flexibility. J. Comb. Optim. **25**, 1–18 (2013). (online)
10. Frazelle, E.H., Sharp, G.P.: Correlated assignment strategy can improve any order-picking operation. Ind. Eng. **21**, 33–37 (1989)
11. Borodin, A., El-yaniv, R.: Online computation and competitive analysis. Cambridge University Press, Cambridge (1998)

Union Closed Tree Convex Sets

Tian Liu[1][(✉)] and Ke Xu[2][(✉)]

[1] Key Laboratory of High Confidence Software Technologies, Ministry of Education,
Peking University, Beijing 100871, China
lt@pku.edu.cn
[2] National Lab of Software Development Environment,
Beihang University, Beijing 100191, China
kexu@nlsde.buaa.edu.cn

Abstract. We show that the union closed sets conjecture holds for tree convex sets. The union closed sets conjecture says that in every union closed set system, there is an element to appear in at least half of the members of the system. In tree convex set systems, there is a tree on the elements such that each subset induces a subtree. Our proof relies on the well known Helly property of hypertrees and an equivalent graph formulation of the union closed sets conjecture given in (Bruhn, H., Charbit, P. and Telle, J.A.: The graph formulation of the union-closed sets conjecture. *Proc. of EuroComb 2013*, 73–78 (2013)).

Keywords: The union closed sets conjecture · Tree convex sets · Hypertree · Tree convex bipartite graphs · Helly property

1 Introduction

A *set system* is a family of subsets of a given finite universe. A set system is called *union closed*, if the union of any two sets in the system is still a member of the system. *The union closed sets conjecture* due to Peter Frankl [6] says that in every union closed set system, there is an element to appear in *at least* half of the members of the system. Although the conjecture dates back to 1979, it is still far from having a complete proof at this moment [3]. For example, we can not even show that there is an element to appear in a constant portion of the member sets. Only some special cases are verified and some equivalent formulations are given. We refer to [3] for an excellent survey on this conjecture.

Recently, a graph formulation of the conjecture is proposed and the conjecture is shown to hold for some special graph classes [2]. Especially, the conjecture is equivalent to say that in every finite bipartite graph with at least one edge, each of the two bipartition classes contains a vertex belonging to *at most* half of *maximal stable sets*, and the conjecture holds for chordal bipartite graphs [2].

A set system is nothing else but a *hypergraph* [1]. The elements in the universe are the vertices and the subsets are the hyperedges. In this way, a *tree convex*

Partially supported by Natural Science Foundation of China (Grant Nos. 61370052 and 61370156).

© Springer International Publishing Switzerland 2015
J. Wang and C. Yap (Eds.): FAW 2015, LNCS 9130, pp. 198–203, 2015.
DOI: 10.1007/978-3-319-19647-3_19

set system is also called a *hypertree*. In a hypertree, there is a tree associated with the vertex set, such that each hyperedge induces a subtree. A well known property of Hypertree is *Helly property*, that is, if every two hyperedges have a non-empty intersection, then there is a vertex to appear in every hyperedge.

A set system also has a *bipartite incidence graph*, where the two bipartition classes are the universe and the set system respectively, and there is an edge between an element in the universe and a subset if the element is in the subset. In this way, a tree convex set system has a *tree convex bipartite graph* as its bipartite incidence graph. For some algorithmic aspects of tree convex sets or tree convex bipartite graphs, we refer to the survey paper [4] and e.g. [8,10–13].

In this paper, we show that the union closed sets conjecture holds for tree convex set systems. First, we show that the union closed sets conjecture holds for tree convex set systems without empty set as a member. This part of the proof is built on the Helly property of hypertrees. Then, we recall the equivalent graph formulation of the union closed sets conjecture in [2] to show that the union closed set conjecture also holds for tree convex set systems with empty set as a member. Putting together, we show that the union closed sets conjecture holds for tree convex bipartite graphs under the graph formulation in [2].

Since chordal bipartite graphs are a proper subset of tree convex bipartite graphs [9], our result extends the validity of the union closed set conjecture from chordal bipartite graphs in [2] to tree convex bipartite graphs. In the language of hypergraphs, we extends the validity of the union closed set conjecture from β-acyclic hypergraphs (as shown in [2]) to hypertrees, since a hypergraph is β-acyclic if and only if its bipartite incidence graph is chordal bipartite, and if and only if every its subhypergraph is a hypertree, and thus itself is also a hypertree (Theorem 8.2.1 to Theorem 8.2.5, pages 125–126, [1]).

We note that in [2], besides the result for chordal bipartite graphs, also there are results for other kinds of bipartite graphs. Their proof for chordal bipartite graphs depends on a sufficient condition for the union closed sets conjecture and a characteristics for chordal bipartite graphs, while our proof depends on Helly property of hypertrees and an equivalent graph formulation of the union closed sets conjecture developed in [2]. Thanks to the graph formulation in [2], we can find a simpler proof while get a stronger result.

This paper is structured as follows. After introducing basic notions and facts (Sect. 2), the union closed sets conjecture is shown to hold for tree convex sets (Sect. 3), and finally are some concluding remarks (Sect. 4).

2 Preliminaries

In this section, we give some definitions and known results.

Definition 1. *A set system $(U, \mathcal{F})$ has a universe U and each member X in $\mathcal{F}$ is a subset of U. $\mathcal{F}$ is union closed if for any X, Y in $\mathcal{F}$, $X \cup Y$ is in $\mathcal{F}$.*

Definition 2. *A set system $(U, \mathcal{F})$ is also a hypergraph, where U is the vertex set and and each member X in $\mathcal{F}$ is a hyperedge. $(U, \mathcal{F})$ is a hypertree or tree* *convex*, if there is a tree $T = (U, E)$ associated on U, such that each X in $\mathcal{F}$ induces a subtree T_X in T.

Definition 3. *Given a set system* $(U, \mathcal{F})$, *if every two subsets* X, Y *in* $\mathcal{F}$ *have a non-empty intersection, then* $\bigcap \mathcal{F} = \{x | x \in X \text{ for all } X \in \mathcal{F}\} \neq \emptyset$, *this is called Helly property. In other words, there is an element* w *to appear in every member set of* $\mathcal{F}$, *if every two member sets have a non-empty intersection.*

It is well known that hypertrees have Helly property (pages 23–24 in [1]). In this paper, we consider the following conjecture [6].

Conjecture 1. For every union closed set system $(U, \mathcal{F})$, there is an element x in U which appears in at least half of members of $\mathcal{F}$, that is, $|\{X | x \in X\}| \geq |\mathcal{F}|/2$.

An equivalent conjecture is the following one [2]. A *stable set* in a graph is a subset of vertices which induce no edge. A maximal stable set is a stable set whose proper supersets are not stable sets any more. Stable sets are also called independent sets, and a maximal stable set is nothing else but an independent dominating set, where a dominating set is a subset of vertices such that every vertex outside it has a neighbor in it.

Conjecture 2. For every finite bipartite graph with at least one edge, each of the two bipartition classes contains a vertex belonging to at most half of maximal stable sets.

We will call both conjectures as the union closed sets conjecture in this paper.

All remaining undefined notions mentioned in this paper, such as chordal bipartite graphs and β-acyclic hypergraphs, can be found in [1].

3   Main Results

In this section, we show our main result that the union closed sets conjecture holds for tree convex sets, or equivalently for tree convex bipartite graphs.

We first show that the union closed sets conjecture holds for tree convex sets without empty set as a member, by using Helly property of hypertrees.

Theorem 1. *The union closed sets conjecture holds for tree convex sets without empty set as a member.*

Proof. Assume that we have a set system $(U, \mathcal{F})$, where $\emptyset \notin \mathcal{F}$ and $\mathcal{F}$ is tree convex. We have a tree $T = (U, E)$ such that for each X in $\mathcal{F}$, X induce a subtree T_X in T. We will build a new tree T' and a new set system $(U', \mathcal{F}')$ as follows.

First, we build T' and U'. For each edge (u, v) in T, we replace it by two edges $(u, uv), (uv, v)$, where uv is a new vertex inserted between u and v, and we treat uv and vu as the same vertex. In this way, we get a new tree $T' = (U', E')$, where $U' = U \cup \{uv | (u, v) \in E\}$ and $E' = \{(u, uv), (uv, v) | (u, v) \in E\}$. An example of tree T and tree T' is given in Fig. 1.

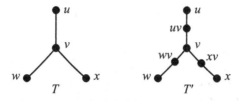

Fig. 1. An example of tree T and tree T'.

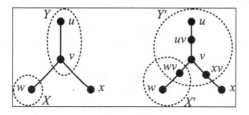

Fig. 2. An example of subsets X, Y and subsets X', Y'.

Second, we build $\mathcal{F}'$. For each X in $\mathcal{F}$, we replace it by $X' = X \cup \{uv | u \in X \text{ or } v \in X\}$. In this way, we get a new set system $(U', \mathcal{F}')$, where $\mathcal{F}' = \{X' | X \in \mathcal{F}\}$. Clearly, $\mathcal{F}'$ is still tree convex, or in other words, it is a hypertree. Indeed, each X' in $\mathcal{F}'$ induces a subtree $T'_{X'}$ in T, where $T'_{X'} = (X', E'_{X'})$ and $E'_{X'} = \{(uv, u) | u \in X\}$. An example of subsets X, Y and subsets X', Y' is given in Fig. 2.

Now the union closed property and the tree convex property of $\mathcal{F}$ will translate to the Helly property of $\mathcal{F}'$. That is, for all $X', Y' \in \mathcal{F}'$, we have that $X' \cap Y' \neq \emptyset$. Indeed, for any $X, Y \in \mathcal{F}$, we have that $X \cup Y \in \mathcal{F}$ and $X \cup Y$ induces a subtree $T_{X \cup Y}$ in T, since $\mathcal{F}$ is union closed and tree convex. But $X \cup Y$ induces a subtree $T_{X \cup Y}$ in T if and only if either there is w in $X \cap Y$ or there is u in X and v in Y and (u, v) in E. In the first case, w is in $X' \cap Y'$, and in the second case, uv is in $X' \cap Y'$. Thus, $X' \cap Y'$ is non-empty for all X', Y' in $\mathcal{F}'$.

Then by Helly property of hypertrees (pages 23–24 in [1]), all members of $\mathcal{F}'$ have a non-empty intersection. Thus, there is a w or uv in $\bigcap \mathcal{F}'$. If w is in $\bigcap \mathcal{F}'$, then w is also in $\bigcap \mathcal{F}$, by the construction of $\mathcal{F}'$. In this case, $\mathcal{F}$ also has the Helly property and the union closed sets conjecture holds for $\mathcal{F}$. If uv is in $\bigcap \mathcal{F}'$, then each member X' in $\mathcal{F}'$ contains uv. By the construction of $\mathcal{F}'$, each member X in $\mathcal{F}$ either contains u or v or both. Therefore, either u or v will appear in at least half of members of $\mathcal{F}$. Say u appears in at least half of members in $\mathcal{F}$. In this case, the union closed sets conjecture also holds for $\mathcal{F}$. □

We next show that the union closed sets conjecture holds for tree convex sets with empty set as a member. To this end, we need the graph formulation of the union closed sets conjecture (Conjecture 2) and the equivalence of Conjectures 1 and 2 in [2].

Theorem 2. *The union closed sets conjecture holds for tree convex sets with empty set as a member.*

Proof. If a union closed tree convex set system $(U, \mathcal{F})$ has empty set as a member, that is, $\emptyset \in \mathcal{F}$, then its bipartite incidence graph $B = (U, \mathcal{F}, \{(u, X) | u \in X\})$ is a tree convex bipartite graph with an isolated vertex $\emptyset$ (the empty set, and we think that the empty set also induces a subtree, the null tree without any vertex).

An isolated vertex will appear in every maximal stable sets. Thus, if we remove this isolated vertex from the bipartite incidence graph B to get a new bipartite graph $B' = (U, \mathcal{F} \setminus \{\emptyset\}, \{(u, X) | u \in X\})$, the number of maximal stable sets will not change, and D' is a maximal stable set of B' if and only if $D \cup \{\emptyset\}$ is a maximal stable set of B.

But the set system with B' as its bipartite incidence graph is a union closed tree convex set system without the empty set as a member. By our above result Theorem 1, the union closed sets conjecture (Conjecture 1) holds for this set system. By the equivalence of Conjectures 1 and 2, we known that the union closed sets conjecture (Conjecture 2) holds for B'. Since B' and B have the same number of maximal stable sets, we known that the union closed sets conjecture (Conjecture 2) holds for B. Again, by the equivalence of Conjectures 1 and 2, we known that the union closed sets conjecture (Conjecture 1) holds for $\mathcal{F}$.   □

Now we have the following two theorems by Theorems 1 and 2.

Theorem 3. *The union closed sets conjecture (Conjecture 1) holds for tree convex sets.*

Theorem 4. *The union closed sets conjecture (Conjecture 2) holds for tree convex bipartite graphs.*

4   Conclusions

We have shown that the union closed sets conjecture holds for tree convex set systems or tree convex bipartite graphs. Thus under the graph formulation in [2], the validity of the union closed sets conjecture can be extended from chordal bipartite graphs to tree convex bipartite graphs, or equivalently, form β-acyclic hypergraphs to hypertrees.

Another proper superset of chordal bipartite graphs is the so-called *perfect elimination bipartite graphs* [1,7], which are incomparable with tree convex bipartite graphs [9]. A proper superset of β-acyclic hypergraphs is the so-called α-*acyclic hypergraphs* [1,5]. Thus the validity of the union closed conjecture for perfect elimination bipartite graphs and for α-acyclic hypergraphs are interesting open problems.

Finally, can our techniques be refined to show that there is an element to appear in a *constant* portion of the member sets for union closed sets?

Acknowledgments. The help of anonymous reviewers has improved our presentation greatly.

References

1. Brandstad, A., Le, V.B., Spinrad, J.P.: Graph Classes - A Survey. Society for Industrial and Applied Mathematics, Philadelphia (1999)
2. Bruhn, H., Charbit, P., Telle, J.A.: The graph formulation of the union-closed sets conjecture. In: Nesetril, J., Pellegrini, M. (eds.) EuroComb 2013. CRM, pp. 73–78. Scuola Normale Superiore, Pisa (2013)
3. Bruhn, H. and Schaudt, O.: The journey of the union-closed sets conjecture. ArXiv 1212.4175v2 (2012)
4. Bao, F.S., Zhang, Y.: A review of tree convex sets test. Comput. Intell. **28**(3), 358–372 (2012)
5. Fagin, R.: Degrees of acyclicity for hypergraphs and relational database schemes. J. ACM **30**, 514–550 (1983)
6. Frankl, P.: Handbook of Combinatorics, vol. 2, pp. 1293–1329. MIT Press, Cambridge (1995)
7. Golumbic, M.C., Goss, C.F.: Perfect elimination and chordal bipartite graphs. J. Graph Theory **2**, 155–163 (1978)
8. Jiang, W., Liu, T., Ren, T., Xu, K.: Two hardness results on feedback vertex sets. In: Atallah, M., Li, X.-Y., Zhu, B. (eds.) FAW-AAIM 2011. LNCS, vol. 6681, pp. 233–243. Springer, Heidelberg (2011)
9. Liu, T.: Restricted bipartite graphs: comparison and hardness results. In: Gu, Q., Hell, P., Yang, B. (eds.) AAIM 2014. LNCS, vol. 8546, pp. 241–252. Springer, Heidelberg (2014)
10. Lu, M., Liu, T., Xu, K.: Independent domination: reductions from circular- and triad-convex bipartite graphs to convex bipartite graphs. In: Fellows, M., Tan, X., Zhu, B. (eds.) FAW-AAIM 2013. LNCS, vol. 7924, pp. 142–152. Springer, Heidelberg (2013)
11. Lu, M., Liu, T., Tong, W., Lin, G., Xu, K.: Set cover, set packing and hitting set for tree convex and tree-like set systems. In: Gopal, T.V., Agrawal, M., Li, A., Cooper, S.B. (eds.) TAMC 2014. LNCS, vol. 8402, pp. 248–258. Springer, Heidelberg (2014)
12. Song, Y., Liu, T., Xu, K.: Independent domination on tree convex bipartite graphs. In: Snoeyink, J., Lu, P., Su, K., Wang, L. (eds.) AAIM 2012 and FAW 2012. LNCS, vol. 7285, pp. 129–138. Springer, Heidelberg (2012)
13. Wang, C., Chen, H., Lei, Z., Tang, Z., Liu, T., Xu, K.: Tree convex bipartite graphs: $cal NP$-complete domination, hamiltonicity and treewidth. In: Proceedings of FAW (2014)

Fast Quantum Algorithms for Least Squares Regression and Statistic Leverage Scores

Yang Liu and Shengyu Zhang[✉]

Department of Computer Science and Engineering, The Institute of Theoretical Computer Science and Communications, The Chinese University of Hong Kong, Sha Tin, Hong Kong
{yliu,syzhang}@cse.cuhk.edu.hk

Abstract. Least squares regression is the simplest and most widely used technique for solving overdetermined systems of linear equations $Ax = b$, where $A \in \mathbb{R}^{n \times p}$ has full column rank and $b \in \mathbb{R}^n$. Though there is a well known unique solution $x^* \in \mathbb{R}^p$ to minimize the squared error $\|Ax - b\|_2^2$, the best known classical algorithm to find x^* takes time $\Omega(n)$, even for sparse and well-conditioned matrices A, a fairly large class of input instances commonly seen in practice. In this paper, we design an efficient quantum algorithm to generate a quantum state proportional to $|x^*\rangle$. The algorithm takes only $O(\log n)$ time for sparse and well-conditioned A. When the condition number of A is large, a canonical solution is to use regularization. We give efficient quantum algorithms for two regularized regression problems, including ridge regression and δ-truncated SVD, with similar costs and solution approximation.

Given a matrix $A \in \mathbb{R}^{n \times p}$ of rank r with SVD $A = U\Sigma V^T$ where $U \in \mathbb{R}^{n \times r}$, $\Sigma \in \mathbb{R}^{r \times r}$ and $V \in \mathbb{R}^{p \times r}$, the statistical leverage scores of A are the squared row norms of U, defined as $s_i = \|U_i\|_2^2$, for $i = 1, ..., n$. The matrix coherence is the largest statistic leverage score. These quantities play an important role in many machine learning algorithms. The best known classical algorithm to approximate these values runs in time $\Omega(np)$. In this work, we introduce an efficient quantum algorithm to approximate s_i in time $O(\log n)$ when A is sparse and the ratio between A's largest singular value and smallest non-zero singular value is constant. This gives an exponential speedup over the best known classical algorithms. Different than previous examples which are mainly modern algebraic or number theoretic, this problem is linear algebraic. It is also different than previous quantum algorithms for solving linear equations and least squares regression, whose outputs compress the p-dimensional solution to a $\log(p)$-qubit quantum state.

1   Introduction

Quantum Algorithms for Solving Linear Systems, and the Controversy. The past two decades witnessed the development of quantum algorithms [Mos09], and one recent discovery is quantum speedup for solving linear systems $Ax = b$ for sparse and well-conditioned matrices $A \in \mathbb{R}^{n \times p}$. Solving linear

© Springer International Publishing Switzerland 2015
J. Wang and C. Yap (Eds.): FAW 2015, LNCS 9130, pp. 204–216, 2015.
DOI: 10.1007/978-3-319-19647-3_20

systems is a ubiquitous computational task, and sparse and well-conditioned matrices form a fairly large class of inputs frequently arising in many practical applications, especially in recommendation systems where the data set can be very sparse [ZWSP08]. The best known classical algorithm for solving linear systems for this class of matrices runs in time $O(\sqrt{\kappa}sn)$ [She94], where κ is the condition number of A (i.e. the ratio between A's largest and smallest singular values), and the sparseness parameter s is the maximum number of non-zero entries in each row of A. Harrow, Hassidim and Lloyd [HHL09] introduced an efficient quantum algorithm, thereafter referred to as HHL algorithm, for the linear system problem, and the algorithm runs in time $O(s^2\kappa^2 \log n)$. The dependence on κ is later improved by Ambainis [Amb12] and the algorithm was used for solving least squares regression (defined next) by Wiebe, Braun and Lloyd [WBL12]. HHL algorithm was also extended in [CJS13] to more general problem specifications.

Though the costs of these quantum algorithms are exponentially smaller than those of the best known classical algorithms, there is a catch that these quantum algorithms do not output the entire solution x^*, but compress $x^* \in \mathbb{R}^n$ (assuming $n = p$) into a $\log n$-qubit quantum state. More precisely, the output is a quantum state $|\overline{x^*}\rangle$ proportional to $\sum_{i=1}^n x_i^*|i\rangle$. This important distinction between outputs of classical and quantum algorithms caused some controversy for these quantum algorithms. After all, one cannot read out the values x_i^* from $|\overline{x^*}\rangle$. Indeed, if outputting all x_i^* is required as classical algorithms, then any quantum algorithm needs $\Omega(n)$ time even for just writing down the answer, thus no exponential speedup is possible.

Despite this drawback, the quantum output $|\overline{x^*}\rangle$ can be potentially useful in certain context where only global information of x^* is needed. For instance, sometimes only the expectation value of some operator associated with x^*, namely $x^{*T}Mx^*$ for some matrix M is needed [HHL09]. Another example is when one desires to compute only the weighted sum $\sum c_i x_i^*$, then SWAP test can be used on $|\overline{c}\rangle = \sum_i \frac{c_i}{\|c\|_2}|i\rangle$ and $|\overline{x^*}\rangle$ to get a good estimate of $\sum c_i x_i^*$ in time $O(\log n)$. As argued in [Amb12], this is impossible for classical algorithms unless P = BQP.

In this paper, we give new quantum algorithms, which also address the controversial issue on two levels. First, we design an efficient quantum algorithm for least squares regression, which runs in time $O(\log n)$ for sparse and well-conditioned A. Same as the one in [WBL12], our quantum algorithm outputs a quantum sketch $|\overline{x^*}\rangle$ only, but our algorithm is simpler, and more efficient with a better dependence on s and κ.

In addition, we consider the case that A is ill-conditioned, or even not full-rank. Classical resolutions for such cases use regularization. We give efficient quantum algorithms for two popular regularized regression problems, including ridge regression and δ-truncated SVD, based on our algorithm for least squares regression.

Second, we also design new efficient quantum algorithms for calculating statistic leverage scores (SLS) and matrix coherence (MC), two quantities playing important roles in many machine learning algorithms [Sar06, DMM08, MD09, BMD09, DMMS11]. Our algorithm has cost $O(\log n)$ for approximately calculating the k-th statistic leverage score s_k for any index $k \in [n]$, exponentially faster than the best known classical algorithms. Repeatedly applying this allows us to approximately calculate all the statistic leverage scores in time $O(n \log n)$ and to calculate matrix coherence in time $O(\sqrt{n} \log n)$, which has a polynomial speedup to their classical counterparts of cost $O(n^2)$ [DMIMW12]. Note that different than all aforementioned quantum algorithm that outputs a quantum sketch only, our algorithms for calculating SLS and MC indeed produce the requested values, same as their classical counterpart algorithms' output. Our algorithms are based on the phase estimation idea as in the HHL algorithm, and the results showcase the usefulness of the HHL algorithm even in the standard computational context without controversial issue any more.

Next we explain our results in more details.

Least Squares Regression. Least squares regression (LSR) is the simplest and most widely used technique for solving overdetermined systems. In its most important application – data fitting, it finds a hyperplane through a set of data points while minimizing the sum of squared errors. The formal definition of LSR is as follows. Given an $n \times p$ matrix A $(n \geq p)$ together with an n-dimensional vector b, the goal of LSR is to compute a p-dimensional vector

$$x^* = \arg \min_{x \in \mathbb{R}^p} \|Ax - b\|_2^2. \tag{1}$$

For well-conditioned problems, i.e. those with the condition number of A being small (which in particular implies that A has full column rank), it is well known that Eq. (1) has a unique and closed-form solution

$$x^* = A^+ b, \tag{2}$$

where A^+ is the Moore-Penrose pseudoinverse of A. If one computes x^* naively by first computing A^+ and then the product $A^+ b$, then the cost is $O(p^2 n + n^2 p)$, which is prohibitively slow in the big data era[1]. Therefore, finding fast approximation algorithms which output a vector $\tilde{x} \approx x^*$ is of great interest. Classically, there are known algorithms that output an $\tilde{x}$ with a relative error bound $\|\tilde{x} - x^*\|_2 \leq \epsilon \|x^*\|_2$ for any constant error $0 < \epsilon < 1$, and run in time $\tilde{O}(\text{nnz}(A) + nr)$ [CW13, NN13], where $\text{nnz}(A)$ is the number of non-zero entries in A, r is the rank of A and the $\tilde{O}$ notation hides a logarithmic factor. These algorithms are much faster than the naive ones for the special case of sparse or low rank matrices, but remain linear in size of A for general cases. Given that it is impossible to have classical approximation algorithms to run in time $o(np)$ for general cases, it would be great if there exist much faster quantum algorithms for LSR. Similar to [HHL09], one can only hope to produce a quantum state close to $|\overline{x^*}\rangle$ fast. [WBL12] gives a quantum algorithm which outputs a quantum state close to $|\overline{x^*}\rangle$ in time $O(\log(n+p)s^3\kappa^6)$. Their algorithm is based

---

[1] Though theoretically more efficient algorithms for matrix multiplication exist [Sto10], in practice they are seldom used due to the complication in implementing, parallelization and non-robustness. Thus in machine learning algorithms matrix multiplication $A_{m \times n} B_{n \times k}$ is assumed to take time $O(mnk)$. In any case it is just a polynomial saving, in contrast to the exponential gap to the quantum algorithm cost.

on the observation that $x^* = A^+b = (A^TA)^{-1}A^Tb$ when A has full column rank, and their main idea is to construct the quantum state $|\overline{x^*}\rangle$ by apply the operator $(A^TA)^{-1}$ to the state $|A^Tb\rangle$. In this paper, we propose another quantum algorithm that outputs a quantum state close to $|\overline{x^*}\rangle$ in time $O(\log(n+p)s^2\kappa^3)$.

We highlight four advantages of our algorithm compared to [WBL12]. First, our algorithm is much simpler since we directly apply the operator A^+ to the state $|b\rangle$, while they first applied A^T to $|b\rangle$ to get $|A^Tb\rangle$, then prepared $(A^TA)^{-1}$ and applied it to $|A^Tb\rangle$. Second, the simplicity also leads to a better dependence on s and κ in our algorithm. Third, [WBL12] assumes that A is Hermitian, which is usually not the case for typical machine learning applications[2]. Our algorithm works for non-Hermitian matrices as well, for which we need to work on singular Hermitian matrices A. Fourth, note that $|\overline{x^*}\rangle$ misses one important information of x^*, namely its ℓ_2 norm, which is actually crucially needed when we want to compute $\sum_i c_i x_i$ by SWAP test. Our algorithm also gives a good estimate to $\|x^*\|_2^2$ without introducing much extra running time.

The more precise description of the performance of our algorithm is stated in the next theorem. As in [HHL09], we assume that the vector b is in a nice form in the sense that each b_i and $\sum_{i_1}^{i_2}|b_i|^2$ can be efficiently computed, which enables us to prepare $|b\rangle$ efficiently [GR02].

Theorem 1. *Let $A \in \mathbb{R}^{n \times p}$ and $b \in \mathbb{R}^n$ be the input of the least squares regression problem and suppose that x^* is its optimal solution. Assume that each row of A has at most s non-zero entries, and all the non-zero singular values of A are in the range $[\frac{1}{\kappa}, 1]$. Then there exists a quantum algorithm that outputs a quantum state proportional to $\tilde{x}$ satisfying $\|\tilde{x} - x^*\|_2 \leq \epsilon \cdot \max\{\|x^*\|_2, \|b\|_2\}$, and outputs a value l satisfying $|l - \|x^*\|_2^2| \leq \epsilon(\|x^*\|_2^2 + \|b\|_2^2)$, in time $O(\log(n+p)s^2\kappa^3/\epsilon^2)$.*

Ridge Regression and Truncated SVD. For ill-conditioned problems, i.e. when the condition number of A is large, the solution given by Eq. (2) becomes very sensitive to errors in A and b. A prevailing solution in practice is to use regularization. Two of the most commonly used regularization methods are ridge regression [GHO99] (a.k.a Tikhonov regularization) and truncated singular value decomposition [Han87].

For ridge regression (RR) problem, we are given an $n \times p$ matrix A, an n-dimensional vector b together with a parameter $\lambda > 0$, and we want to compute

$$x^* = \arg\min_{x \in \mathbb{R}^p} \|Ax - b\|_2^2 + \lambda \|x\|_2^2. \tag{3}$$

The unique minimizer of Eq. (3) is $x^* = (A^TA + \lambda I_p)^{-1}A^Tb$ [Tik63], which takes $O(np^2 + p^3)$ time to compute in the naive way. An alternative solution uses the dual space approach by computing an equivalent expression $x^* = A^T(AA^T + \lambda I_n)^{-1}b$ [SGV98], which takes $O(n^2p + n^3)$ time to compute in the naive way,

---

[2] Although they mentioned a standard pre-processing technique to deal with the non-Hermitian case, but they seem to have overlooked the fact that after the pre-processing, the new input matrix is not full (column) rank (unless $n = p$, which is hardly the case in machine learning settings).

and is faster than the original one when $p \gg n$. When approximation is allowed, the best known classical algorithm for ridge regression outputs an approximation solution $\tilde{x}$ satisfying $\|\tilde{x} - x^*\|_2 \leq \epsilon \|x^*\|_2$ in time $\tilde{O}(nnz(A) + n^2 r)$ [CLL+14]. This algorithm has an significant speedup over the previous algorithms when A is sparse or of low rank, but still slow for general cases. Based on the algorithm in Theorem 1, we design a quantum algorithm to solve ridge regression problem efficiently (in the sense of generating quantum sketch of the solution).

Theorem 2. *Let $A \in \mathbb{R}^{n \times p}$, $b \in \mathbb{R}^n$ and λ be the input of the ridge regression problem and suppose that x^* is its optimal solution. Assume that each row of A has at most s non-zero entries, and all the non-zero singular values of A are in the range $[\frac{1}{\kappa}, 1]$. Then there exists a quantum algorithm that generates a quantum state proportional to $|\tilde{x}\rangle$ satisfying $\|\tilde{x} - x^*\|_2 \leq \epsilon \cdot \max\{\|x^*\|_2, \|b\|_2\}$, and outputs a value l satisfying $|l - \|x^*\|_2^2| \leq \epsilon(\|x^*\|_2^2 + \|b\|_2^2)$, in time $O(\log(n+p)s^2\kappa'^3/\epsilon^2)$, for $\kappa' = \frac{\max\{1, \sqrt{\lambda}\}}{\min\{\frac{1}{\kappa}, \sqrt{\lambda}\}}$.*

Next we discuss truncated singular value decomposition (truncated SVD). In this problem we are given an $n \times p$ matrix A, an n-dimensional vector b together with a parameter $k < \mathtt{rank}(A)$, and we want to compute

$$x^* = \arg\min_{x \in \mathbb{R}^p} \|A_k x - b\|_2^2, \tag{4}$$

where A_k is the best rank-k approximation of A obtained through SVD. More specifically, let $A = \sum_i^r \lambda_i u_i v_i^T$ be the SVD of A, where r is the rank of A, λ_i is the i-th largest singular value of A, and $u_i \in \mathbb{R}^n, v_i \in \mathbb{R}^p$ are the corresponding left and right singular vectors for $i = 1, ..., r$. Then A_k is defined to be $A_k = \sum_{i=1}^k \lambda_i u_i v_i^T$.

The basic idea of truncated SVD is to impose an additional requirement that the ℓ_2 norm of the solution x^* should be small by removing the large influence from the small non-zero singular values of A. If the number k is chosen properly, the ratio between λ_1 and λ_k is small and then the solution $x^* = A_k^+ b$ to Eq. (4) becomes not sensitive to errors in A and b. A more direct way to remove the influence by the small non-zero singular values of A is to set a gap δ on the singular values and neglect all those singular values smaller than δ. Define the δ-truncated singular value decomposition (δ-TSVD) problem as follows.

Given an $n \times p$ matrix A, an n-dimensional vector b together with a parameter $\delta > 0$ and we want to find

$$x^* = \arg\min_{x \in \mathbb{R}^p} \|A_\delta x - b\|_2^2, \tag{5}$$

where $A_\delta = \sum_{i:\lambda_i \geq \delta} \lambda_i u_i v_i^T$, assuming $A = \sum_i^r \lambda_i u_i v_i^T$ is the SVD of A.

A naive algorithm to solve δ-TSVD needs to first compute the matrix A_δ and then solve the least squares regression problem with the new input A_δ and b in $O(n^2 p + np^2)$ time. Our algorithm in Theorem 1 can be also adapted to solve δ-TSVD efficiently (again in the sense of generating quantum sketch of the optimal solution).

Theorem 3. *Let* $A \in \mathbb{R}^{n \times p}$, $b \in \mathbb{R}^n$ *and* δ *be the input of the* δ-*truncated singular value decomposition problem and let* x^* *be the optimal solution of this problem. Assume that each row of* A *has at most* s *non-zero entries, and that the largest singular value of* A *is at most* 1. *Let* $\Lambda_1 = \max\{\lambda_i : \lambda_i < \delta, i \in [n]\}$, $\Lambda_2 = \min\{\lambda_i : \lambda_i \geq \delta, i \in [n]\}$ *where* λ_i *is the* i-*th largest singular value of* A. *Let* $\Lambda = \Lambda_2 - \Lambda_1$. *Then there exists a quantum algorithm that generates a quantum state proportional to* $|\tilde{x}\rangle$ *satisfying* $\|\tilde{x} - x^*\|_2 \leq \epsilon \cdot \max\{\|x^*\|_2, \|b\|_2\}$, *and outputs a value* l *satisfying* $|l - \|x^*\|_2^2| \leq \epsilon(\|x^*\|_2^2 + \|b\|_2^2)$, *in time* $O(\log(n + p)s^2/(\min\{\Lambda, \delta\epsilon\}\delta^2\epsilon))$.

Calculating Statistic Leverage Scores and Matrix Coherence. The definition of statistic leverage scores (SLS) and matrix coherence (MC) are as follows. Though the definition uses A's SVD, which is not necessarily unique, it is not hard to see that each s_i depends on A only, not on any specific SVD decomposition of A.

Definition 1. *Given an* $n \times p$ *matrix* A *of rank* r *with SVD* $A = U\Sigma V^T$ *where* $U \in \mathbb{R}^{n \times r}$, $\Sigma \in \mathbb{R}^{r \times r}$ *and* $V \in \mathbb{R}^{p \times r}$, *the statistic leverage scores of* A *are defined as* $s_i = \|U_i\|_2^2$, $i \in \{1, ..., n\}$, *where* U_i *is the* i-*th row of* U. *The matrix coherence of* A *is defined as* $c = \max_{i \in \{1,...,n\}} s_i$, *the largest statistic leverage score of* A.

Statistic leverage scores measure the correlation between the singular vectors of a matrix and the standard basis and they are very useful in large-scale data analysis and randomized matrix algorithms [MD09, DMM08]. These quantities have been used in statistical data analysis since a long time ago. Actually they are equal to the diagonal entries of the "hat-matrix" which interprets the influence associated with the data points and so they are widely used to indicate possible outliers in regression diagnostics [HW78, CH+86]. They have also been found useful in many theoretical computer science and machine learning problems. Many random sampling algorithms for matrix problems like least-squares regression [Sar06, DMMS11] and low-rank matrix approximation [Sar06, DMM08, MD09, BMD09] use them as an important indicator to design the sampling distribution which are used to sample the input data matrix.

The related parameter, matrix coherence, has also been of interest recently in problems like Nystrom-based low-rank matrix approximation [TR10] and matrix completion [CR09].

A naive algorithm to compute the statistic leverage scores and matrix coherence first performs SVD or QR decomposition to get an orthogonal basis of A, and then calculates the squared ℓ_2 norm of the rows of the basis matrix to get the statistic leverage scores. This takes $O(np^2)$ time (assuming $n \geq p$), which is extremely slow when n and p are large. The best known classical approximation algorithm for the problem of calculating statistic leverage scores runs in time $O((np + p^3) \log n)$ [DMIMW12], which is much faster than the naive algorithm for the case when $n >> p >> 1$. This algorithm is also the best for calculating matrix coherence for now.

An important difference of calculating SLS/MC than previously mentioned LSR related problems is that each SLS s_i (and the MC c) is a scalar instead of a vector. Thus the previous barrier of outputting a vector does not exist, which makes it possible to design efficient quantum algorithm for complete solution to the problem rather than generating a quantum sketch as before. In this work, we design a fast quantum algorithm to approximate the statistic leverage score s_i for any index $i \in \{1, ..., n\}$ in time $O(\log n)$ when A is sparse and the ratio between A's largest singular value and smallest non-zero singular value is small. And thus we can approximate all the statistic leverage scores in time $O(n \log n)$ by running the algorithm for index $i = 1, ..., n$. And we can approximate the matrix coherence in time $O(\sqrt{n} \log n)$ by using amplitude amplification. More specifically, we have the following theorem and corollary.

Theorem 4. *Let $A \in \mathbb{R}^{n \times p}$, let s_i be the i-th statistic leverage score of A for $i = 1, ..., n$. Assume that each row of A has at most s non-zero entries, and all the non-zero singular values of A are in the range $[\frac{1}{\kappa}, 1]$. Then there exists a quantum algorithm that, on any requested $i \in [n]$, returns $\tilde{s}_i$ satisfying $|\tilde{s}_i - s_i| \leq \epsilon$ in time $O(\log(n + p)s^2 \kappa/\epsilon)$.*

Corollary 5. *Let $A \in \mathbb{R}^{n \times p}$, let c be the coherence of A. Assume that each row of A has at most s non-zero entries, and all the non-zero singular values of A are in the range $[\frac{1}{\kappa}, 1]$. Then there exists a quantum algorithm that returns $\tilde{c}$ satisfying $|\tilde{c} - c| \leq \epsilon$ in time $O(\sqrt{n} \log(n + p)s^2 \kappa/\epsilon)$.*

2   Preliminaries

Given a matrix $A \in \mathbb{R}^{n \times p}$ with rank $r \leq \min\{n, p\}$, let A_i denote the transpose of the i-th row of A, namely take the i-th row and view it as a column vector $\in \mathbb{R}^p$. Let $nnz(A)$ denote the number of non-zero entries of A. Let λ_i denote the i-th largest singular value of A, and let $\lambda_{\max}$ denote the largest singular value of A, unless specified otherwise. Let I_r denote the identity matrix of dimension $r \times r$, e_i the unite vector with the i-th coordinate being 1 and all the rest being 0, and 0_n the zero vector of dimension n.

For a rank-r matrix $A \in \mathbb{R}^{n \times p}$, its thin SVD is $A = U \Sigma V^T$ where $U \in \mathbb{R}^{n \times r}$, $V \in \mathbb{R}^{p \times r}$ satisfy $U^T U = V^T V = I_r$, and $\Sigma = \text{diag}(\lambda_1, ..., \lambda_r)$ with the λ_i's being the singular values of A. The Moore-Penrose pseudoinverse of A is defined to be $A^+ = V \Sigma^{-1} U^T$. The full SVD of A is $A = U_F \Sigma_F V_F^T$ where $U_F \in \mathbb{R}^{n \times n}, \Sigma_F \in \mathbb{R}^{n \times p}, V \in \mathbb{R}^{p \times p}$ and $U^T U = UU^T = I_n, V^T V = VV^T = I_p$. When a matrix $A \in \mathbb{R}^{n \times n}$ is full rank, the thin SVD and full SVD are the same.

Quantum phase estimation [Kit95, CEMM97, BDM99] or quantum eigenvalue estimation allows one to estimate the eigenphase of an eigenvector of a unitary operator. It has been widely used as subroutine in other algorithms. In the Phase Estimation problem, we are given a unitary matrix U by black-boxes of controlled-U, controlled-U^2, controlled-U^{2^2}, $\cdots$, controlled-$U^{2^{t-1}}$ operations, and an eigenvector $|u\rangle$ of U with eigenvalue $e^{2\pi i \varphi}$ with the value of $\varphi \in [0, 1)$ unknown. The task is to output an n-bit estimation of φ.

Theorem 6. *There is a quantum algorithm that, on input $|0^t\rangle|u\rangle$ where $t = \log \frac{1}{\epsilon\delta} + O(1)$, output $|\tilde{\varphi}\rangle|u\rangle$ in time $O(t)$ using each controlled-U^{2^i} once, and $|\varphi - \varphi'| \leq \delta$ with probability at least $1 - \epsilon$.*

Two comments are in order. First, if we do not have controlled-U^{2^i}, and need to implement them, then the total time becomes $O(\frac{1}{\delta}\log\frac{1}{\epsilon})$ assuming that implementing controlled-U takes unit time. Second, when the input is $|0^t\rangle|b\rangle$ where $b = \sum_i \beta_i|u_i\rangle$, then the output is $b = \sum_i \beta_i|\tilde{\varphi}_i\rangle|u_i\rangle$ where each $\tilde{\varphi}_i$ approximates φ_i.

The next Amplitude Estimation theorem estimates the success probability of an algorithm.

Theorem 7 [BHMT00]. *Suppose that an algorithm A has success probability $p < 1 - \Omega(1)$, then there exists an algorithm B running A exactly M times to output a number p' satisfying that*

$$|p' - p| \le O\Big(\frac{\sqrt{p}}{M} + \frac{1}{M^2}\Big).$$

Algorithm QLSR

Input: $A \in \mathbb{R}^{n\times n}$, $b \in \mathbb{R}^n$. A is Hermitian with spectral decomposition $A = \sum_{i=1}^n \lambda_i|v_i\rangle\langle v_i|$, where all the eigenvalues $\lambda_1, ..., \lambda_n$ satisfy $\frac{1}{\kappa} \le |\lambda_i| \le 1$ for $i = 1, ..., r$ for some known value κ and $\lambda_i = 0$ for $i = r+1, ..., n$. Suppose that $b = \sum_{i=1}^n \beta_i|v_i\rangle$.

Output: A quantum state proportional to $|\tilde{x}\rangle$ where $\tilde{x} \approx x^* \stackrel{\text{def}}{=} A^+ b$, and a value $\ell \approx \|x^*\|_2^2$.

Algorithm:

1. Prepare the quantum state $|b\rangle = \frac{1}{\|b\|_2}\sum_{i=1}^n \beta_i|v_i\rangle$.

2. Perform phase estimation to create the state $\frac{1}{\|b\|_2}\sum_{i=1}^n \beta_i|v_i\rangle|\tilde{\lambda}_i\rangle$, where $\tilde{\lambda}_i$ is the estimated value of λ_i satisfying $|\tilde{\lambda}_i - \lambda_i| \le \delta_{\text{PE}} \stackrel{\text{def}}{=} \frac{\epsilon}{2\kappa}$ for $i = 1, ..., n$.

3. Add a qubit $|0\rangle$ to the state and perform a controlled rotation as follows. If $\tilde{\lambda}_i \ge \frac{1}{2\kappa}$, rotate the qubit to $(\frac{1}{2\kappa\tilde{\lambda}_i}|1\rangle + \sqrt{1 - \frac{1}{4\kappa^2\tilde{\lambda}_i^2}}|0\rangle)$; otherwise do nothing. The resulting state is

$$\frac{1}{\|b\|_2}\sum_{i=1}^r \beta_i|v_i\rangle|\tilde{\lambda}_i\rangle\left(\frac{1}{2\kappa\tilde{\lambda}_i}|1\rangle + \sqrt{1 - \frac{1}{4\kappa^2\tilde{\lambda}_i^2}}|0\rangle\right) + \frac{1}{\|b\|_2}\sum_{i=r+1}^n \beta_i|v_i\rangle|\tilde{\lambda}_i\rangle|0\rangle.$$
(6)

4. Use amplitude amplification (by repeating the previous steps $O(\kappa^2/\epsilon)$ times) to boost the amplitude squared for $|1\rangle$ (in the last qubit) to be at least 0.99.

5. Measure the last qubit.

6. **if** we observe $|1\rangle$,
   (a) The remaining state is proportional to $\sum_{i=1}^r \frac{\beta_i}{\tilde{\lambda}_i}|v_i\rangle|\tilde{\lambda}_i\rangle$.
   (b) Reverse the phase estimation process and get the state proportional to $\sum_{i=1}^r \frac{\beta_i}{\tilde{\lambda}_i}|v_i\rangle = |\tilde{x}\rangle$ as our output.
   else output 0 as an estimate to $|x^*\rangle$.

7. Use amplitude estimation to get an estimate p' to the probability p of observing $|1\rangle$ when measuring the state in Eq. (6), to precision $\delta = \epsilon/(4\kappa^2)$ and with success probability 0.99. Output $\ell = p' \cdot 4\|b\|_2^2\kappa^2$.

3   Quantum Algorithm for LSR

In this section, we present our quantum approximation algorithm QLSR for Least Squares Regression, then analyze its error rate and running time.

Without loss of generality, we can assume that $A \in \mathbb{R}^{n \times n}$ with rank r is Hermitian and $b \in \mathbb{R}^n$. See the full version of this paper for discussions on how to deal with the non-Hermitian case. Recall that our goal is to compute $x^* = A^+ b$. Now we will analyze the precision, error probability and the cost. For convenience, we summarize the parameters here: $\delta_{PE} = \frac{\epsilon}{2\kappa}$ and $\delta = \frac{\epsilon}{4\kappa^2}$. We first analyze the quality of the solution $|\tilde{x}\rangle$.

Lemma 8. *With probability at least 0.99, the outputted vector $\tilde{x}$ satisfies*

$$\|\tilde{x} - x^*\|_2 \leq \epsilon \cdot \max\{\|x^*\|_2, \|b\|_2\}.$$

Proof. First we will show that if we observe $|1\rangle$ in Step 6, then the outputted state (normalized) $|\tilde{x}\rangle$ satisfies $\|\tilde{x} - x^*\|_2 \leq \epsilon \|x^*\|_2$. Indeed, if we observe $|1\rangle$, then the remaining state is proportional to $|\tilde{x}\rangle = \sum_{i=1}^r \frac{\beta_i}{\tilde{\lambda}_i} |v_i\rangle$. Recall that $|x^*\rangle = \sum_{i=1}^r \frac{\beta_i}{\lambda_i} |v_i\rangle$. Thus

$$
\begin{aligned}
\|\tilde{x} - x^*\|_2^2 &= \sum_{i=1}^r \left(\frac{\beta_i}{\tilde{\lambda}_i} - \frac{\beta_i}{\lambda_i} \right)^2 = \sum_{i=1}^r \frac{\beta_i^2}{\lambda_i^2} \left(1 - \frac{\lambda_i}{\tilde{\lambda}_i} \right)^2 = \sum_{i=1}^r \frac{\beta_i^2}{\lambda_i^2} \frac{(\tilde{\lambda}_i - \lambda_i)^2}{\tilde{\lambda}_i^2} \\
&\leq \left(\frac{\delta_{PE}}{\frac{1}{\kappa} - \delta_{PE}} \right)^2 \|x^*\|_2^2 \leq \epsilon^2 \|x^*\|_2^2.
\end{aligned}
\tag{7}
$$

Next we show that if we do not observe $|1\rangle$, then the outputted 0 vector is still a good estimation to $|x^*\rangle$, because $|x^*\rangle$ itself is too short. More precisely, define $\rho = \frac{1}{\|b\|_2^2} \sum_{i=1}^r \beta_i^2$, the fraction of $|b\rangle$ falling into the non-zero eigenspace of A. Note that the probability of observing $|1\rangle$ in Step 6 is

$$
p = \frac{1}{\|b\|_2^2} \sum_{i=1}^r \frac{\beta_i^2}{4\kappa^2 \tilde{\lambda}_i^2} = \frac{1}{4\kappa^2 \|b\|_2^2} \sum_{i=1}^r \frac{\beta_i^2}{\tilde{\lambda}_i^2} \geq \frac{1}{4\kappa^2 \|b\|_2^2} \sum_{i=1}^r \beta_i^2 = \frac{1}{4\kappa^2} \rho.
$$

If $\rho \geq \epsilon^2/\kappa^2$, then $p \geq \epsilon^2/(4\kappa^4)$, thus the amplitude amplification already boosts the probability to 0.99 with $O(\kappa^2/\epsilon)$ repetitions, enabling us to observe $|1\rangle$ almost for sure. When $\rho < \epsilon^2/\kappa^2$, if we observe $|1\rangle$, then Eq. (7) still holds. If we observe $|0\rangle$ and output 0 as an estimate to x^*, then the error is

$$
\|0 - x^*\|_2^2 = \|x^*\|_2^2 = \sum_{i=1}^r \frac{\beta_i^2}{\lambda_i^2} \leq \kappa^2 \rho \|b\|_2^2 < \epsilon^2 \|b\|_2^2.
$$

Next we analyze the estimated norm.   □

Lemma 9. *With probability at least 0.99, the outputted value ℓ satisfies*

$$\left| \ell - \|x^*\|_2^2 \right| \leq \epsilon \left(\|x^*\|_2^2 + \|b\|_2^2 \right).$$

Proof. Recall that $\ell = p' \cdot 4\|b\|_2^2 \kappa^2$, and $|p - p'| \leq \delta$.

$$\left|\ell - \|x^*\|_2^2\right| \leq \left|p \cdot 4\|b\|_2^2 \kappa^2 - \|x^*\|_2^2\right| + \delta \cdot 4\|b\|_2^2 \kappa^2$$

$$= \left|\sum_{i=1}^{r} \left(\frac{\beta_i^2}{\tilde{\lambda}_i^2} - \frac{\beta_i^2}{\lambda_i^2}\right)\right| + \delta \cdot 4\|b\|_2^2 \kappa^2$$

Using the fact that $\lambda_i \geq 1/\kappa$ and that $|\lambda_i - \tilde{\lambda}_i| \leq \delta_{\text{PE}}$, it is not hard to see that

$$\left|\sum_{i=1}^{r} \left(\frac{\beta_i^2}{\tilde{\lambda}_i^2} - \frac{\beta_i^2}{\lambda_i^2}\right)\right| \leq 2\kappa\delta_{\text{PE}} \sum_{i=1}^{r} \frac{\beta_i^2}{\lambda_i^2} = 2\kappa\delta_{\text{PE}}\|x^*\|_2^2 = \epsilon\|x^*\|_2^2.$$

Since $\delta = \epsilon/4\kappa^2$, we have $\delta \cdot 4\|b\|_2^2 \kappa^2 = \epsilon\|b\|_2^2$. Thus $\left|\ell - \|x^*\|_2^2\right| \leq \epsilon\left(\|x^*\|_2^2 + \|b\|_2^2\right)$. □

The error probability is a small constant as guaranteed by the error rate of phase estimation, amplitude amplification, and amplitude estimation. Finally let us analyze the cost. For Step 1, we can efficiently prepare $|b\rangle$ in time $O(\log n)$ provided that b_i $(i = 1, ..., n)$ and $\sum_{i_1}^{i_2} |b_i|^2 (1 \leq i_1 < i_2 \leq n)$ are efficiently computable by using the procedure of [GR02].

For Step 2, we perform quantum phase estimation by simulating e^{iAt}, which takes time $O((\log n)s^2)$ if A has at most s non-zero entries each row and is efficiently row computable by the results in [BACS07]. In order that the eigenvalue estimation has error at most $\delta_{\text{PE}} = \frac{\epsilon}{2\kappa}$, the phase estimation algorithm needs $O(\kappa/\epsilon)$ time.

For Step 4, the original probability of seeing $|1\rangle$ is $p = \frac{1}{\|b\|_2^2} \sum_{i=1}^{r} \frac{\beta_i^2}{4\kappa^2 \lambda_i^2}$. Recall that $p \geq \rho/(4\kappa^2)$. To boost this probability to a constant (say, 0.99) it needs to repeat the previous procedure $\sqrt{1/p} \leq \sqrt{4\kappa^2/\rho} = O(\kappa^2/\epsilon)$ when $\rho \geq \epsilon^2/\kappa^2$. Therefore, if we do not need to estimate the norm $\|x^*\|_2$, then the algorithm can just stop before Step 7. The total time cost is $O((\log n) \cdot s^2 \cdot \frac{\kappa}{\epsilon} \cdot \frac{\kappa^2}{\epsilon}) = O((\log n)s^2\kappa^3/\epsilon^2)$.

To estimate the norm $\|x^*\|_2$, the Amplitude Estimation needs to repeat Steps 1 to 3 at most $O(1/\delta) = O(\kappa^2/\epsilon)$ times. So the total cost is $O((\log n) \cdot s^2 \cdot \frac{\kappa}{\epsilon} \cdot \frac{\kappa^2}{\epsilon}) = O((\log n)s^2\kappa^3/\epsilon^2)$.

This completes the proof of Theorem 1.

Our quantum algorithms for the two extensions, Ridge Regression and Truncated SVD problem can be found in the full version of the paper.

4 Quantum Algorithm for Calculating Statistic Leverage Scores and Matrix Coherence

In this section, we present quantum algorithms for calculating statistic leverage scores and matrix coherence, and analyze their performance. Given an $n \times p$ matrix A of rank r with SVD $A = U\Sigma V^T$ where $U \in \mathbb{R}^{n \times r}$, $\Sigma \in \mathbb{R}^{r \times r}$

and $V \in \mathbb{R}^{p \times r}$, the statistic leverage scores of A are defined as $s_i = \|U_i\|_2^2$, $i \in \{1, ..., n\}$, and the matrix coherence c is the largest statistic leverage score of A.

Without loss of generality, we assume that A is Hermitian. (See the full version of this paper for the detailed technique to deal with the non-Hermitian case.) We have the following quantum algorithm for calculating the k-th statistic leverage score of A, s_k for any index $k \in [n]$. Denote the k-th computational basis by $|e_k\rangle$, which has the form $e_k = \sum_{i=1}^n \beta_i |v_i\rangle$ as a decomposition into A's eigenvectors $|v_i\rangle$.

Algorithm QSLS

Input: $A \in \mathbb{R}^{n \times n}$, $k \in [n]$. A is Hermitian with rank r and spectral decomposition $A = \sum_{i=1}^n \lambda_i |v_i\rangle\langle v_i|$. The eigenvalues $\lambda_1, ..., \lambda_n$ satisfy $\frac{1}{\kappa} \leq |\lambda_i| \leq 1$ for $i \leq r$ and $\lambda_i = 0$ for $i > r$. Suppose that $e_k = \sum_{i=1}^n \beta_i |v_i\rangle$. **Output**: A value $\tilde{s}_k \approx s_k$.

Algorithm:

1. Prepare the quantum state proportional to $e_k = \sum_{i=1}^n \beta_i |v_i\rangle$.
2. Perform phase estimation to create the state proportional to $\sum_{i=1}^n \beta_i |v_i\rangle |\tilde{\lambda}_i\rangle$, where $\tilde{\lambda}_i$ is the estimated value of λ_i satisfying $|\tilde{\lambda}_i - \lambda_i| \leq \epsilon_{\mathrm{PE}} \overset{\text{def}}{=} \frac{1}{3\kappa}$ for $i = 1, ..., n$.
3. Add one qubit $|0\rangle$ to the state and perform a controlled rotation as follows. If $\tilde{\lambda}_i \geq \frac{1}{2\kappa}$, rotate the qubit to $|1\rangle$; otherwise do nothing. The resulting state is proportional to

$$\sum_{i=1}^r \beta_i |v_i\rangle |\tilde{\lambda}_i\rangle |1\rangle + \sum_{i=r+1}^n \beta_i |v_i\rangle |\tilde{\lambda}_i\rangle |0\rangle. \tag{8}$$

4. Measure the last qubit.
5. Use Amplitude Estimation to get an estimate p' to the probability p of observing $|1\rangle$ when measuring the state in Eq. (8) to precision ϵ. Output $\tilde{s}_k = p'$.

Now we analyze the running time and performance. For Step 1, the state needed is now $|e_k\rangle$ which is trivially easy to prepare. For Step 2, we perform quantum phase estimation by simulating e^{iAt}, which takes time $O((\log n)s^2)$ if A has at most s non-zero entries each row and is efficiently row computable by the results in [BACS07]. In order that the eigenvalue estimation has error $\leq \epsilon_{\mathrm{PE}} = \frac{1}{3\kappa}$, i.e. $|\tilde{\lambda}_i - \lambda_i| \leq \epsilon_{\mathrm{PE}}, \forall i = 1, ..., n$, we need to run the procedure $O(\kappa)$ times.

To estimate the value p, the Amplitude Estimation needs to repeat Steps 1 to 4 at most $O(1/\epsilon)$ times. So the total cost is $O((\log n)s^2) \times O(\kappa) \times O(\frac{1}{\epsilon}) = O(\log(n)s^2\kappa/\epsilon)$.

The probability of seeing $|1\rangle$ in Step 4 of Algorithm QSLS is $p = \sum_{i=1}^r \beta_i^2$, and we have that

$$s_k = \|U_k\|_2^2 = \|e_k^T U\|_2^2 = e_k^T U U^T e_k = e_k^T A A^\dagger e_k = \sum_{i=1}^r \beta_i^2 = p.$$

Since $\tilde{s}_k = p'$, and $|p' - p| \leq \epsilon$, we have that $|\tilde{s}_k - s_k| \leq \epsilon$. This proves Theorem 4. Using a simple amplitude amplification, we easily get Corollary 5.

References

[Amb12] Ambainis, A.: Variable time amplitude amplification and quantum algorithms for linear algebra problems. In: Proceedings of the 29th International Symposium on Theoretical Aspects of Computer Science, pp. 636–647 (2012)

[BACS07] Berry, D.W., Ahokas, G., Cleve, R., Sanders, B.C.: Efficient quantum algorithms for simulating sparse hamiltonians. Commun. Math. Phys. **270**(2), 359–371 (2007)

[BDM99] Bužek, V., Derka, R., Massar, S.: Optimal quantum clocks. Phys. Rev. Lett. **82**(10), 2207 (1999)

[BHMT00] Brassard, G., Hoyer, P., Mosca, M., Tapp, A.: Quantum amplitude amplification and estimation. arXiv preprint quant-ph/0005055 (2000)

[BMD09] Boutsidis, C., Mahoney, M.W., Drineas, P.: An improved approximation algorithm for the column subset selection problem. In: Proceedings of the twentieth Annual ACM-SIAM Symposium on Discrete Algorithms, pp. 968–977. Society for Industrial and Applied Mathematics (2009)

[CEMM97] Cleve, R., Ekert, A., Macchiavello, C., Mosca, M.: Quantum algorithms revisited (1997). arXiv preprint quant-ph//9708016

[CH+86] Chatterjee, S., Hadi, A.S., et al.: Influential observations, high leverage points, and outliers in linear regression. Stat. Sci. **1**(3), 379–393 (1986)

[CJS13] Clader, B.D., Jacobs, B.C., Sprouse, C.R.: Preconditioned quantum linear system algorithm. Phys. Rev. Lett. **110**, 250–504 (2013)

[CLL+14] Chen, S., Liu, Y., Lyu, M., King, I., Zhang, S.: Fast relative-error approximation algorithm for ridge regression. (Submitted 2014)

[CR09] Candès, E.J., Recht, B.: Exact matrix completion via convex optimization. Found. Comput. Math. **9**(6), 717–772 (2009)

[CW13] Clarkson, K.L., Woodruff, D.P.: Low rank approximation and regression in input sparsity time. In: Proceedings of the 45th Annual ACM Symposium on Symposium on Theory of Computing, pp. 81–90. ACM (2013)

[DMIMW12] Drineas, P., Magdon-Ismail, M., Mahoney, M.W., Woodruff, D.P.: Fast approximation of matrix coherence and statistical leverage. J. Mach. Learn. Res. **13**(1), 3475–3506 (2012)

[DMM08] Drineas, P., Mahoney, M.W., Muthukrishnan, S.: Relative-error cur matrix decompositions. SIAM J. Matrix Anal. Appl. **30**(2), 844–881 (2008)

[DMMS11] Drineas, P., Mahoney, M.W., Muthukrishnan, S., Sarlós, T.: Faster least squares approximation. Numer. Math. **117**(2), 219–249 (2011)

[GHO99] Golub, G.H., Hansen, P.C., O'Leary, D.P.: Tikhonov regularization and total least squares. SIAM J. Matrix Anal. Appl. **21**(1), 185–194 (1999)

[GR02] Grover, L., Rudolph, T.: Creating superpositions that correspond to efficiently integrable probability distributions. arXiv preprint quant-ph/0208112 (2002)

[Han87] Hansen, P.C.: The truncated SVD as a method for regularization. BIT Numer. Math. **27**(4), 534–553 (1987)

[HHL09] Harrow, A.W., Hassidim, A., Lloyd, S.: Quantum algorithm for linear systems of equations. Phys. Rev. Lett. **103**(15), 150502 (2009)

[HW78] Hoaglin, D.C., Welsch, R.E.: The hat matrix in regression and anova. Am. Stat. **32**(1), 17–22 (1978)

[Kit95] Kitaev, A.Y.: Quantum measurements and the abelian stabilizer problem. arXiv preprint quant-ph/9511026 (1995)

[MD09] Mahoney, M.W., Drineas, P.: CUR matrix decompositions for improved data analysis. Proc. Nat. Acad. Sci. **106**(3), 697–702 (2009)

[Mos09] Mosca, M.: Quantum algorithms. In: Meyers, R.A. (ed.) Encyclopedia of Complexity and Systems Science. Springer, New York (2009)

[NN13] Nelson, J., Nguyên, H.L.: Osnap: Faster numerical linear algebra algorithms via sparser subspace embeddings. In: 2013 IEEE 54th Annual Symposium on Foundations of Computer Science (FOCS), pp. 117–126. IEEE (2013)

[Sar06] Sarlos, T.: Improved approximation algorithms for large matrices via random projections. In: 47th Annual IEEE Symposium on Foundations of Computer Science, FOCS 2006, pp. 143–152. IEEE (2006)

[SGV98] Saunders, C., Gammerman, A., Vovk, V.: Ridge regression learning algorithm in dual variables. In: (ICML-1998) Proceedings of the 15th International Conference on Machine Learning, pp. 515–521. Morgan Kaufmann (1998)

[She94] Shewchuk, J.R.: An introduction to the conjugate gradient method without the agonizing pain (1994)

[Sto10] Stothers, A.J.: On the complexity of matrix multiplication (2010)

[Tik63] Tikhonov, A.: Solution of incorrectly formulated problems and the regularization method. Soviet Math. Dokl. **5**, 1035–1038 (1963)

[TR10] Talwalkar, A., Rostamizadeh, A.: Matrix coherence and the nystrom method. arXiv preprint arXiv:1004.2008 (2010)

[WBL12] Wiebe, N., Braun, D., Lloyd, S.: Quantum algorithm for data fitting. Phys. Rev. Lett. **109**, 050505 (2012)

[ZWSP08] Zhou, Y., Wilkinson, D., Schreiber, R., Pan, R.: Large-scale parallel collaborative filtering for the netflix prize. In: Fleischer, R., Xu, J. (eds.) AAIM 2008. LNCS, vol. 5034, pp. 337–348. Springer, Heidelberg (2008)

A New Distributed Algorithm for Computing a Dominating Set on Grids

Photchchara Pisantechakool[1]([✉]) and Xuehou Tan[2]

[1] School of Science and Technology, Tokai University, 4-1-1 Kitakaname,
Hiratsuka 259-1292, Japan
3btad008@mail.tokai-u.jp
[2] School of Information Science and Technology, Tokai University, 4-1-1 Kitakaname,
Hiratsuka 259-1292, Japan

Abstract. This paper presents a new distributed algorithm that computes a dominating set of size $\lfloor \frac{(m+2)(n+2)}{5} \rfloor - 3$ on an $m \times n$ grid, $m, n \geq 8$. This improves upon the previous distributed algorithm of Fata et al. by 4 on the size of the found dominating set. Our result is obtained by exploring new distributed techniques for corner handling. Also, we point out an error in the termination stage of Fata et al.'s algorithm and give a corrected termination method. Our algorithm finds applications in robotics and sensor networks.

Keywords: Grid domination problem · Dominating set · Distributed grid domination problem · Sensor network

1 Introduction

The problem of finding a dominating set for a graph is a well-studied problem in graph theory, and has many potential applications in sensor networks and swarm robots, as well as routing problems in mobile networks. The dominating set problem [6] is a graph problem where every vertex of a given graph $G = (V, E)$ must be either in a dominating set $U \subseteq V$ or adjacent to a member of the dominating set, and the goal is to find a smallest set U in the graph G. For path graphs and trees, a linear-time algorithm to find a dominating set has been given [3].

Finding a domination number (i.e., the size of a smallest dominating set) of an arbitary graph is NP-hard [6], and planar graph is also proven to be NP-hard. Grid graphs, which lie in a class of planar graph, have a special structure that allows their domination number to be determined optimally. For $m \times n$ grid, the size of the optimal dominating set was unknown until recently, but the upper bound of $\lfloor \frac{(m+2)(n+2)}{5} \rfloor - 4$ was shown in [4]. It has also been shown that the lower bound of domination number is equal to the upper bound for $m, n \geq 16$, thus characterizing the domination number of grids [7].

This work was partially supported by the Grant-in-Aid (MEXT/JSPS KAKENHI 15K00023).

© Springer International Publishing Switzerland 2015
J. Wang and C. Yap (Eds.): FAW 2015, LNCS 9130, pp. 217–228, 2015.
DOI: 10.1007/978-3-319-19647-3_21

Previous efforts were focused on the problem of computing the dominating numbers for grids [1,4,6,7]. Two previous works for computing a dominating set were Chang's doctoral thesis [4] and Fata et al.'s conference paper [5]. Chang's method is constructive, and one can simply derive from his method to give a centralized algorithm so as to find a dominating set of optimal size $\lfloor \frac{(m+2)(n+2)}{5} \rfloor -$ 4. A distributed algorithm was given in [5], which computes a dominating set of size $\lceil \frac{(m+2)(n+2)}{5} \rceil$.

It should be pointed out that the algorithm of Fata et al. [5] is incomplete in its termination stage. A set of agents is initially located at vertices of the grid. The number of agents may be larger than $\lceil \frac{(m+2)(n+2)}{5} \rceil$, and some agents may even be at the same grid vertex. The agents have three modes: (a) sleep, (b) active, and (c) settled. All the agents, in the sleep mode at the beginning, will activate in a randomized or previously scheduled manner. The very first agent becomes settled just at its original vertex. Each active agent can communicate with the settled agents so as to find the place (vertex) where it becomes settled. As soon as settled agents no longer have to communicate, each settled agent goes back to sleep mode. The remaining non-activated agents are required to leave the grid afterwards, but the final operation is NOT a distributed one (see page 5 of [5]). Following their algorithm, the agent being active after the dominating set is found will simply restart the algorithm again, due to the fact that she cannot know whether the dominating set has already been found.

The goal of this paper is to give a new algorithm to compute a dominating set on grids in a distributed manner (that can terminate correctly). We first define a distributed system model. In particular, each agent is equipped with an answering machine that can record a broadcast message at a time, which is the most updated message. This makes it possible to let the remaining non-activated agents leave the grid. Next, we explore the techniques of Chang's corner handling so that they can work in the distributed system. Our distributed algorithm can produce a dominating set of size $\lfloor \frac{(m+2)(n+2)}{5} \rfloor - 3$, which improves upon the previous result [5] by 4. This is the best result to our knowledge.

Distributed grid domination algorithms can be adopted by distributed systems for many applications. For example, swarm robots equipped with short-ranged landmine detection devices can be deployed from an airplane into a designated area (considered as a grid). These robots can move to align themselves in optimized formation to maximize the coverage on their own without having to manually control them or using the centralized system.

In Sect. 2 of this paper, we introduce the essential definitions and notation for the dominating set problem on grids. In Sect. 3 we briefly revisit Chang's centralized constructive method. The tools in Sect. 3 are explored in Sect. 4 to give a distributed algorithm. Finally, Sect. 5 concludes the paper and poses some open problems.

2 Preliminaries

A graph $G = (V, E)$ is defined as a set of vertices V connected by a set of edges $E \subseteq V \times V$. We assume the graph is undirected, i.e., $(v, u) \in E \leftrightarrow (u, v) \in E, \forall v, u \in V$.

A vertex $u \in V$ is defined as a neighbour of vertex $v \in V$, if $(u, v) \in E$. The set of all neighbours of vertex v is denoted by $N(v)$. For a subset $U \subseteq V$, we define $N(U)$ as $\bigcup_{u \in U} N(u)$. For a subset $U \subseteq V$, we say the vertices in $N(U)$ are *dominated* by the vertices in U. For graph G, a set of vertices $S \subseteq V$ is a dominating set if each vertex $v \in V$ is either in S or is dominated by S.

A dominating set with minimum cardinality is called an optimal dominating set of a graph G; its cardinality is called the domination number of G and is denoted by $\gamma(G)$. Note that although the domination number of a graph, $\gamma(G)$, is unique, there may be different optimal dominating sets [1]. An $m \times n$ grid graph $G = (V, E)$ is defined as a graph with vertex set $V = \{v_{i,j} | 1 \leq i \leq m, 1 \leq j \leq n\}$ and edge set $E = \{(v_{i,j}, v_{i,j'}) ||j - j'| = 1\} \bigcup \{(v_{i,j}, v_{i',j}) ||i - i'| = 1\}$ [2]. For ease of presentation, we will fix an orientation and labelling of the vertices, so that vertex $v_{0,0}$ is the lower-left vertex and vertex $v_{m-1,n-1}$ is the upper-right vertex of the grid. In this paper we will include super-grid in grid indices. We denote the domination number of an $m \times n$ grid G by $\gamma_{m,n} = \gamma(G)$.

Theorem 1 *(Gonçalves et al., [7]). For an $m \times n$ grid with $16 \leq m \leq n, \gamma_{m,n} = \lfloor \frac{(m+2)(n+2)}{5} \rfloor - 4$*

Definition 1 *(Grid Boundary). For an $m \times n$ grid $G = (V, E)$, we define the boundary of G, denoted by $B(G)$, as the set of vertices with less than 4 neighbours.*

Definition 2 *(Sub-Grids and Super-Grids). An $m \times n$ grid $G = (V, E)$ is called a sub-grid of an $m' \times n'$ grid $G' = (V', E')$ if G is induced by vertices $v'_{i,j} \in V'$, where $1 \leq i \leq m' - 2$ and $1 \leq j \leq n' - 2$. If G is a sub-grid of G', G' is called the super-grid of G (see Fig. 1).*

Definition 3 *(Optimal Grid Pattern). A subset $U \subseteq V$ constitutes an optimal grid pattern on grid $G = (V, E)$ if there exists a fixed $r \in 0, 1, 2, 3, 4$ such that for any vertex $v_{x,y} \in U$ we have $x - 2y \equiv r (mod\ 5)$.*

One can also define an optimal grid pattern as a set of vertices whose (x, y) coordinates satisfy $y - 2x \equiv r(\mathrm{mod}\ 5)$, for some fixed r. This corresponds to swapping the x and y axes. For the proofs we only analyze the case mentioned above; the other case can be treated similarly.

Definition 4 *(Grid Optimization). A subset $U \subseteq V$ optimizes grid $G = (V, E)$ if it constitutes an optimal grid pattern and there exists no vertex $v \in V \setminus U$ that can be added to U so that U remains an optimal grid pattern. See Fig. 1(a).*

Definition 5 *(Orphans). Let $U \subseteq V$ be a set of vertices that optimizes grid $G = (V, E)$. A vertex $v \in V$ that has no neighbour in U is called an orphan (see Fig. 1(a)).*

Definition 6 *(Projection). Consider a grid $G = (V, E)$ and its super-grid $G' = (V', E')$. For a set $U' \subseteq V'$, its projection is defined as the set $U'' = (N(U' \setminus V) \cup U') \cap V$. Similarly, we say a vertex $v \in U' \setminus V$ is projected if it is mapped to its neighbour in V. See Fig. 1.*

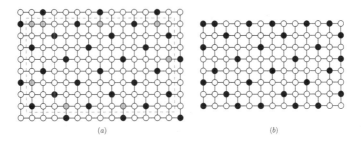

Fig. 1. In (a), a 15×10 grid G' is demonstrated and its 13×8 sub-grid G is highlighted in dashed square. Orphans are shown in grey vertices. G' is optimized by a set of U' of 30 vertices. In (b), vertices in $U' \setminus V$ are projected onto their neighbors in G

Definition 7 *(Slot). Given a $m \times n$ grid $G = (V, E)$, let $v_{i,j} \in V$ be a vertex occupied by a settled agent. The four vertices $\{v_{i+2,j+1}, v_{i-1,j+2}, v_{i-2,j-1}, v_{i+1,j-2}\}$ from $v_{i,j}$ within the boundary of G are called slots of the settled agent occupying $v_{i,j}$.*

Definition 8 *(Pseudo-Slot). When an agent settles at corner points, the agent may assign a vertex as pseudo-slot. Pseudo-slots have the same priority as slots when an active agent seeks a location to occupy, but the agent who settles there will not calculate its slots and instead go into sleep mode. See Fig. 6.*

3 Chang's Centralized Constructive Method Revisited

In this section we revisit Chang's centralized constructive method that can produce a dominating set of size $\lfloor \frac{(m+2)(n+2)}{5} \rfloor - 4$ for an $m \times n$ grid, $m, n \geq 8$. Chang's constructive method consists of the following three main ideas:

(i) Initialization: At this step, a subset $U' \subseteq V'$ that optimizes the super-grid G' is provided. Basically, it can select the smaller between two permutations (of super-grid) of the optimal grid pattern. See Fig. 1(a).

(ii) Corner Handling: Each corner (i.e., a 5×5 portion) of the super-grid has one vertex removed from U', and the vertices around four corners of the super-grid are moved into the original grid. See Fig. 3.

(iii) Projection: Using a process called projection, the vertices in $U' \setminus V$ except for four corners are characterized and put into the original grid G.

Initialization: As stated in Definition 3 and proven in [4], for a given $m \times n$ grid graph, there exist some r in $x - 2y \equiv r \pmod 5$ such that $|S| \leq \lfloor \frac{(m+2)(n+2)}{5} \rfloor$. Careful observation showed that the optimal grid pattern repeats itself every 5×5 block. It is known that when the pattern is shifted around in one super-grid, it produces different number of the dominating vertices. There are five disjoint permutations of the pattern, based on $r \in \{0, 1, 2, 3, 4\}$ in Definition 3. For one grid size, some permutations produce smaller number of dominating vertices, but when the size changed, others may produce smaller number.

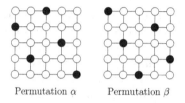

Permutation α   Permutation β

Fig. 2. Permutation α and Permutation β

For a known size of grid, picking a suitable permutation can further reduce the number of elements. We can simply use two permutations whose maximum sizes of produced sets in all possible grid size do not overlap.[1] That is, one can always choose smaller number of dominating vertices between the two and get the minimum size. We refer to these two permutation as Permutation α and β (shown in Fig. 2). The derived algorithm uses these two disjoint permutations in Initialization step.

Corner Handling and **Projection:** One can further reduce the number of dominating vertices around the corners of a grid. Recall that using Projection (Definition 6) from super-grid, we can dominate the orphans (as stated in Definition 5). For grid with large size, the vertices on the boundary that are not a part of a corner must be dominated by projected vertices. However, at each corner, some elements overlap and the placement is not ideal.

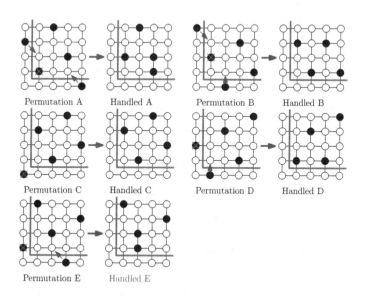

Permutation A   Handled A   Permutation B   Handled B

Permutation C   Handled C   Permutation D   Handled D

Permutation E   Handled E

Fig. 3. Handling each corner's permutation

---

[1] Note also that Chang's constructive method has to choose from five different permutations based on input grid size.

By performing Corner Handling step before Projection step, we can reduce one vertex at each corner before moving all the vertices at boundary of super-grid and while doing so, move vertices around the corner to the original grid. Chang's case-based handling method considers each corner as a 5×5 block and handles each permutation differently, as shown in Fig. 3.

Lastly, as previously stated in Definition 6, we use projection process to move the remaining vertices on the boundary of super-grid to its sub-grid (the original grid), and dominating the remaining orphans.

4   Distributed Grid Domination

In this section, we first introduce the distributed system model. (A prototype model can be found in [5].) Next, we show that Chang's centralized constructive method, especially the Corner Handling Step, can be extended so as to compute a dominating set of size $\lfloor \frac{(m+2)(n+2)}{5} \rfloor - 3$ for a given $m \times n$ grid in a distributed manner, $m, n \geq 8$.

4.1   Model and Notation

Assume that the environment is an $m \times n$ grid $G = (V, E)$ with $m, n \in \mathbb{N}$. The goal is to dominate the grid environment in a distributed fashion using several robots (or agents) without any knowledge of environment size. At the start, there exist k agents in the environment, where k can be smaller or greater than the number of agents needed to dominate the grid. The following assumptions are made for the grid and agents.

Assumption A: Agents can be located only on the vertices of the grid, and can move between grid vertices only on the edges of the grid. More than one agent can be at the same vertex at any given time. We refer to the vertices using standard Cartesian coordinates defined in Sect. 2[2].

Assumption B: The agents, denoted by $a_1, ..., a_k$ are initially located at arbitrary vertices on the grid. The agents have three modes; (a) sleep, (b) active, and (c) settled. The sleep mode in this algorithm means that the agent will not contribute to the distributed algorithm, but can still perform other unrelated functions, such as detecting intruders or landmines.

Assumption C: Each agent is equipped with an answering machine that can record a broadcast message at a time when the agent is in sleep mode. At the beginning of the procedure, all agents are in sleep mode. During each epoch, or time interval with specific length, one agent goes into active mode. The activation sequence of agents is arbitrary; it can be scheduled in advance, or randomized, but each agent will activate at fix length of time after the previous one.

Assumption D: Agents in active mode and settled mode can communicate. The active agent can communicate with the settled agents to perform the distributed

---

[2] We include vertices of super-grid in labelling for the ease of presentation.

dominating set algorithm. Once an agent activates and performs its parts, it goes into settled mode. After the settled agents form a dominating set, all of them go back to sleep mode and will not activate again.

Assumption E: Each agent is equipped with suitable bearing sensor (incoming direction) and range sensor to help computing the location of the sender of signal it receives from, in its own local coordinates with itself as origin and an arbitrary orientation. Additionally, agents are equipped with short-ranged proximity sensors to sense the environment boundary. Agents are able to sense the boundary only if they are on a vertex v whose neighbor is a boundary vertex of the grid.

4.2 Outline of Our Distributed Algorithm

The idea of our algorithm is to implement the optimal grid pattern used in the centralized algorithm in a distributed manner, using communications among active and settled agents.

Agents will keep track of surrounding four locations (vertices) that are correctly aligned in optimal grid pattern. These locations are called slots (Definition 7), and any new agents becoming active later will communicate and attempt to occupy these slots to contribute to the optimal grid pattern. If there exists a slot outside the boundary, an agent will keep a location of neighbor vertex whose position is on the boundary instead. These locations are called orphans as defined in Definition 5, and each agent outside the corner area will have at most one orphan tracked at a time. Orphans are occupied by agents after all slots are occupied, which allow us to move around some agents in similar manner to that of centralized grid domination algorithm.

Corollary 1 *(Orphan Number). Given a $m \times n$ grid $G = (V, E)$, for any vertices outside 4 corner points (Fig. 5) of each corner, an agent will have at most one orphan when aligned with optimal grid pattern.*

When an agent activates, it checks the most recent message to see whether there exists a message or not. Initially, there is no message in any agents' short-memory storage in message receptor, the agent then concludes that itself is the first agent to activate. After, the first agent checks for boundary to see whether this initial location is around the corner or not before settle. If an agent finds itself at the corner, it moves to ideal location around the corner instead. If there exists a message but not termination message, the agent then sends a broadcast signal to find the settled agents on the grid who have slots available, and waits for some specified time for response. The new active agent then contributes to optimal grid pattern using information received from settled agents.

We break down the distributed algorithm into three main steps for easier presentation.

(i) Initialization: How the very first activated agent works.
(ii) Settlement: How active and settled agents communicate, and how an active agent gets settled.

(iii) Termination: How the algorithm finishes. Particularly, how the termination condition is verified in the case that the number of agents is larger than necessary to dominate the grid.

4.3   Initialization

As stated in *Assumption C*, the algorithm starts with all agents being in sleep mode. Each agent will activate at certain time apart from one another.

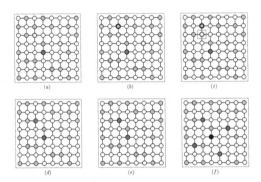

Fig. 4. (a) The first agent settles with four slots. (b) An agent becomes activate. (c) The closest slot is computed. (d) The active agent occupies that slot. (e) The lists of slots are updated. (f) An agent with 4 slots occupied goes into sleep mode.

Each agent acts similarly when it first activates. The first thing each agent does is checking most recently stored message to see if there exists a termination message. We will describe about termination message later in Sect. 4.5. Since at the beginning, no message has been broadcast yet, the agent will not see a single saved message and then conclude that it is the *First Agent* to activate in the system.

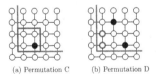

(a) Permutation C   (b) Permutation D

Fig. 5. (a) If the first agent activates at corner points (black square), it will move to designated location (black circle). (b) A termination condition with two orphans.

The first agent has special action to take before entering settled mode. First, the agent must check whether it is at one of the four corner points or not, using functions described in *Assumption E*. If it finds itself in one of the four corner points, as illustrated in Fig. 5(a), it will move to the specified location of that corner. By doing so, we can assume that for any corners, the agents are forming optimal grid pattern from any locations outside the four corner points, or starting a Permutation C corner.

If the first agent activates outside corner points, it will enter settled mode normally. Before an agent enters settled mode, it will make a list of unoccupied locations around itself that aligned in optimal grid pattern (Fig. 4(a)). After an agent enters settled mode, the list of unoccupied slots and orphan is updated when another active agent occupies any of them.

4.4 Settlement

For other agents activating after the first one, they take the following actions; check the recorded message, broadcast for slots, compute for closest location to occupy, then move to occupy and settle at computed location.

Since there is at least one settled agent after the first, the active agents will receive response signal from settled agents. Active agents can contribute to the algorithm by either settle in unoccupied slots, or settle in pseudo-slots or orphans. An active agent will compute for closest settled agent among those that responded and send request for slot list. Chosen settled agent sends out its whole list to active agent, who then computes for closest location of eligible slots. After determining the location to settle using $L^1 norm$ distance, active agent sends out notification that the location will be taken to all settled agents, and travels to the location. Settled agents whose lists hold such location in slot list then remove the location from their own's list. Once a list is empty and no orphan in the surrounding, a settled agent will go into sleep mode (Fig. 4(f)).

Once the active agent reaches the chosen location, it will first check whether it is at corner points or not, in similar fashion to that of the first agent. If an active agent finds itself in one of the corner point, it will take special action, called corner settlement, as shown in Fig. 6. This is the step similar to corner handling in centralized algorithm, but performed in a distributed fashion. For example, if an active agent chooses a corner point $v_{1,1}$, shown as Permutation A in Fig. 6, it will move to new location $v_{1,2}$ and create a list with only one location, called pseudo-slot, shown as $v_{3,1}$ in Fig. 6. A pseudo-slot is considered as a slot by settled agents when responding to request signal, thus both slots and pseudo-slots will be occupied before orphans.

If the chosen location for an active agent is not corner point, the active agent will create a list of slots and orphan normally, similar to that of first agent.

When an active agent receives a location marked as pseudo-slot and chooses it to settle, it will notify settled agents of its choice then move to the location. However, it will not create a list of slots and orphan, and instead go directly into sleep mode.

Eventually all slots and pseudo-slots are occupied, and active agents will not receive any response signal when requesting for slots. Active agents then send out request signal for orphans, and settled agents with orphans in their lists will respond by sending orphan locations. An active agent then computes for closest

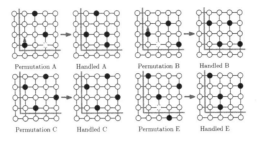

Permutation A   Handled A   Permutation B   Handled B

Permutation C   Handled C   Permutation E   Handled E

Fig. 6. Distributed algorithm's case-based method on how to handle each corner point

location then notifies all settled agents of its choice. Like pseudo-slot, active agents settling at orphans will go directly into sleep mode.

4.5   Termination

Termination of the distributed algorithm normally happens in the following cases.

Case 1: The number of agents is not enough to dominate the grid. After settled agents receive no broadcast signal for a fixed amount of time (longer than an interval of activation sequence), all the settled agents will enter sleep mode, making themselves a subset of dominating set of the grid.

Case 2: The number of agents in the grid is more than enough to dominate the grid. The distributed algorithm will produce a complete dominating set for the grid. In this case, algorithm has three different conditions for last active agent to check.

(a) Only one settled agent responds to an active agent with only one orphan.
(b) Two settled agents respond to an active agent, but both have the same orphan.
(c) Two settled agents respond to an active agent, but both orphans are in the corner (Fig. 6(b)).

If one of the conditions is satisfied, the last active agent will send out the termination message after notifying settled agents that it will occupy the location, and then go directly into sleep mode. The termination message will be the last broadcast message, and stored in answering machine's memory of every agent, including agents that have yet to activate.

Any agent activates after the broadcast will see termination message as the most updated message, and leave the grid without contributing to the distributed algorithm.

4.6   The Algorithm

We now provide a complete algorithm and prove that it is correct and creates dominating set for grid correctly according to initializaion step in centralized algorithm.

Algorithm: DistributedGridDomination

Initialization

1. First agent activates.
2. First agent concludes that it is the first agent because there is no stored message.
3. First agent checks for corner points as described in Sect. 4.3, then moves to designated location as necessary.
4. First agent settles at its current location.

Settlement

1. Other agents activate one by one in uniformly distributed interval. Activated agent checks stored message, making sure that it is not a termination message.
2. Active agents communicate with settled agents to find the closest slot to settle.
   (a) Active agent sends out request signal for slots. Settled agent responds if it has unoccupied slots.
   (b) Active agent computes the closest settled agent, then sends request for a list of all slot locations. Settled agent sends list of unoccupied slots.
   (c) Active agent computes the closest slot then notifies other settled agents. Settled agents remove to-be-occupied location.
   (d) Active agent moves to closest slot, then checks for corner points and proceeds as described in Sect. 4.4.
   (e) **If** active agent is at pseudo-slot location, **then** it goes directly into sleep mode.
   (f) Active agent settles at current location and becomes settled agent, then updates its list of slots.
3. During step 2.(a), **If** active agent receives no response, **then** it sends out request signal for orphans and occupy orphan as stated in Sect. 4.4.

Termination

1. During Settlement step 3., **If** active agent detects any of the following conditions; (i) only one settled agent responds, or (ii) two settled agents respond with the same orphan or two different orphans which are at corner points, **then** it sends out termination message.
2. During Settlement step 2.(a), **If** settled agents receives no request signal for a fixed period of time, **then** the algorithm terminates.
3. During Settlement step 1., **If** active agent sees termination message, **then** it leaves the grid.

Lemma 1. *During the algorithm, the agents occupying non-pseudo, non-orphan slots contribute to the optimal grid pattern correctly.*

Proof. An active agent always settles in slots computed by other settled agents. Since the slots computed by an settled agent are aligned with the settled agent in optimal grid pattern according to Definition 3, as long as the new active agent settles at non-pseudo, non-orphan slots of previously settled agents, it will contribute to the optimal grid pattern correctly.

Theorem 2. *The number of agent used to dominate the grid in our algorithm is bounded to* $\lfloor \frac{(m+2)(n+2)}{5} \rfloor - 3$ *for any* $m \times n$ *grid such that* $m, n \geq 8$.

Proof. Our distributed grid domination algorithm computes a dominating set correctly with the exception that the smaller of two permutations cannot be chosen in the distributed manner. So, the size of the computed dominating agent set is upper-bounded to $\lfloor \frac{(m+2)(n+2)}{5} \rfloor - 3$.

Note that agents may perform their tasks such as traversing the grid in at most $m + n$ steps, and number of agents required to dominate the grid is bounded to $\lfloor \frac{(m+2)(n+2)}{5} \rfloor - 3$, the running time of algorithm can be upper-bounded polynomial time of $O(mn(m + n))$ steps to contruct a dominating set algorithm on an $m \times n$ grid.

5   Conclusions

We presented an algorithm to the problem of finding dominating sets on an $m \times n$ grid where $m, n \geq 8$ in the distributed manner. First, we briefly revisited Chang's centralized algorithm that obtains a dominating set of size $\lfloor \frac{(m+2)(n+2)}{5} \rfloor - 4$. We then presented our distributed algorithm that computes a dominating set of size $\lfloor \frac{(m+2)(n+2)}{5} \rfloor - 3$ using similar methods under the restrictions of the distributed system models.

References

1. Alanko, S., Crevals, S., Isopoussu, A., Östergård, P., Pettersson, V.: Computing the domination number of grid graphs. Electron. J. Comb. **18**(1), no. P141 (2011)
2. Bondy, A., Murty, U.: Graph Theory. Series: Graduate Texts in Mathematics. Springer, London (2008)
3. Chang, G.J.: Algorithmic aspects of domination in graphs. In: Pardalos, P.M., Du, D.Z., Graham, R.L. (eds.) Handbook of Combinatorial Optimization, pp. 221–282. Springer, New York (2013)
4. Chang, T.Y.: Domination numbers of grid graphs. Ph.D. Dissertation, University of South Florida (1992)
5. Fata, E., Smith, S.L., Sundaram, S.: Distributed dominating sets on grids. In: 2013 American Control Conference, pp. 211–216. IEEE Press, New York (2013)
6. Garey, M., Johnson, D.: Computers and intractability: a guide to the theory of NP-completeness. In: Klee, V. (ed.) A Series of Books in the Mathematical Sciences. W. H. Freeman and Company, New York (1979)
7. Gonçalves, D., Pinlou, A., Rao, M., Thomassé, S.: The Domination number of grids. SIAM J. Discrete Math. **25**, 1443–1453 (2011). SIAM

Approximate Model Counting
via Extension Rule

Jinyan Wang[1,2]([✉]), Minghao Yin[3], and Jingli Wu[1,2]

[1] Guangxi Key Lab of Multi-source Information Mining & Security,
Guangxi Normal University, Guilin, China
{wangjy612,wjlhappy}@gxnu.edu.cn
[2] College of Computer Science and Information Technology,
Guangxi Normal University, Guilin, China
[3] School of Computer Science and Information Technology,
Northeast Normal University, Changchun, China
ymh@nenu.edu.cn

Abstract. Resolution principle is an important rule of inference in theorem proving. Model counting using extension rule is considered as a counterpart to resolution-based methods for model counting. Based on the exact method, this paper proposes two approximate model counting algorithms, and proves the time complexity of the algorithms. Experimental results show that they have good performance in the accuracy and efficiency.

Keywords: Propositional satisfiability · Model counting · Resolution principle · Extension rule

1 Introduction

In recent years, there are tremendous improvements in the field of propositional satisfiability (SAT). Many hard combinatorial problems in artificial intelligence and computer science, such as planning problems, have been compiled into SAT instances, and solved effectively by SAT solvers [1–3]. On the other hand, the problem of counting the number of models or satisfying assignments for a Boolean formula (#SAT), is an important extension of satisfiability testing [4]. Recent researches have also shown that model counting corresponds to numerous #P-complete problems such as performing inference in Bayesian networks, probabilistic planning and diagnosis [4–9]. Resolution principle is the rule of inference at the basis of most procedures for both SAT and #SAT [10]. In [11], Birnbaum and Lozinskii directly extended Davis-Putnam (DP) procedure [12] to solve #SAT problems, and proposed a model counter CDP. Based on CDP, Bayardo and Pehoushek introduced the idea of connected component analysis to enhance Relsat's model counting ability [13]. Furthermore, the introduction of component caching and clause learning accelerated greatly the model counting procedure [14]. Since the space requirement was an important concern in

ⓒ Springer International Publishing Switzerland 2015
J. Wang and C. Yap (Eds.): FAW 2015, LNCS 9130, pp. 229–240, 2015.
DOI: 10.1007/978-3-319-19647-3_22

implementing component caching, sharpSAT solver employed advanced component caching mechanism to make components be stored more succinctly [15]. In addition, Bacchus et al. [16] utilized backtracking search to solve #SAT problem. The run time and memory usage of these algorithms often increase exponentially with problem size. Consequently, they are limited to relatively small formulas. In 2005, Wei and Selman presented an approximate model counting algorithm ApproxCount [17] based on SampleSat [18], which sampled from the set of solutions of a Boolean formula near-uniformly. Following the scheme outlined by Jerrum et al. [19], ApproxCount used an exact model counter like Relsat [13] or Cachet [14] to count models of the residual formula after some variables had been set by SampleSat. There are other model counting techniques based on sample method, such as SampleCount [20], SampleMinisat [21]. Kroc et al. [22] utilized belief propagation and MiniSat to design approximate model counting algorithm.

In [23,24], Lin et al. used extension rule to solve SAT problem. The key idea is to use inclusion-exclusion principle to solve the problem. We proposed a model counting algorithm using extension rule [25,26]. We compared the method with resolution-based methods. The more pairs of clauses with complementary literals are, the more efficient the method is, and the less efficient resolution-based methods are. So the method is considered as a counterpart to resolution-based methods for model counting. Furthermore, Bennett and Sankaranarayanan [27] presented a pruning technique to model counting using the inclusion-exclusion principle. Also, Linial et al. [28,29]considered approximate inclusion-exclusion, when intersection sizes are known for only some of the subfamilies, or when these quantities are given to within some error, or both. In this paper, we use extension rule to present two approximate model counting algorithms: ULBApprox and SampleApprox. For ULBApprox, we firstly analyze that under what condition a upper or lower bound of S can be viewed as the approximate value of S (the number of different maximum terms, i.e., the number of unsatisfying assignments), then count the approximate number of models. SampleApprox is an approximate algorithm which combines extension rule with SampleSat algorithm. Experimental results indicate the two approximate algorithms have good performance in the accuracy and efficiency.

The paper is organized as follows. We review the extension rule and describe the model counting method based on extension rule in the next section. In Sect. 3, we show how to count the approximate number of models using extension rule. Some experimental results are reported in Sect. 4. In the last section, we summarize this paper.

2   Model Counting Using Extension Rule

We begin by specifying the notations that will be used in the rest of this paper. Σ denotes a set of clauses or a Boolean formula in conjunctive normal form (CNF), C denotes a single clause, V denotes the set of all variables appearing in Σ, and $M(\Sigma)$ denotes the number of models of Σ.

2.1 Extension Rule

We review the extension rule in brief. The readers are referred to [23] for more details. Given a clause C and a set of variables V, $D = \{C \vee a, C \vee \neg a | a \in V$ and a does not appear in $C\}$. The operation proceeding from C to D is called extension rule on C, and D is the result of extension rule. A clause C is logically equivalent to the result of extension rule D. So extension rule is a legal inference rule.

A clause is a maximum term on a set V if it contains all variables in V in either positive form or negative form. For example, given a set of variables $V = \{a, b, c\}$, $a \vee b \vee c$ is a maximum term on V, but $a \vee b$ is not. Given a set of clauses Σ with its set of variables $V(|V| = v)$, if clauses in Σ are all maximum terms on V, Σ is unsatisfiable if it contains 2^v clauses; otherwise, Σ is satisfiable. Therefore, we can decide the satisfiability of a set of maximum terms.

2.2 Model Counting Using Extension Rule

According to the definition of maximum terms, we can count the number of models of a set of maximum terms. Given a set of clauses Σ with its set of variables $V(|V| = v)$, if the clauses in Σ are all maximum terms on V, the number of models of Σ is $2^v - S$ if Σ contains S distinct clauses, where $S \leq 2^v$.

For example, given a set of clauses $\Sigma = \{a \vee b, a \vee \neg b, \neg a \vee b\}$ with its set of variables $V = \{a, b\}$, the clause $\neg a \vee \neg b$ does not appear in Σ. It is clear the assignment, which a is assigned 1 and b is assigned 1, is a model of Σ. Actually, The number of maximum terms is equal to the number of unsatisfying assignments in a Boolean formula. Therefore, if we want to count models of a set of clauses, we should proceed by finding an equivalent set of clauses such that all the clauses in it are maximum terms by using extension rule. Then, we can know how many models there are. For example, given a set of clauses $\Sigma = \{a \vee \neg b, \neg a, \neg b \vee c\}$ with its set of variables $V = \{a, b, c\}$, the sets of maximum terms $\{a \vee \neg b \vee c, a \vee \neg b \vee \neg c\}$, $\{\neg a \vee b \vee c, \neg a \vee b \vee \neg c, \neg a \vee \neg b \vee c, \neg a \vee \neg b \vee \neg c\}$ and $\{a \vee \neg b \vee c, \neg a \vee \neg b \vee c\}$ are the extended results of the clauses $a \vee \neg b$, $\neg a$ and $\neg b \vee c$ by using extension rule, respectively. Consequently, the union of the three maximum terms $\{a \vee \neg b \vee c, a \vee \neg b \vee \neg c, \neg a \vee b \vee c, \neg a \vee b \vee \neg c, \neg a \vee \neg b \vee c, \neg a \vee \neg b \vee \neg c\}$ is the set of maximum terms generated from Σ by using extension rule. Therefore, we can get the number of models of Σ is $2^3 - 6 = 2$.

In fact, it is sufficient to count the number of all the maximum terms generated by Σ rather than to list them. We can use inclusion-exclusion principle to compute the sum of maximum terms, which generate from Σ by using extension rule. Given a set of clauses $\Sigma = \{C_1, C_2, \ldots, C_n\}$, let V be the set of variables which appear in Σ ($|V| = v$), P_i be the set of all the maximum terms generated from C_i by using extension rule, and S be the sum of all different maximum terms generated from Σ, then we have

$$S = \sum_{i=1}^{n} |P_i| - \sum_{1 \leq i < j \leq n} |P_i \cap P_j| + \sum_{1 \leq i < j < l \leq n} |P_i \cap P_j \cap P_l| - \ldots$$
$$+ (-1)^{n+1} |P_1 \cap P_2 \cap \ldots \cap P_n|. \tag{1}$$

where $|P_i| = 2^{v-|C_i|}$, and $|P_i \cap P_j| = 0$, if there are complementary literals in $C_i \cup C_j$; otherwise, $|P_i \cap P_j| = 2^{v-|C_i \cup C_j|}$. $|C_i \cap C_j|$ denotes the number of variables appearing in $C_i \cup C_j$. In Eq. (1), we call the computation for an absolute value as a term, and the computation for a continuous sum as a sum term.

For example, we count the number of models of the set of clauses $\Sigma = \{a \vee \neg b, \neg a, \neg b \vee c\}$. The number of distinct maximum terms generated from Σ is $S = 2^1 + 2^2 + 2^1 - 0 - 2^0 - 2^0 + 0 = 6$, so the number of models of Σ is $2^3 - 6 = 2$. This process is called model counting using extension rule (MCER) [25,26].

Clearly, the higher the complementary factor is, the more efficient MCER algorithm is. However, the worst-case time complexity of the algorithm is exponential. Furthermore, we note that the value of the first $2k - 1(k = 1, 2, \ldots, \lfloor (n+1)/2 \rfloor)$ sum terms, denoted by S_{2k-1}, is an upper bound of S, and the value of the first $2k(k = 1, 2, \ldots, \lfloor n/2 \rfloor)$ sum terms, denote by S_{2k}, is a lower bound of S in Eq. (1). In this sense, $2^v - S_{2k-1}$ acts as a lower bound of the result; $2^v - S_{2k}$ acts as an upper bound of the result. This enables us to design an approximate model counting algorithm. In the next section, we give the detailed description.

3   Approximate Model Counting Using Extension Rule

In this section, we present two approximate model counting algorithms based on extension rule. At first, we give a method to measure the dispersion between the approximate and exact value, so it is a benchmark to measure the approximate model counting algorithms.

Definition 1. Let S_{accur}, S_{appr} be the exact and approximate value, respectively. We call $\sigma = \frac{|S_{accur} - S_{appr}|}{S_{accur}}$ approximate dispersion[1].

The smaller approximate dispersion σ is, the nearer S_{appr} is to S_{accur}. Otherwise, the farther S_{appr} is to S_{accur}.

3.1   ULBApprox

We analyze firstly in what condition the value of the first $2k - 1$ or $2k$ sum terms in Eq. (1) can be viewed as the approximate value of S. Then we present the approximate model counting algorithm ULBApprox.

---

[1] If σ is very small, then $S_{accur} \approx S_{appr}$. Consequently, $\sigma \approx \sigma' = \frac{|S_{accur} - S_{appr}|}{S_{appr}}$. So σ' is also considered as approximate dispersion. In this paper, we use σ' to get the approximate number of unsatisfying assignments of Σ, because it is difficult to use σ in the condition that we do not know the exact number of unsatisfying assignments of Σ. However, in the experimental results, we obtain the approximate value of models and know the exact number of models, so we use σ to measure the two approximate algorithms.

Theorem 1. *Given a set of clauses $\Sigma = \{C_1, C_2, \ldots, C_n\}$ and a threshold r of approximate dispersion, let P_i be the set of all the maximum terms generated from C_i by using extension rule, where $i = 1, 2, \ldots, n$, S be the number of distinct maximum terms generated from Σ, S_{2k-1} and S_{2k} be the value of the first $2k - 1(k = 1, 2, \ldots, \lfloor (n+1)/2 \rfloor)$ sum terms and the value of the first $2k(k = 1, 2, \ldots, \lfloor n/2 \rfloor)$ sum terms in Eq. (1), respectively. Then we have that*

(1) if $\dfrac{\displaystyle\sum_{1 \leq i < j < \ldots < l < o \leq n} \overbrace{|P_i \cap P_j \cap \ldots P_l \cap P_o|}^{2k}}{S_{2k-1}} \leq r$, then $\dfrac{S}{S_{2k-1}} \in [1-r, 1]$;

(2) if $\dfrac{\displaystyle\sum_{1 \leq i < j < \ldots < o < q \leq n} \overbrace{|P_i \cap P_j \cap \ldots P_o \cap P_q|}^{2k+1}}{S_{2k}} \leq r$, then $\dfrac{S}{S_{2k}} \in [1, 1+r]$.

Proof. We prove only (1). The proof of (2) is similar.

$$\frac{S}{S_{2k-1}} = 1 - \frac{S_{2k-1} - S}{S_{2k-1}} = 1 - \sigma'$$

$$= 1 - [\sum_{1 \leq i < j < \ldots < l < o \leq n} \overbrace{|P_i \cap P_j \cap \ldots P_l \cap P_o|}^{2k} -$$

$$(\sum_{1 \leq i < j < \ldots < o < q \leq n} \overbrace{|P_i \cap P_j \cap \ldots P_o \cap P_q|}^{2k+1} - \ldots$$

$$+ (-1)^{n+1}|P_1 \cap P_2 \cap \ldots \cap P_n|)]/S_{2k-1}.$$

Since S_{2k} is a lower of S, we have

$$S - S_{2k} = \sum_{1 \leq i < j < \ldots < o < q \leq n} \overbrace{|P_i \cap P_j \cap \ldots P_o \cap P_q|}^{2k+1} - \ldots$$

$$+ (-1)^{n+1}|P_1 \cap P_2 \cap \ldots \cap P_n|) \geq 0.$$

Furthermore, by $\dfrac{\displaystyle\sum_{1 \leq i < j < \ldots < l < o \leq n} \overbrace{|P_i \cap P_j \cap \ldots P_l \cap P_o|}^{2k}}{S_{2k-1}} \leq r$, we obtain $\sigma' \leq r$.

Consequently, $\frac{S}{S_{2k-1}} \geq 1 - r$; S_{2k-1} is a upper of S, that is, $\frac{S}{S_{2k-1}} \leq 1$. So we get $\frac{S}{S_{2k-1}} \in [1-r, 1]$.

From Theorem 1, it is evident that if we consider S_{2k-1} or S_{2k} as the approximate value of S, we have to judge whether its condition is true. This leads to the ULBApprox algorithm described in algorithm 1. Let $\Sigma = \{C_1, C_2, \cdots, C_n\}$ be a set of clauses, $V(|V| = v)$ be its set of variables, and r is a threshold of approximate dispersion.

We use *number*, *svalue* and *sumterm* to denote the value of a term, an impermanent value about S during process of computation, and the value of a sum term, respectively. From line 3 to 11, we get the value *sumterm* of a sum term. If *sumterm* $= 0$, the algorithm turns to 24 line to get the exact number of models; however, if $\frac{sumterm}{svalue} <= r$, it turns to 24 line to get the approximate number of models[2]; otherwise, we update *svalue* according to Eq. (1), and continue to count the value of next sum term.

In the worst case, the run time of ULBApprox algorithm is the same as the exact model counting algorithm MCER, but it returns the exact number of models. In other words, the algorithm is equivalent with the MCER algorithm in the worst case. Therefore, we analyze only the complexity of ULBApprox algorithm when it returns an approximate value.

Algorithm 1. ULBApprox (CNF: Σ, r)

1: $i = 1; number = 0; svalue = 0; sumterm = 0;$
2: **while** $i <= n$ **do**
3:    **for** every sets L that contains i clauses **do**
4:       $union=$ the union of all the clauses in L;
5:       **if** there are no complementary literals in *union* **then**
6:          $number = 2^{v-|union|};$
7:       **else**
8:          $number = 0;$
9:       **end if**
10:       $sumterm = sumterm + number;$
11:    **end for**
12:    **if** $sumterm = 0$ **then**
13:       printf ("Exact value!"); skip to 24;
14:    **else if** $i >= 2 \&\& \frac{sumterm}{svalue} <= r \&\& svalue < 2^v$ **then**
15:       printf ("Approximate value!"); skip to 24;
16:    **else if** $i \bmod 2 = 1$ **then**
17:       $svalue = svalue + sumterm;$
18:    **else**
19:       $svalue = svalue - sumterm;$
20:    **end if**
21:    $sumterm = 0;$
22:    $i = i + 1;$
23: **end while**
24: return $2^v - svalue;$

Theorem 2. *Given a set of clauses $\Sigma = \{C_1, C_2, \ldots, C_n\}$, if ULBApprox algorithm returns an approximate value and svalue is the value of first l sum terms $(1 \leq l < n)$, then the time complexity of ULBApprox algorithm is $O(n^{l+1})$.*

Proof. For the set of clauses Σ, if ULBApprox algorithm returns an approximate value, we know that the result of the value of the $l + 1th$ sum term divided by

---

[2] When *sumterm* is the value of the first sum term, $svalue = 0$. In this case, we can not run the division operator, so we restrict $i \geq 2$; because model counting is generally to count the number of models of a Boolean formula, the number of unsatisfying assignments is less than 2^v. So we restrict its approximate value is less than 2^v.

the value of the first l sum terms, is not more than r according to Theorem 1. So it counts $C_n^1 + C_n^2 + \ldots + C_n^l + C_n^{l+1}$ times, and the time complexity is $O(n^{l+1})$.

ULBApprox algorithm firstly counts the approximate number of unsatisfying assignments, and then counts the approximate number of models. It is necessary to analyze that the proportion covered by the number of models in the total assignments is how to affect the approximate number of models. We use S_{num} to denote the number of assignments, S_{false} and S_{fappro} to denote the exact and approximate value of the unsatisfying assignments, respectively, S_{true} and S_{tappro} to denote the exact and approximate value of the satisfying assignments, respectively. Clearly, $S_{false} + S_{true} = S_{num}, S_{fappro} + S_{tappro} = S_{num}$, and the dispersion $S_{div} = |S_{false} - S_{fappro}| = |S_{true} - S_{tappro}|$. When the approximate dispersion of the number of unsatisfying assignments $\sigma' = S_{div}/S_{fappro} \leq r$, we analyze the change about the approximate dispersion of the number of satisfying assignments $\sigma = S_{div}/S_{true}$. If $S_{true} \geq S_{fappro}$, then $\sigma \leq r$; otherwise, $\sigma > r$. In other words, when $S_{true} > S_{false}$, ULBApprox algorithm can get better approximate value of the number of models.

3.2   SampleApprox

In this subsection, we propose the other approximate model counting algorithm SampleApprox. Firstly, we give a sub-algorithm SubMCER to count the number of models of a restricted subspace. Given a set of clauses Σ with its set of variables $V(|V| = v)$, when we use inclusion-exclusion principle to compute the sum of all different maximum terms generated from Σ, we are actually searching through the entire space of maximum terms, and the size is 2^v. The entire space can be divided into two subspaces: one contains all maximum terms in which a appears, and the other contains all maximum terms in which $\neg a$ appears, where $a \in V$. Also, the entire space can be divided into four subspaces: the first one contains all maximum terms in which a and b appear, and the second one contains all maximum terms in which a and $\neg b$ appear, the three one contains all maximum terms in which $\neg a$ and b appear, and the fourth one contains all maximum terms in which $\neg a$ and $\neg b$ appear, where $a, b \in V$. When the restricted clause C is given, we restrict the subspace of all maximum terms, in which the literals of C are all appearing. Then we use algorithm SubMCER to count the number of maximum terms in the subspace, which are not generated from Σ. That is the number of models of the restricted subspace.

The SubMCER algorithm is given in algorithm 2. The input of the algorithm is a set of clauses Σ with its set of variables $V(|V| = v)$ and any clause C defined in V. c is the number of variables included by C.

In 4 line, we eliminate the clauses, which do not generate any maximum terms in the restricted subspace by clause C. In 6 line, in order to use the MCER algorithm, if $L = Literal(C') \cap Literal(C) \neq \varnothing$, we eliminate the literals L from C', where $Literal(C)$ is the set of literals of clause C. For example, if C is $a \vee \neg b$, we are actually searching through the space of all the maximum terms in which a appears positively and b appears negatively. Namely, we are searching

Algorithm 2. SubMCER (CNF: Σ, C)

1: $\Sigma_1 = \Sigma$;
2: **for** every clause C' in Σ_1 **do**
3:    **if** C' and C have complementary literals **then**
4:       eliminate C' from Σ_1;
5:    **else if** $L = Literal(C') \cap Literal(C) \neq \varnothing$ **then**
6:       eliminate the literals L from C';
7:    **end if**
8: **end for**
9: return MCER(Σ_1); // Σ_1 contains n_1 clauses; the number of variables in Σ_1 is $v - c$.

---

through the assignment space in which a is assigned 0 and b is assigned 1 to get the number of models of the subspace.

Theorem 3. *Given the set of clauses $\Sigma = \{C_1, C_2, \ldots, C_n\}$ with its set of variables $V(|V| = v)$ and any clause C defined in V, the best time complexity is $O(n_1)$ and the worst time complexity is $O(2^{n_1})$ for algorithm SubCER, where n_1 denotes the number of residual clauses after Σ is filtrated by C, and $n_1 < n$.*

Proof. We assume that the set of residual clauses is Σ_1 after Σ is filtrated by C, and the number of clauses in Σ_1 is n_1. In the best case, each pair of clauses in Σ_1 contains complementary literals, so we only need to compute the value of the first sum term in Eq. (1), namely, compute $C_{n_1}^1$ times, and time complexity is $O(n_1)$. In the worst case, SubMCER computes all sum terms, namely, computes $C_{n_1}^1 + C_{n_1}^2 + \ldots + C_{n_1}^{n_l}$ times, and time complexity is $O(2^{n_1})$.

Like Approxcount algorithm, SampleApprox algorithm uses also SampleSat algorithm to get the approximate value of the total number of models divided by the number of models of the restricted subspace. SampleSat algorithm is based on Walksat [30] widely used in solving SAT problems, and is injected with MCMC (Markov Chain Monte Carlo) moves to make it sample from the set of solutions of a Boolean formula near-uniformly. We describe the SampleApprox algorithm in detail as follows.

Given a set of clauses Σ, we can obtain the approximate value of $M(\Sigma)$ by taking the ratio β, that is the ratio of the number of samples, in which variable x is assigned *true* (or *false*), in the total sample size. Namely, $M(\Sigma) \approx (1/\beta) \cdot M(\Sigma|_{x=true})$ (or $M(\Sigma) \approx (1/\beta) \cdot M(\Sigma|_{x=false})$), where $M(\Sigma|_{x=true})$ (or $M(\Sigma|_{x=false})$) can be counted by using SubMCER algorithm. We can reduce further the subspace. If we choose q variables and assign randomly to every variable x_i denoted by t_i ($i = 1, 2, \ldots, q$), then $M(\Sigma) \approx (1/\beta') \cdot M(\Sigma|_{x_1=t_1 \wedge \ldots x_q=t_q})$, where β' is the ratio of the number of samples, in which variable x_i is assigned t_i ($i = 1, 2, \ldots, q$), in the total sample size. SampleApprox algorithm is given in algorithm 3. Let $\Sigma = \{C_1, C_2, \ldots, C_n\}$ be a set of clauses, $V(|V| = v)$ be its set of variables, and K be the sample size.

From 3 to 6 lines, we call SampleSat algorithm to get K models of Σ and put them into *Modelsamples*. From 7 to 11 lines, we count the number of models

---

Algorithm 3. SampleApprox (CNF: Σ)

1: Choose q variables from V and assign randomly to every variable x_i denoted by
   $t_i(i = 1, 2, \ldots, q)$;
2: $j = 1; S_q = 0$;
3: **while** $j <= K$ **do**
4:   $Modelsamples \leftarrow SampleSat(\Sigma)$;
5:   $j = j + 1$;
6: **end while**
7: **for** every sample in $Modelsamples$ **do**
8:   **if** $x_1 = t_1 \wedge \ldots \wedge x_q = t_q$ **then**
9:     $S_q = S_q + 1$;
10:   **end if**
11: **end for**
12: $Multiplier = K/S_q$;
13: $S'_q \leftarrow$ SubMCER (Σ, C_q);
14: $M(\Sigma) = Multiplier * S'_q$;
15: **return** $M(\Sigma)$;

S_q in the sample set in which variable x_i is assigned t_i $(i = 1, 2, \ldots, q)$. In 13 line, we call SubMCER(Σ, C_q) to compute $M(\Sigma|_{x_1=t_1 \wedge \ldots \wedge x_q=t_q})$, where C_q is the clause composed with these literals, which are negative forms of assignments of q variables (e.g., provided that variable x_i is assigned true, C_q contains the literal $\neg x_i$). So we can get the approximate value of $M(\Sigma)$ by taking $Multiplier * S'_q$.

4   Experimental Results

We utilize Microsoft Visual C++ 6.0 to achieve MCER, ULBApprox and SampleApprox algorithms, and all experiments are conducted on a computer with a four-core 2.10 GHz CPU and 1.95 GB RAM. Experimental purpose is to show the accuracy and efficiency of the two approximate algorithms. The instances are obtained by a random generator [23]. It takes three parameters $< v, n, k >$ as input, where v is the number of variables, n is the number of clauses, k is the length of each clause, and each clause is obtained by choosing randomly k variables from the set of m variables and by determining the polarity of each literal with probability $p = 0.5$.

We test these instances: $< 30, n, 7 >$ and $< v, 150, 7 >$, where n lies between 100 and 150, and v lies between 24 and 29. The results given from Figs. 1, 2, 3 and 4 are mean values on 50 experiments for each point. The abscissa denotes the number of clauses or the number of variables, and the ordinate denotes the approximate dispersion σ or run time. For ULBApprox algorithm, we give two different values of r, i.e., 0.05 and 0.01, in order to show that it is how to influence the performance of ULBApprox algorithm.

From Figs. 1 and 3, we can see that ULBApprox algorithm has good performance in the accuracy. For these instances, their approximate dispersions are almost all less than the given threshold r. The smaller r is, the higher accuracy

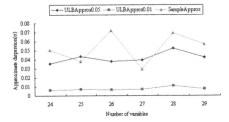

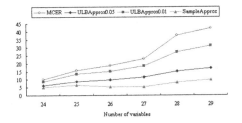

Fig. 1. The approximate dispersion for $< 30, n, 7 >$

Fig. 2. The run time for $< 30, n, 7 >$

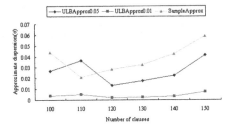

Fig. 3. The approximate dispersion for $< v, 150, 7 >$

Fig. 4. The run time for $< v, 150, 7 >$

of ULBApprox algorithm is. Namely, the smaller σ is. However, the more time is used. On the whole, the run time of ULBApprox algorithm is less than MCER algorithm. When $r = 0.05$, for the instances $< 30, 110, 7 >$, the efficiency of ULBApprox algorithm is 4.98 times that of MCER algorithm.

For SampleApprox algorithm, their approximate dispersions of these instances are almost all larger than that got by ULBApprox algorithm ($r = 0.05$). For these instances, the maximum approximate dispersion is 0.072. The run time of SampleApprox algorithm increases slowly as the number of clauses or variables become larger. Namely, the larger the number of clauses or variables is, the higher the efficiency of SampleApprox is. For the instances $< 30, 150, 7 >$, the efficiency of SampleApprox algorithm is 6.02 times that of MCER algorithm.

5   Conclusions

This paper proposes two approximate model counting algorithms by using extension rule. The key characteristic of ULBApprox algorithm is that it firstly counts the approximate number of unsatisfying assignments, then counts the approximate number of models. And SampleApprox is an approximate algorithm by combining extension rule with SampleSat. Experiment results indicate the two approximate algorithms have good performance in the efficiency and accuracy.

Acknowledgments. This paper was supported by the National Key Basic Research Program of China (973 Program, No. 2012CB326403), National Natural Science Foundation of China (Nos. 61272535, 61370156, 61363035, 61165009), Guangxi "Bagui Scholar" Teams for Innovation and Research Project, Guangxi Natural Science Foundation (No. 2013GXNSFBA019263), Science and Technology Research Projects of Guangxi Higher Education (Nos. 2013YB028, 2013YB029), Scientific Research Foundation of Guangxi Normal University for Doctors, and Guangxi Collaborative Innovation Center of Multisource Information Integration and Intelligent Processing.

References

1. Rintanen, J.: Planning as satisfiability: heuristics. Artif. Intell. **193**, 45–86 (2012)
2. Cai, D., Yin, M.: On the utility of landmarks in SAT based planning. Knowl.-Based Syst. **36**, 146–154 (2012)
3. Nakhost, H., Mller, M.: Towards a theory of random walk planning: regress factors. fair homogeneous graphs and extensions. AI Communications **27**(4), 329–344 (2014)
4. Bacchus, F., Dalmao, S., Pitassi, T.: Algorithms and complexity results for # SAT and bayesian inference. In: 44th Symposium on Foundations of Computer Science (FOCS), pp. 340–351 (2003)
5. Li, W., Poupart, P., Beek, P.V.: Exploiting structure in weighted model counting approaches to probabilistic inference. J. Artif. Intell. Res. **40**, 729–765 (2011)
6. Domshlak, C., Hoffmann, J.: Probabilistic planning via heuristic forward search and weighted model counting. J. Artif. Intell. Res. **30**, 565–620 (2007)
7. Satish Kumar, T.K.: A model counting characterization of diagnoses. In: 13th International Workshop on Principles of Diagnosis, pp. 70–76 (2002)
8. Zhang, S., Zhang, C., Yan, X.: Post-mining: maintenance of association rules by weighting. Inf. Syst. **28**(7), 691–707 (2003)
9. Wu, X., Zhang, S.: Synthesizing high-frequency rules from different data sources. IEEE Trans. Knowl. Data Eng. **15**(2), 353–367 (2003)
10. Gomes, C.P., Sabharwal, A., Selman, B.: Model counting, handbook of satisfiability. In: Biere, A., et al. (eds.) Frontiers in Artificial Intelligence and Applications, vol. 185, pp. 633–654. IOS Press, Amsterdam (2009)
11. Birnbaum, E., Lozinskii, E.: The good old davis-putnam procedure helps counting models. J. Artif. Intell. Res. **10**, 457–477 (1999)
12. Davis, M., Logemann, G., Loveland, D.: A machine program for theorem-proving. Commun. ACM **5**(7), 394–397 (1962)
13. Bayardo Jr., R.J., Pehoushek, J.D.: Counting models using connected components. In: Seventeenth National Conference on Artificial Intelligence (AAAI), pp. 157–162 (2000)
14. Sang, T., Bacchus, F., Beame, P., Kautz, H., Pitassi, T.: Combining component caching and clause learning for effective model counting. In: Seventh International Conference on Theory and Applications of Satisfiability Testing (SAT), pp. 20–28 (2004)
15. Thurley, M.: SharpSAT – counting models with advanced component caching and implicit bcp. In: Biere, A., Gomes, C.P. (eds.) SAT 2006. LNCS, vol. 4121, pp. 424–429. Springer, Heidelberg (2006)
16. Bacchus, F., Dalmao, S., Pitassi, T.: Solving # SAT and bayesian inference with backtracking search. J. Artif. Intell. Res. **34**, 391–442 (2009)

17. Wei, W., Selman, B.: A new approach to model counting. In: Bacchus, F., Walsh, T. (eds.) SAT 2005. LNCS, vol. 3569, pp. 324–339. Springer, Heidelberg (2005)

18. Wei, W., Erenrich, J., Selman, B.: Towards efficient sampling: exploiting random walk strategies. In: Nineteenth National Conference on Artificial Intelligence (AAAI), pp. 670–676 (2004)

19. Jweeum, M.R., Valiant, L.G., Vazirani, V.V.: Random generation of combinatorial structures from a uniform distribution. Theoret. Comput. Sci. **43**, 169–188 (1986)

20. Gomes, C.P., Hoffmann, J., Sabharwal, A., Selman, B.: From sampling to model counting. In: Twentieth International Joint Conference on Artificial Intelligence (IJCAI), pp. 2293–2299 (2007)

21. Gogate, V., Dechter, R.: Approximate counting by sampling the backtrackfree search space. In: Twenty-Second National Conference on Artificial Intelligence (AAAI), pp. 198–203 (2007)

22. Kroc, L., Sabharwal, A., Selman, B.: Leveraging belief propagation, backtrack search, and statistics for model counting. Ann. Oper. Res. **184**(1), 209–231 (2011)

23. Lin, H., Sun, J.: Zhang, Y.: Theorem proving based on extension rule. J. Autom. Reasoning **31**(1), 11–21 (2003)

24. Iwama, K.: CNF satisfiability test by counting and polynomial average time. SIAM J. Comput. **18**(2), 385–391 (1989)

25. Yin, M., Sun, J.: Counting models using extension rules. In: Twenty-Second National Conference on Artificial Intelligence (AAAI), pp. 1916–1917 (2007)

26. Yin, M., Lin, H., Sun, J.: Solving # SAT using extension rules. J. Softw. **20**(7), 1714–1725 (2009)

27. Bennett, H., Sankaranarayanan, S.: Model counting using the inclusion-exclusion principle. In: Sakallah, K.A., Simon, L. (eds.) SAT 2011. LNCS, vol. 6695, pp. 362–363. Springer, Heidelberg (2011)

28. Linial, N., Nisan, N.: Approximate inclusion-exclusion. Combinatorica **10**(4), 349–365 (1990)

29. Kahn, J., Linial, N., Samorodnitsky, A.: Inclusion-exclusion: exact and approximate. Combinatorica **16**(4), 465–477 (1996)

30. Selman, B., Kautz, H., Cohen, B.: Local search strategies for satisfiability testing. In: Trick, M., Johnson, D.S (eds.) Second DIMACS Challenge on Cliques, Coloring and Satisfiability, pp. 521–532 (1993)

Improved Information Set Decoding for Code-Based Cryptosystems with Constrained Memory

Maoning Wang[1]($\boxtimes$) and Mingjie Liu[2]

[1] Key Laboratory of Cryptologic Technology and Information Security, Ministry of Education, School of Mathematics, Shandong University, Jinan 250100, China
wangmaoning@mail.sdu.edu.cn
[2] Beijing International Center for Mathematical Research, Peking University, Beijing 100871, China
liumj9705@gmail.com

Abstract. The decoding of random linear codes is one of the most fundamental problems in both computational complexity theory and algorithmic cryptanalysis. Specifically, the best attacks known against existing code-based cryptosystems, such as McEliece, are unstructural, i.e., these attacks directly use generic decoding algorithms that treat the hidden binary codes as random linear codes. This topic is also attracting increasing interest in a post-quantum context as this area becomes increasingly active. In an attempt to solve this problem, several algorithms and their variants have recently been proposed, with increasingly lower time complexities. However, their memory complexities, which are even more important in practice for real attacks, are neglected.

In this paper, we consider the performance of information set decoding (ISD) algorithms for the problem of syndrome decoding for random binary linear codes with restricted memory. Using Finiasz and Sendrier's standard framework for ISD algorithms, we propose an exact algorithm that performs better when the memory is constrained; also this improvement can be mathematically proven. From a practical standpoint, our approach can yield good time complexities for any given space bound, hence providing a good measure of the effectiveness of a cryptanalytic attack on code-based cryptosystems. Our method can also be seen as an extended application of the dissection technique proposed by Dinur et al. at CRYPTO 2012 [11].

Keywords: Random linear codes · Syndrome decoding · Information set decoding · McEliece · Constrained memory · Dissection technique

This work was supported by National Natural Science Foundation of China (Grant No. 61133013), China's 973 Program (Grant No. 2013CB834205), China Postdoctoral Science Foundation (Grant No. 2013M540786) and Nature Science Foundation of Shandong Province (Grant No. ZR2012FM005).

J. Wang and C. Yap (Eds.): FAW 2015, LNCS 9130, pp. 241–258, 2015.
DOI: 10.1007/978-3-319-19647-3_23

1   Introduction

We first introduce the following problem, which is essentially intrinsic to all existing code-based cryptosystems:

The set $\mathcal{C} = \{\boldsymbol{m}^T G : \boldsymbol{m} \in \mathbb{F}_2^k\}$ is called a linear code of length n and dimension k, specifically, the linear code generated by $G \in \mathbb{F}_2^{k \times n}$. Here, $\boldsymbol{m}$ is a column vector and $\boldsymbol{m}^T$ represents the transpose of $\boldsymbol{m}$. The matrix G is called a generator matrix for this code. The elements of $\mathcal{C}$ are called codewords. If the linear code $\mathcal{C}$ is equal to $\{\boldsymbol{c} \in \mathbb{F}_2^n : H\boldsymbol{c} = \boldsymbol{0}\}$, then the matrix $H \in \mathbb{F}_2^{(n-k) \times n}$ is called a parity-check matrix for the code. The number of non-zero elements in a vector is its so-called Hamming weight.

Now, our task is to solve the following.

Problem 1 (Syndrome Decoding Problem for Binary Linear Codes). Find an error vector $\boldsymbol{e} \in \mathbb{F}_2^n$ whose Hamming weight is equal to w (w is prescribed) given the matrix H and $\boldsymbol{x} = \boldsymbol{c} + \boldsymbol{e}$ for an unknown codeword $\boldsymbol{c}$.

Equivalently, we know $H\boldsymbol{e} \in \mathbb{F}_2^{n-k}$, the so-called syndrome of $\boldsymbol{e}$, because $H\boldsymbol{e} = H(\boldsymbol{x} - \boldsymbol{c}) = H\boldsymbol{x} - \boldsymbol{0} = H\boldsymbol{x} \in \mathbb{F}_2^{n-k}$, which can be immediately obtained from the inputs. In addition, this problem can be understood as the attempt to discover the w of n columns of H that sum to $\boldsymbol{s} = H\boldsymbol{x} \in \mathbb{F}_2^{n-k}$.

Throughout this paper, we consider only random linear codes, namely, codes whose parity-check matrices H are random over $\mathbb{F}_2^{(n-k) \times n}$. Under this condition, the distance d of code $\mathcal{C}$, which is defined as the minimum Hamming weight of its non-zero codewords, asymptotically satisfies

$$\frac{k}{n} = 1 - \mathrm{h}(\frac{d}{n}) \tag{1}$$

(the proverbial Gilbert-Varshamov bound) when n becomes large. Here, the function h is the binary entropy function

$$\mathrm{h}(x) = -x \log_2(x) - (1 - x) \log_2(1 - x). \tag{2}$$

In addition, for relevancy to the case of public-key encryption schemes or identification schemes, we focus on the problem when the weight w of e is set to $\lfloor \frac{d-1}{2} \rfloor$ (the error correction capacity of $\mathcal{C}$), which is the largest value that guarantees the uniqueness of the decoding.

Theoretically, this problem is known to be NP-hard [2,5], and thus, its difficulty can serve as a foundation for cryptographic constructions. It continues to attract increasing interest because to date, it has remained hard even when quantum computing algorithms are considered. From another viewpoint, for certain existing code-based cryptosystems, the best attacks appear to be unstructural, i.e., to directly use generic decoding algorithms [7,12]. Taking McEliece [23] as an example, an attacker who merely knows the public-key matrix G, which is indistinguishable from a random one (and similarly for H), may succeed in recovering a message vector $\boldsymbol{m}$ after receiving its ciphertext $\boldsymbol{x} = \boldsymbol{c} + \boldsymbol{e} = (\boldsymbol{m}^T G)^T + \boldsymbol{e}$ if he can utilize an algorithm that is sufficiently efficient to complete the task of syndrome decoding. This approach also has several potential variants that are useful in algorithmic cryptanalysis [10,14,20,25]. Therefore, a series of studies has appeared in the literature, proposing algorithms to solve this problem with increasingly lower time complexities [3,4,6,13,15,18,21,22,26].

As the results of such studies have become increasingly advanced, they have primarily focused on time complexity. However, in practice, the memory complexity also plays an important role. Specifically, all the records reported in real-world problem solving in cryptanalysis, such as integer factorization [19], collision-seeking [28] and other types of problems, as summarized in [16,27], use parallelization methods because these attacks must be performed on either a large number of workstations or a large number of inexpensive, specialized processors. Because a smaller-space algorithm can make better use of fast cache memories and, in particular, because a smaller-space algorithm often enables easier and more efficient large-scale parallelization, as mentioned in [1], we must focus on memory-constrained algorithms when considering real attacks.

It is therefore not very reasonable to simply ignore the storage constraints when devising or analyzing actual algorithms. When we wish to run an algorithm, we are obliged to provide a cost estimation, which requires considering questions such as "what is the concrete number of PCs and workstations that we need to access if the budget is $1 million?" or "do we have sufficient funds to construct some type of fixed-scale fast cache memories for specialized processors?" Hence, the true measure of the effectiveness of a cryptanalytic attack is the time required to complete the attack given certain resource limitations, as discussed in [27]. Therefore, it is important to develop an algorithm that is feasible in practice, i.e., given the typical processing speed and memory resources of modern computers, and to analyze its efficiency.

Moreover, from a theoretical point of view, a high memory complexity might be considered to be a serious drawback for any exact algorithms using exponential memory. Therefore, in recent years, numerous papers in the field of computer science and information theory have discussed better time-memory tradeoffs and low-memory algorithms applied to various problems, especially those concerning NP-hard problems [1,8,9,11,17,29]. It is the lack of such research on coding-theoretical problems that inspired us to write this paper.

The contribution of this paper is the proposal of a more efficient ISD algorithm under memory-restricted conditions. As shown in Table 1, this algorithm offers an improved time complexity for any prescribed space bound. Our approach substantially benefits from the new type of algorithm proposed by Dinur et al., which they call dissection [11], for composite problems such as the cryptanalysis of multiple-encryption schemes. Furthermore, the improvements achieved using our algorithm can be confirmed through a formal mathematical proof. Hence, our method is valuable in both theory and practice.

2 FS-ISD

We describe the FS-ISD framework [13], where the abbreviation "FS-ISD" denotes the ISD algorithm used by Finiasz and Sendrier. We choose this

Table 1. Comparison of worst-case time complexities under different memory constraints

method ╲ memory constraint[a]	0.003n	0.006n	0.009n	0.012n	0.015n
[18] (memoryless)	0.05751n	*[b]	*	*	*
FS-ISD [13]	-[c]	-	-	-	0.05558n
plain FS-ISD with parameter constraint eq (7)	0.056619n	0.056089n	0.055763n	0.0556n	0.05558n
ours	0.05609n	0.0556n	0.055569n	*	*

[a] For instance, the first data cell of the 4-th row indicates that the algorithm presented in this paper operates on a time scale of $\tilde{O}(2^{0.05609n})$ under the condition that the memory that can be accessed is at most $O(2^{0.003n})$.

[b] * denotes that the value is identical to its neighbor to the left.

[c] − denotes that this case is not mentioned in the literature.

algorithm as our starting point because it is the most efficient algorithm that is guaranteed to find a solution and it serves as a general structure for subsequent proposals, including our method.

Recall that our objective is to recover the weight w (column) vector $e \in \mathbb{F}_2^n$ given the random matrix $H \in \mathbb{F}_2^{(n-k)\times n}$ and the syndrome of e, $s = He \in \mathbb{F}_2^{n-k}$. The FS-ISD algorithm sequentially executes the following two steps until the solution e is found: First, a random permutation of n columns of H together with the corresponding elements of e is performed. Second, a partial match-seeking within certain columns of the matrix transformed from the rearranged H is performed, followed naturally by a compensation of the bits that are different from the current syndrome.

Specifically, the changes to the column order in the first phase can be performed through multiplication by a random permutation matrix $P \in \mathbb{F}_2^{n\times n}$. Let us denote the resulting matrix by $\tilde{H} = HP \in \mathbb{F}_2^{(n-k)\times n}$. This also transforms the target solution to $\tilde{e}$, which has the same weight as the original e, but its 1s and 0s are arranged in a different order. It follows that $\tilde{e} = eP$.

Next, we transform $\tilde{H}$ into the form $\begin{bmatrix} Q & 0 \\ & I_{n-k-l} \end{bmatrix}$ by applying elementary row operations or, equivalently, by multiplying it by an invertible matrix $C \in \mathbb{F}_2^{(n-k)\times(n-k)}$, where 0 is an $l\times(n-k-l)$ zero matrix, I_{n-k-l} is an $(n-k-l)\times(n-k-l)$ identity matrix, and Q is in $\mathbb{F}_2^{(n-k)\times(k+l)}$. This form is always achievable because H (or, likewise, $\tilde{H}$) is of full rank with probability exponentially close to 1 in the case that H is random. In addition, the corresponding processing must also be performed for s, which then becomes s'.

We introduce p as a parameter to optimize the final complexity expressions. Then, it is sufficient to find, in $k+l$ columns of Q, those p that sum up to s' in the first l coordinates and to then complete the remaining $n-k-l$ positions by selecting $w-p$ columns of $\begin{bmatrix} 0 \\ I_{n-k-l} \end{bmatrix}$. A simple method of accomplishing this is to check each of the $\binom{k+l}{p}$ linear combinations of p columns of Q to determine whether that combination satisfies the condition on the first l rows and whether its difference from s' has a weight of $w - p$ because we can only choose columns from a partial identity matrix to eliminate this difference. However, [13] applies the birthday algorithm to decrease the time complexity. Specifically, we search for collisions on the elements' first l of $n - k$ bits in two lists, one of which contains all possible sums of $\frac{p}{2}$ columns taken from the first $\frac{k+l}{2}$ columns of Q and the other of which contains all the s's subtracting a sum of $\frac{p}{2}$ columns taken from the remaining $\frac{k+l}{2}$ columns of Q. These lists (stored in $\mathcal{L}_1$ or $\mathcal{L}_2$) can be computed by multiplying all the length-$(k+l)$ binary vectors with $\frac{p}{2}$ 1s in the first or last $\frac{k+l}{2}$ positions by the matrix Q respectively. If a collision of this type satisfies the weight requirements, then adding the other $w - p$ columns will yield the solution following reverse permutation. Therefore, the running time cost in the searching step is $\mathcal{T}_{\text{FS-ISD}} =$ $\tilde{O}(\max\{|\mathcal{L}_1|, |\mathcal{L}_2|, \frac{|\mathcal{L}_1| \cdot |\mathcal{L}_2|}{2^l}\}) = \tilde{O}(\max\{\binom{(k+l)/2}{p/2}, \binom{(k+l)/2}{p/2}, \frac{\binom{(k+l)/2}{p/2}^2}{2^l}\})$,[1] and the memory cost is $\mathcal{S}_{\text{FS-ISD}} = O(\min\{|\mathcal{L}_1|, |\mathcal{L}_2|\}) = O(\binom{(k+l)/2}{p/2})$.

It remains to estimate the expected number of repetitions of the procedures above, where one repetition refers to the act discussed above of retrieving $\tilde{e} \in \mathbb{F}_2^n$ divided into two components, $\tilde{e} = [\tilde{e}_1, \tilde{e}_2] \in \mathbb{F}_2^{k+l} \times \mathbb{F}_2^{n-k-l}$, where $\tilde{e}_1$ is obtained via the birthday algorithm and $\tilde{e}_2$ is collected thereafter. This process succeeds as long as the weight distribution of $\tilde{e}$ is as desired, i.e., $\tilde{e}_1$ is of weight p and $\tilde{e}_2$ is of weight $w - p$; then, we can obtain the original solution e via inverse permutation. Moreover, the probability that an $\tilde{e}$ with this type of distribution will appear is $\mathcal{P} = \frac{\binom{(k+l)/2}{p/2}^2 \cdot \binom{n-k-l}{w-p}}{\binom{n}{w}}$ because of the randomness of the initial permutation. Hence, it is anticipated that the solution will be reached within $\mathcal{P}^{-1}$ repetitions.

In conclusion, the entire algorithm runs in time $\mathcal{T}_{\text{FS-ISD}} \cdot \mathcal{P}^{-1}$ and with memory $\mathcal{S}_{\text{FS-ISD}}$, where p and l are parameters. Minimizing the expression for the time complexity requires solving the following optimization problem (OPT1) under the quite natural specified constraint conditions:

$$\min \quad T_{\text{FS-ISD}}(c_k, c_l, c_p) - P(c_k, c_w, c_l, c_p) \tag{3}$$

$$\text{s.t.} \begin{cases} 0 \le c_l \le 1 - c_k \\ \max\{0, c_k + c_l + c_w - 1\} \le c_p \le \min\{c_k + c_l, c_w\}, \end{cases} \tag{4} \tag{5}$$

where c_k, c_w, c_l and c_p denote $\frac{k}{n}$, $\frac{w}{n}$, $\frac{l}{n}$ and $\frac{p}{n}$, respectively, and $T_{\text{FS-ISD}}(c_k, c_l, c_p)$ is a term such that $\mathcal{T}_{\text{FS-ISD}} = \tilde{O}(2^{T_{\text{FS-ISD}}(c_k, c_l, c_p)n})$ (and, likewise, $\mathcal{P} =$

---

[1] The polynomial factors hidden by the notation $\tilde{O}$ originate from the operations of merging the two lists to obtain collisions that are described in Algorithm 2, such as computing the indexes and sorting, and their concrete coefficients and powers are determined by the data structures. In addition, later in this paper, we will often omit the notations $\tilde{O}$ and O for simplicity of presentation when there is no ambiguity arising from the context.

$\tilde{O}(2^{P(c_k,c_w,c_l,c_p)n}))$; the latter can be calculated using the expressions obtained above and Stirling's formula

$$\binom{An}{Bn} = \tilde{O}(2^{Ah(B/A)n}), \tag{6}$$

where h(x) is the binary entropy function (2). In addition, note that the constraint inequalities (4) and (5) originate from the fact that we must choose each submatrix or vector to be of non-negative dimension. In this manner, we find that for $c_k \in [0, 1]$, the worst-case time complexity for solving Problem 1 using the FS-ISD algorithm is $\tilde{O}(2^{0.05558n})$ with memory of $O(2^{0.0147n})$.

We use the following pseudo-codes to summarize the entire flow. Note that the abbreviation "MitM" in the title of Algorithm 2 stands for Meet-in-the-Middle. Here, the notation $|_t$ denotes either the first t bits of some vector of length greater than t or the first t rows of a matrix, depending on the context.

Algorithm 1. FS-ISD

Input: parity check matrix $H \in \mathbb{F}_2^{(n-k)\times n}$, syndrome $s = He$, parameters p and l
Output: error vector $e \in \mathbb{F}_2^n$ of known weight w
1: **loop**
2:    generate a random permutation matrix P;
3:    $\tilde{H} \leftarrow HP$;
4:    $\left[Q \begin{array}{c|c} 0 \\ I_{n-k-l} \end{array} \middle| s' \right] \leftarrow C \left[\tilde{H} \middle| s \right]$; // perform row operations to obtain the desired shape
5:    set $t = l$;
6:    $\mathcal{L} \leftarrow$ FS-MitM$(Q|_t, s'|_t, k + l, t, p)$; // see Algorithm 2
7:    **for all** $\tilde{e}_1$ such that $\tilde{e}_1 \in \mathcal{L}$ **do**
8:       **if** wt$(s' - Q\tilde{e}_1) = w - p$ **then**
9:          $\tilde{e}_2 \leftarrow s' - Q\tilde{e}_1$;
10:         $\tilde{e} \leftarrow [\tilde{e}_1, \mathbf{0}] + [\mathbf{0}, \tilde{e}_2]$;
11:         $e^T \leftarrow \tilde{e}^T P^{-1}$;
12:         output e and terminate the program;
13:      **end if**
14:   **end for**
15: **end loop**

A note about Algorithm 2: when it is called by Algorithm 1 (the FS-ISD), the output $\mathcal{L}$ is constructed on the fly, which causes the memory complexity to become $|\mathcal{L}_1|$ (or $|\mathcal{L}_2|$) instead of max$\{|\mathcal{L}_1|, |\mathcal{L}|\}$; similarly, this also holds for the generation of $\mathcal{L}_2$, where the memory complexity becomes $|\mathcal{L}_1|$ instead of max$\{|\mathcal{L}_1|, |\mathcal{L}_2|\}$.

Considered to be a universal model for analyzing ISD-type algorithms, FS-ISD has been carefully and extensively examined by numerous researchers, among whom the author of [24] provides very clear insight into the parameters. Concretely, he rigorously proved that for a given code rate c_k, the parameters c_l^* and c_p^* that minimize $T_{\text{FS-ISD}}(c_k, c_w, c_l, c_p) - P(c_k, c_w, c_l, c_p)$ satisfy

---

Algorithm 2. FS-MitM

Input: matrix $Q \in \mathbb{F}_2^{t \times (k+l)}$; vector $s \in \mathbb{F}_2^t$; parameters $k+l$, t and p
Output: list $\mathcal{L}$
1: // first, initialize the lists
2: $\mathcal{L} \leftarrow$ empty list;
3: $\mathcal{L}_1 \leftarrow \{x_i\}$, where $\{x_i\}$ denotes all vectors of length $(k+l)$ with $\frac{p}{2}$ 1s in the first $\frac{k+l}{2}$ positions;
4: $\mathcal{L}_2 \leftarrow \{y_i\}$, where $\{y_i\}$ denotes all vectors of length $(k+l)$ with $\frac{p}{2}$ 1s in the last $\frac{k+l}{2}$ positions;
5: // next, search for collisions
6: // put $\mathcal{L}_1$ into some data structure (possible choices: hash table or multimap) DS
7: **for** $i = 1$ to $|\mathcal{L}_1|$ **do**
8:      $a \leftarrow$ first t bits of Qx_i; // in fact, Qx_i is of length t; however, for consistency with the following algorithms, we write "first t bits" here
9:      $DS[a] \leftarrow x_i$; // store x_i in DS at index a
10: **end for**
11: // then, for each element in $\mathcal{L}_2$, search for the matching elements in DS
12: **for** $j = 1$ to $|\mathcal{L}_2|$ **do**
13:     $b \leftarrow$ first t bits of $s - Qy_j$;
14:     $\mathcal{L}' \leftarrow \{DS[b]\} = \{x_{i_1}, \cdots, x_{i_{|\mathcal{L}'|}}\}$; // fetch all elements in DS at index b and put them in list $\mathcal{L}'$
15:     **for** $m = 1$ to $|\mathcal{L}'|$ **do**
16:         insert element $x_{i_m} + y_j$ into $\mathcal{L}$;
17:     **end for**
18: **end for**
19: output $\mathcal{L}$;

$0 < c_l^* < 1 - c_k$, $0 < c_p^* < c_w, c_w - c_p^* < \frac{1 - c_k - c_l^*}{2}$, and $c_p^* < \frac{c_k + c_l^*}{2}$. Intuitively, these conclusions indicate that the optimal parameters c_l^* and c_p^* are not located at the corners of the parameter space. In addition, as a corollary, the author also showed the following.

Lemma 1 (Corollary 5.2.23. in [24]). *For optimal c_l^* and c_p^*, it holds that $\mathcal{S}_{FS\text{-}ISD}(c_l^*, c_p^*) \le c_l^*$.*

Here, $\mathcal{S}_{\text{FS-ISD}}(c_l^*, c_p^*)$ is shorthand for $\mathcal{S}_{\text{FS-ISD}}(c_k, c_w, c_l, c_p)$ such that $\mathcal{S}_{\text{FS-ISD}} = O(2^{\mathcal{S}_{\text{FS-ISD}}(c_k, c_w, c_l, c_p)n})$ for a given c_k and its corresponding c_w that satisfy the Gilbert-Varshamov bound (1).

Because there is no straightforward literature to follow for the consideration of memory constraints, the only available method of obtaining a memory-conditioned algorithm is to adjust the parameters. Specifically, when we solve the optimization problem above, one more restriction should be added:

$$\mathcal{S}_{\text{FS-ISD}}(c_k, c_w, c_l, c_p) \le b, \tag{7}$$

where 2^{bn} is the memory bound, which depends on the real resources we are using. The numerical results are presented in Table 1, line 3. This method of addressing the parameters can be regarded as a way to facilitate a time-memory tradeoff; however, because its focus is not on memory concerns, we might obtain further improvements if we were to use techniques that do focus on memory.

3   The New Algorithm

3.1   The Motivation

In light of the explanation above, intuitively, we should be able to improve the memory efficiency of the FS-ISD algorithm by replacing the subroutine FS-MitM with a better one. Our new algorithm also benefits from the concept underlying Shamir et al.'s dissection algorithm [11] for composite problems with memory concerns. However, [11] primarily discuss such algorithms in the context of analyzing the security of multiple-encryption schemes; therefore, we begin by introducing the concept in this scenario and then show how to apply it to our decoding problem.

A multiple-encryption scheme is a concatenation of R perfect N-bit block ciphers with independent round keys, each of which is also N bits, and the objective of cryptanalyzing such schemes is to recover all the round keys, given R plaintext-ciphertext pairs. The dissection concept involves guessing the middle values, which are "ciphertexts" obtained through sequential encryptions of the first several block ciphers. This makes it such that in each iteration of the algorithm, several middle-value blocks are fixed such that non-satisfactory key candidates are filtered out to avoid unnecessary computation. All possible keys are enumerated by looping over all possible values of those middle blocks. Furthermore, a more flexible tradeoff between the time and memory costs is obtained by properly selecting u and m according to the value of R for the current layer. This makes it feasible to minimize the time complexity given limited memory resources.

The pseudo-code below represents the process of the algorithm, where each X_i^u denotes the u-th intermediate value from P_i to C_i, i.e., the "ciphertext" of the sequential encryptions of P_i obtained using $k_1, k_2, \ldots, k_u$.

For syndrome decoding, or, more specifically, to search for vectors whose products with the matrix Q match a certain vector s within the first l bits, namely for the task performed by FS-MitM, we modify the scenario such that the dissection procedures discussed above still function. The desired vector $e \in \mathbb{F}_2^{k+l}$ ensures that $0 + Qe = s$ holds for at least the first l bits; therefore, informally, we can regard 0 as the plaintexts; s, the vector input to FS-MitM, as the ciphertexts; and e as the sum of the R round keys to be determined, i.e., $e = k_1 + k_2 + \cdots + k_R$. Each of the R initial candidate key lists contains all vectors of some desired shape consistent with k_i. Moreover, we define encryption as the addition of vectors in $\mathbb{F}_2$ and decryption as the inverse.

In addition, the smallest units into which the "key" e can be divided are bits instead of integral multiples of fixed-length blocks, which provides greater flexibility. Therefore, similar to splitting the R round keys $k_1, k_2, \ldots, k_R$ into two sets $k_1, k_2, \ldots, k_u$ and $k_{u+1}, k_{u+2}, \ldots, k_R$ for a single middle-value guess in

---

Algorithm 3. Dissection for Mult-Enc

Input: R plaintext-ciphertext   pairs $(P_1, C_1), (P_2, C_2), \ldots, (P_R, C_R)$,   parameters
  u and m

Output: round keys $k_1, k_2, \ldots, k_R$

 1: **for all** possible values of $X_1^u, X_2^u, \ldots, X_{u-m}^u$ **do**
 2:    obtain the 2^{mN} candidate keys for $k_1, k_2, \ldots, k_u$ given the plaintext-"ciphertext"
        pairs $(P_1, X_1^u), (P_2, X_2^u), \ldots, (P_{u-m}, X_{u-m}^u)$;
 3:    **for all** such candidate keys **do**
 4:       compute the u-th intermediate values of $P_{u-m+1}, P_{u-m+2}, \ldots, P_R$ and store
           the outputs in a table as the index of their candidate keys;
 5:    **end for**
 6:    obtain the $2^{(R-u-(u-m))N}$ candidate keys $k_{u+1}, k_{u+2}, \ldots, k_R$ given "plaintext"-
        ciphertext pairs $(X_1^u, C_1), (X_2^u, C_2), \ldots, (X_{u-m}^u, C_{u-m})$;
 7:    **for all** such candidates **do**
 8:       decrypt $C_{u-m+1}, C_{u-m+2}, \ldots, C_R$ to obtain the u-th intermediate values and
           check whether the corresponding values exist in the table;
 9:       **if** there is an index that matches **then**
10:          obtain the output round keys by the candidates $k_1, k_2, \ldots, k_u$ for this index
              together with the candidates $k_{u+1}, k_{u+2}, \ldots, k_R$;
11:       **end if**
12:    **end for**
13: **end for**

Algorithm 3, we divide a $(k + l)$-bit vector e into an $\alpha(k + l)$-bit component and
a $(1 - \alpha)(k + l)$-bit component.

3.2   The New Algorithm

The pseudo-code presented below in Algorithm 4 provides an overview of our
procedures for solving the same problem addressed by FS-MitM (Algorithm 2).

   The complete computation is performed recursively, and its depth is a function of the input value of σ. Specifically, at some inner layer, we guess the t'-bit
middle value, where t' and namely, β are parameters that affect the complexity,
which are discussed in the next section. In addition, for each guess, we produce
two unequally sized subproblems of smaller scale, solve them by either the next
level of recursion or through Basic Dissection, according to their σs, and then
combine the two solution sets, which are also partial solutions to the problem of
current layer, using operations similar to those of Algorithm 2, namely, checking
the $t - t'$ bits that have not yet been addressed to diagnose collisions.

   At the bottom, we reach Basic Dissection, whose complexity can be sum-
marized as $T_{\text{B-Diss}} = 2^{t'} \cdot \max\{T_1, T_2, \frac{|\mathcal{L}_1| \cdot |\mathcal{L}_2|}{2^{t-t'}}\}$, where T_1 and T_2 are the times
required for the construction of $\mathcal{L}_1$ and $\mathcal{L}_2$, respectively, via FS-MitM. Hence,
we have $T_1 = T_2 = \max\{\binom{(k+l)/4}{p/4}, \frac{\binom{(k+l)/4}{p/4}^2}{2^{t'}}\}$. By Stirling's formula (6), we
obtain $T_1 = \binom{(k+l)/4}{p/4}$ because we set $t' = \frac{1}{4}(c_k + c_l)\text{h}(\frac{c_p}{c_k+c_l})n$, i.e., $\tilde{O}(2^{t'}) =$
$\binom{(k+l)/4}{p/4}$. We also have $|\mathcal{L}_1| = |\mathcal{L}_2| = \frac{\binom{(k+l)/2}{p/2}}{2^{t'}} = \binom{(k+l)/4}{p/4}$. Thus, when

---

Algorithm 4. Dissection for FS-ISD

Input: matrix $Q \in \mathbb{F}_2^{t \times (k+l)}$; vector $s \in \mathbb{F}_2^t$; parameters $k+l$, t, p, σ; external parameters α, β depending on σ

Output: list $\mathcal{L}$ containing all vectors e of length $k+l$ and weight p such that $Qe = s$

1: $\mathcal{L}, \mathcal{L}_1, \mathcal{L}_2 \leftarrow$ empty lists;
2: **if** $\sigma \geq 1/4$ **then**
3:    $\mathcal{L} \leftarrow$ Basic Dissection$(Q, s, k+l, t, p)$;
4: **else**
5:    // set $t' = \beta n'$, where $n' = (c_k + c_l)\mathrm{h}(\frac{c_p}{c_k + c_l})n$, i.e., $\tilde{O}(2^{n'}) = \binom{k+l}{p}$
6:    **for** each vector x of length t' **do**
7:       $Q_l \leftarrow$ column 1 to $\alpha(k+l)$ of matrix Q;
8:       $Q_r \leftarrow$ column $\alpha(k+l) + 1$ to $k+l$ of matrix Q;
9:       $\mathcal{L}_1 = \{z_i\} \leftarrow$ Dissection for FS-ISD$(Q_l|_{t'}, x, \alpha(k+l), t', \alpha p, \frac{\sigma}{\alpha})$;
10:     $\mathcal{L}_2 = \{y_i\} \leftarrow$ Dissection for FS-ISD$(Q_r|_{t'}, s|_{t'} - x, (1-\alpha)(k+l), t', (1-\alpha)p, \frac{\sigma}{1-\alpha})$;
11:     // then, merge $\mathcal{L}_1$ and $\mathcal{L}_2$
12:     **for** $i = 1$ to $|\mathcal{L}_1|$ **do**
13:       $a \leftarrow$ last $t - t'$ bits of $Q_l z_i$;
14:       $DS[a] \leftarrow z_i$;
15:     **end for**
16:     **for** $j = 1$ to $|\mathcal{L}_2|$ **do**
17:       $b \leftarrow$ last $t - t'$ bits of $s|_t - Q_r y_j$;
18:       $\mathcal{L}' \leftarrow \{DS[b]\} = \{z_{i_1}, \cdots, z_{i_{|\mathcal{L}'|}}\}$;
19:       **for** $m = 1$ to $|\mathcal{L}'|$ **do**
20:          add the concatenation of z_{i_m} and y_j, i.e. $\begin{bmatrix} z_{i_m} \\ y_j \end{bmatrix}$ to $\mathcal{L}$;
21:       **end for**
22:     **end for**
23:    **end for**
24: **end if**
25: output $\mathcal{L}$;

---

t satisfies $t \geq \frac{1}{2}(c_k + c_l)\mathrm{h}(\frac{c_p}{c_k + c_l})n$, namely, $\frac{|\mathcal{L}_1| \cdot |\mathcal{L}_2|}{2^{t-t'}}$ is exponentially smaller than $\binom{(k+l)/4}{p/4}$, we obtain the following time complexity for Basic Dissection:

$$\mathcal{T}_{\text{B-Diss}} = 2^{t'} \cdot \max\{\mathcal{T}_1, \mathcal{T}_2, \frac{|\mathcal{L}_1| \cdot |\mathcal{L}_2|}{2^{t-t'}}\} = \binom{(k+l)/4}{p/4} \cdot \binom{(k+l)/4}{p/4} = \binom{(k+l)/2}{p/2},$$

where the last equality also holds by Stirling's formula (6) and the memory complexity is $\mathcal{S}_{\text{B-Diss}} = |\mathcal{L}_1| = \binom{(k+l)/4}{p/4}$ because both $\mathcal{L}_2$ and $\mathcal{L}$ are constructed and applied on the fly.

3.3 The Complexity

In this subsection, we analyze the entire recursive algorithm and present the time and memory complexities. First, we recall two theoretically important functions introduced in [11] and [1]:

Algorithm 5. Basic Dissection

Input: matrix $Q \in \mathbb{F}_2^{t \times (k+l)}$; vector $s \in \mathbb{F}_2^t$; parameters $k + l$, t and p
Output: list $\mathcal{L}$ containing all vectors e of length $k + l$ and weight p such that $Qe = s$;
1: $\mathcal{L}, \mathcal{L}_1, \mathcal{L}_2 \leftarrow$ empty lists;
2: // set $t' = \frac{1}{4}n'$, where $n' = (c_k + c_l)\mathrm{h}(\frac{c_p}{c_k+c_l})n$, i.e., $\tilde{O}(2^{n'}) = \binom{k+l}{p}$
3: **for** each vector x of length t' **do**
4:     $Q_l \leftarrow$ left $\frac{k+l}{2}$ columns of matrix Q;
5:     $Q_r \leftarrow$ right $\frac{k+l}{2}$ columns of matrix Q;
6:     $\mathcal{L}_1 = \{z_i\} \leftarrow$ FS-MitM$(Q_l|_{t'}, x, \frac{k+l}{2}, t', \frac{p}{2})$;
7:     $\mathcal{L}_2 = \{y_i\} \leftarrow$ FS-MitM$(Q_r|_{t'}, s|_{t'} - x, \frac{k+l}{2}, t', \frac{p}{2})$;
8:     // then, merge $\mathcal{L}_1$ and $\mathcal{L}_2$
9:     **for** $i = 1$ to $|\mathcal{L}_1|$ **do**
10:         $a \leftarrow$ last $t - t'$ bits of $Q_l z_i$;
11:         $DS[a] \leftarrow z_i$;
12:     **end for**
13:     **for** $j = 1$ to $|\mathcal{L}_2|$ **do**
14:         $b \leftarrow$ last $t - t'$ bits of $s|_t - Q_r y_j$;
15:         $\mathcal{L}' \leftarrow \{DS[b]\} = \{z_{i_1}, \cdots, z_{i_{|\mathcal{L}'|}}\}$;
16:         **for** $m = 1$ to $|\mathcal{L}'|$ **do**
17:             add element $\begin{bmatrix} z_{i_m} \\ y_j \end{bmatrix}$ to $\mathcal{L}$;
18:         **end for**
19:     **end for**
20: **end for**
21: output $\mathcal{L}$;

Definition 1. *For $\sigma \in (0, 1/2]$, let $\mathfrak{l}$ be the solution to $1/\rho_{\mathfrak{l}+1} < \sigma \leq 1/\rho_{\mathfrak{l}}$, where $\rho_{\mathfrak{l}} = 1 + \mathfrak{l}(\mathfrak{l}+1)/2$ for $\mathfrak{l} = 1, 2, \ldots$ (named the magic sequence in [11]). Then,*

$$\tau(\sigma) = 1 - \frac{1}{\mathfrak{l}+1} - \frac{\rho_{\mathfrak{l}} - 2}{\mathfrak{l}+1}\sigma.$$

If there is no such $\mathfrak{l}$, that is, if $\sigma > 1/2$, then define $\tau(\sigma) = 1/2$.

Definition 2. *Define $F : (0, 1] \to (0, 1)$ by the following recurrence for $\sigma < 1/4$:*

$$F(\sigma) = \beta + \max\{\alpha F(\frac{\sigma}{\alpha}), (1 - \alpha)F(\frac{\sigma}{1-\alpha})\},$$

where $\alpha = 1 - \tau(\sigma)$ and $\beta = \alpha - \sigma$. The base case is $F(\sigma) = 1/2$ for $\sigma \geq 1/4$.

The following lemma, given by [1], which can be verified by induction on each interval between two adjacent points in the magic sequence, provides the relationship between the two functions.

Lemma 2. *For every $\sigma \in (0, 1]$, it holds that $F(\sigma) = \tau(\sigma)$.*

We now return to our objective of analyzing Algorithm 4. Obviously, according to the description above, we have

$$\mathcal{T}_{\text{Diss}} = 2^{t'} \cdot \max\{\mathcal{T}_l, \mathcal{T}_r, \mathcal{M}\},$$

where $\mathcal{T}_l$ and $\mathcal{T}_r$ denote the time cost for the construction of lists $\mathcal{L}_1$ and $\mathcal{L}_2$, namely, the time required to solve the left and right subproblems, and $\mathcal{M}$ denotes the time required for the merging of $\mathcal{L}_1$ and $\mathcal{L}_2$, as shown in lines 12 to 23 of Algorithm 4.

We first focus on a much easier case such that in each recursive call, $\mathcal{M}$ is not the dominant term of the three in the argument of $\max\{\}$ of $\mathcal{T}_{\text{Diss}}$. Thus, we have $\mathcal{T}_{\text{Diss}} = 2^{t'} \cdot \max\{\mathcal{T}_l, \mathcal{T}_r\}$, and we can then prove the following lemma, which shows that although Algorithm 4 has many parameters, such as p, k, l and t, the expression for the time cost is rather simple.

Lemma 3. *In the simple case that the parameters in Algorithm 4 are set to $\alpha = 1 - \tau(\sigma)$ and $\beta = \alpha - \sigma$, we have*

$$\mathcal{T}_{Diss} = \tilde{O}(2^{F(\sigma)n'}),$$

where $n' = (c_k + c_l)\text{h}(\frac{c_p}{c_k+c_l})n$, i.e., $\tilde{O}(2^{n'}) = \binom{k+l}{p}$.

Proof. The proof can be performed by induction on the value of σ, in which we use a property of $\tau(\sigma)$: for $\sigma = 1/\rho_{l+1}$, we have $\sigma/\tau(\sigma) = 1/\rho_l$. $\square$

Now, we must address the term $\mathcal{M}$. The time cost of the procedure to merge lists $\mathcal{L}_1$ and $\mathcal{L}_2$ is

$$\mathcal{M} = \max\{|\mathcal{L}_1|, |\mathcal{L}_2|, \frac{|\mathcal{L}_1||\mathcal{L}_2|}{2^{t-t'}}\}.$$

In fact, the sizes of $\mathcal{L}_1$ and $\mathcal{L}_2$ correspond to the numbers of solutions to the two subproblems, which are implicitly included in the time costs of the upper layers, $\mathcal{T}_l$ and $\mathcal{T}_r$ (not, however, in the easier case discussed above), and the term $\frac{|\mathcal{L}_1||\mathcal{L}_2|}{2^t} = 2^{t'} \cdot \frac{|\mathcal{L}_1||\mathcal{L}_2|}{2^{t-t'}}$ that appears in the expansion of the full expression of $\mathcal{T}_{\text{Diss}}$ is the number of solutions of the problem input to the current layer. If, therefore, in each recursive layer, the number of solutions cannot exceed the time cost given above for the easier case, then we can resort to the analysis above. Supported by the following lemma, we can achieve this by appropriately setting the parameters α, β and t that are input to the top layer.

Lemma 4. *Let us set $\alpha = 1 - \tau(\sigma)$, $\beta = \alpha - \sigma$ and $t \geq (1 - \tau(\sigma))n'$, where n' denotes $(c_k + c_l)\text{h}(\frac{c_p}{c_k+c_l})n$, i.e., $\tilde{O}(2^{n'}) = \binom{k+l}{p}$. Then, in each recursive layer, the number of solutions is no greater than $\mathcal{T}_{Diss}$ in Lemma 3.*

Proof. See Appendix A. The proof is achieved through induction on both the left and right subproblems. $\square$

Moreover, it is somewhat easier to analyze the memory complexity. We can verify by induction that the algorithm runs with memory $2^{\sigma n'}$ because the costs for solving the two subproblems are $2^{\sigma/\alpha \cdot \alpha n'} = 2^{\sigma n'}$ and $2^{\sigma/(1-\alpha)\cdot(1-\alpha)n'} = 2^{\sigma n'}$ and the size of the data structure that is used to store $\mathcal{L}_1$ corresponds to the number of solutions of the left subproblem, which is equal to $2^{\alpha n'}/2^{\beta n'} = 2^{\sigma n'}$, given the chosen parameters α and β.

In summary, we can write the following theorem.

Theorem 1. *Algorithm 4 solves an input problem of scale $(k + l, t, p)$ for parameter σ in time $\tilde{O}(2^{\tau(\sigma)n'})$ and with memory $O(2^{\sigma n'})$ if $t \geq (1 - \tau(\sigma))n'$, where n' denotes $(c_k + c_l)\mathrm{h}(\frac{c_p}{c_k + c_l})n$, i.e., $\tilde{O}(2^{n'}) = \binom{k+l}{p}$, and if, in each recursive layer, α and β are set to $\alpha = 1 - \tau(\sigma')$ and $\beta = \alpha - \sigma'$, respectively, in accordance with the σ' of that layer.*

The final part of this subsection describes the entire new version of FS-ISD obtained by replacing FS-MitM with our Algorithm 4. The following lemma shows that $\mathcal{P}$, the probability of successfully obtaining an error vector e of the desired form, remains the same as the original FS-ISD.

Lemma 5. $\mathcal{P}_{Diss\text{-}ISD} = \mathcal{P}$ *as regards the exponents.*

Proof. This can be proven by using the fact that dividing the problem into two parts of unequal size and uniform distribution does not affect the probability in the exponents and that the depth of recursion is polynomial.

To obtain the numerical result of our new ISD algorithm, we must still determine the optimal parameters c_p^*, c_l^* and σ^* of the following optimization problem (OPT2):

$$\min \quad T_{\text{Diss}}(c_k, c_l, c_p, \sigma) - P(c_k, c_w, c_l, c_p) \tag{8}$$

$$\text{s.t.} \begin{cases} 0 \leq c_l \leq 1 - c_k & (9) \\ \max\{0, c_k + c_l + c_w - 1\} \leq c_p \leq \min\{c_k + c_l, c_w\} & (10) \\ (1 - \tau(\sigma))(c_k + c_l)\mathrm{h}(\frac{c_p}{c_k + c_l}) \leq c_l & (11) \\ 0 \leq S_{\text{Diss}}(c_k, c_l, c_p, \sigma) \leq b, & (12) \end{cases}$$

where T_{Diss}, S_{Diss} and P are the exponents of $\mathcal{T}_{\text{Diss}}$, $\mathcal{S}_{\text{Diss}}$ and $\mathcal{P}$ divided by n, i.e., $T_{\text{Diss}}(c_k, c_l, c_p, \sigma) = \tau(\sigma)(c_k + c_l)\mathrm{h}(\frac{c_p}{c_k + c_l})$, $S_{\text{Diss}}(c_k, c_l, c_p, \sigma) = \sigma(c_k + c_l)\mathrm{h}(\frac{c_p}{c_k + c_l})$, and $P(c_k, c_w, c_l, c_p) = (c_k + c_l)\mathrm{h}(\frac{c_p}{c_k + c_l}) + (1 - c_k - c_l)\mathrm{h}(\frac{c_w - c_p}{1 - c_k - c_l}) - c_w \mathrm{h}(c_w)$.

The constraint inequalities in OPT2 can be explained as follows: (9) and (10) are the same as (4) and (5) in OPT1, (11) originates from the condition $t \geq (1 - \tau(\sigma))n'$ in Theorem 1 and the fact that the initial value of t is l, and (11) is the space complexity constraint.

3.4   Comparison

In view of the discussion of complexities in the previous subsection, we can now show that when memory is considered, the new algorithm performs better than plain FS-ISD with the addition of only one more space constraint inequality to OPT1.

First, we present the following lemma.

Lemma 6. *If (c_l^*, c_p^*) is the optimal parameters for OPT1 for FS-ISD plus constraint (7) for a given c_k and the corresponding c_w that satisfy the Gilbert-Varshamov bound (1), then (c_l^*, c_p^*) and $\sigma = 1/4$ is a group of valid parameters for OPT2 for the new Diss-ISD.*

Proof. This lemma states that when $\sigma = 1/4$, the plain FS-ISD can be modified into a more memory-efficient algorithm by using Basic Dissection instead of FS-MitM, hence checking the parameters are valid for OPT2 is enough. See Appendix B.

Moreover, we also have the following.

Lemma 7. *Denote $F_2(\sigma)$ by $F(\sigma)$ when $\alpha = 1/2$ as a constant. Then, $F(\sigma) < F_2(\sigma)$ for all $\sigma < 1/4$.*

Proof. Computing the concrete formula for $F_2(\sigma)$, as done for $F(\sigma)$, will yield the proof. See Appendix C for details.

One conclusion implied by Lemma 7 is that it is better to precisely set the external parameters α and β based on the σ input to the current layer than to fix α at $1/2$. In other words, splitting into two equally sized subproblems, as in FS-ISD, or even further [4, 21] always produces worse results compared with our algorithm.

Intuitively, if we were to draw the memory-time curves for FS-ISD and our algorithm, which requires plotting every possible memory and corresponding time value, then for each point $(S_{\text{FS-MitM}}, T_{\text{FS-MitM}} - P)$, there would be a point $(S_{\text{Diss}}, T_{\text{Diss}} - P)$ lying strictly to the left. Because these curves are continuous and non-intersecting, there must be a point on the Diss-ISD curve that lies below each point on the FS-ISD curve. Thus, our algorithm that outperforms the plain FS-ISD for every memory constraint. See Table 1, lines 3 and 4 for the numerical results of the comparisons.

4   Conclusions

In this paper, we consider the performance of information set decoding (ISD) algorithms for the problem of syndrome decoding for random binary linear codes under memory restrictions. In the FS-ISD framework, we present an exact algorithm that demonstrates superior performance when the memory is constrained; this improvement can be proven using Theorem 1. From a practical standpoint, our algorithm offers good time complexity for any space bounds, thereby providing a good measure of the effectiveness of a cryptanalytic attack on code-based cryptosystems. Moreover, our method can also be regarded as an extended application of the dissection technique of [11].

A   Proof of Lemma 4

Proof (Proof of Lemma 4). To prove this lemma, we use certain properties of the function $\tau(\sigma)$. First, at the top layer, we have a number of solutions $\mathcal{N} = \frac{\binom{k+l}{p}}{2^l} \leq \frac{2^{n'}}{2^{(1-\tau(\sigma))n'}} = 2^{\tau(\sigma)n'}$[2] depending on the chosen range of t. By Lemma 2, the result holds.

Next, we state that the result also holds for both the left and right subproblems. Specifically, for the left subproblem, the number of solutions is $\mathcal{N}_l = \frac{\binom{\alpha(k+l)}{\alpha p}}{2^{l'}} = \frac{2^{\alpha n'}}{2^{\beta n'}} = 2^{(\alpha-\beta)n'}$. Definition 1 indicates that $\tau(\sigma) = 1 - \frac{1}{l+1} - \frac{\rho_l - 2}{l+1}\sigma$, and it is easy to verify the middle term $\frac{1}{l+1} = \frac{\rho_l}{l+1} \cdot \frac{1}{\rho_l} = (\frac{1}{2}l + \frac{1}{l+1})\frac{1}{\rho_l} > (\frac{1}{2}l + \frac{1}{l+1})\sigma$ because $\sigma \in [1/\rho_{l+1}, 1/\rho_l)$; hence, $\tau(\sigma) < 1 - (\frac{1}{2}l + \frac{1}{l+1})\sigma - (\frac{1}{2}l - \frac{1}{l+1})\sigma = 1 - l\sigma \leq 1 - 2\sigma$. Equivalently, we have $1 - \tau(\sigma) - \sigma > \frac{1}{2}(1 - \tau(\sigma))$, which implies $\beta > \frac{1}{2}\alpha$ in light of the chosen parameters. Thus, $\mathcal{N}_l = 2^{(\alpha-\beta)n'} < 2^{\frac{1}{2}\alpha n'}$. However, Lemma 3 states that $\mathcal{T}_l = 2^{\tau(\frac{\sigma}{\alpha})\alpha n'} \geq 2^{\frac{1}{2}\alpha n'}$ because $\tau(\sigma')$ is not smaller than $1/2$ for all σ' in its domain. Thus, it is proven that $\mathcal{N}_l \leq \mathcal{T}_l$.

It remains to show that the result holds for the right subproblem. The proof of Lemma 2, which requires simply computing and simplifying $F(\sigma)$, implicitly contains the property that $\beta + (1-\alpha)\tau(\sigma/(1-\alpha)) = \tau(\sigma)$. Therefore, $\mathcal{N}_r = \frac{\binom{(1-\alpha)(k+l)}{(1-\alpha)p}}{2^{\beta n'}} = 2^{(1-\alpha-\beta)n'} = 2^{(\tau(\sigma)-\beta)n'} = 2^{(1-\alpha)\tau(\sigma/(1-\alpha))n'} = \mathcal{T}_r$.

Thus, the expressions for the time complexity in all layers remain the same as that in the easier case, which completes our proof.   □

B   Proof of Lemma 6

Proof (Proof of Lemma 6). Fix $\sigma = 1/4$; then, the new Diss-ISD becomes the original FS-ISD with FS-MitM replaced with Basic Dissection. We already know that $T_{\text{Diss}}(c_k, c_w, c_l, c_p, 1/4) = T_{\text{FS-MitM}}(c_k, c_w, c_l, c_p)$ and $S_{\text{Diss}}(c_k, c_w, c_l, c_p, 1/4) = 1/2S_{\text{FS-MitM}}(c_k, c_w, c_l, c_p)$. In addition, constraints 1, 2 and 4 in OPT2 naturally hold for (c_l^*, c_p^*). Therefore, only constraint 3 remains to be checked, i.e., $1/2(c_k + c_l)\text{h}(\frac{c_p}{c_k+c_l}) \leq c_l$.

Recall Lemma 1, which was proved rigorously in [24] without memory conditions (7). The result also holds even when constraint (7) is added because the main technique the author used was to obtain a contradiction from $(T_{\text{FS-MitM}} - P)(c_l^* + \epsilon, c_p^*) - (T_{\text{FS-MitM}} - P)(c_l^*, c_p^*) < 0$ and the optimality of $(T_{\text{FS-MitM}} - P)(c_l^*, c_p^*)$, and in fact, $S_{\text{FS-MitM}}(c_l, c_p)$ is a continuous function in c_l (because $\text{h}(x)$ can be expanded using equation (2)), which implies the existence of an ϵ such that $(c_l^* + \epsilon, c_p^*)$ is also in the parameter set (i.e., $S_{\text{FS-MitM}}(c_l^* + \epsilon, c_p^*) \leq b$ also holds). Hence, from the generalization of Lemma 1, we have $S_{\text{FS-MitM}}(c_l^*, c_p^*) = 1/2(c_k + c_l^*)\text{h}(\frac{c_p^*}{c_k+c_l^*}) \leq c_l*$, which completes our proof.   □

---

[2] For convenience and clarity, we continue to use the notations $<, >, \leq$ and $\geq$; however, when they appear in an expression that contains script characters such as $\mathcal{N}$ and $\mathcal{T}$, these symbols denote that the inequality relations hold for the exponents.

C   Proof of Lemma 7

Proof (Proof of Lemma 7). According to the definition of $F(\sigma)$, we have

$$F_2(\sigma) = (1/2 - \sigma) + \max\{1/2F_2(2\sigma)1/2F_2(2\sigma)\} = 1/2 - \sigma + 1/2F_2(2\sigma)$$

for a constant $\alpha = 1/2$, which implies $\beta = \alpha - \sigma = 1/2 - \sigma$. Through inductive calculations, we can easily obtain

$$F_2(\sigma) = 1 - \frac{1}{2^m} - \frac{2^{m-1} - 1}{2^{m-2}}\sigma$$

for $\sigma \in [\frac{1}{2^{m+1}}, \frac{1}{2^m}], m = 2, 3, 4, \cdots$.

A given σ will lie in the interval $[\frac{1}{\rho_{l+1}}, \frac{1}{\rho_l}]$ for some integer l and in the interval $[\frac{1}{2^{m+1}}, \frac{1}{2^m}]$ for some m. Then, one of the following two cases will hold:

$$(1)\frac{1}{\rho_{l+1}} \leq \frac{1}{2^{m+1}} \leq \sigma \leq \frac{1}{\rho_l} \leq \frac{1}{2^m}; \qquad (2)\frac{1}{2^{m+1}} \leq \frac{1}{\rho_{l+1}} \leq \sigma \leq \frac{1}{\rho_l} \leq \frac{1}{2^m}$$

because the interval $[\frac{1}{2^{m+1}}, \frac{1}{2^m}]$ is larger than $[\frac{1}{\rho_{l+1}}, \frac{1}{\rho_l}]$ for any set of $m, l \geq 2$, i.e., for any $\sigma \leq 1/4$. In addition, for $l > 4$, we have $l + 1 < 1/4l(l + 1) < 1/4l(l+1) + 1/2 = \frac{\rho_l}{2}$, and thus, $\frac{1}{l+1} > \frac{2}{\rho_l} \geq \frac{2}{2^{m+1}} = \frac{1}{2^m}$ holds. Moreover, $\frac{\rho_l - 2}{l+1} = \frac{1/2l(l+1)-1}{l+1} = 1/2l - \frac{1}{l+1}$ is monotonically increasing on the integer l and hence is strictly greater than 2 for $l \geq 5$. This also implies that $\frac{\rho_l - 2}{l+1} > 2 > 2 - \frac{1}{2^{m-2}} = \frac{2^{m-1}-1}{2^{m-2}}$. This is now sufficient to conclude that $\frac{1}{l+1} + \frac{\rho_l - 2}{l+1}\sigma > \frac{1}{2^m} + \frac{2^{m-1}-1}{2^{m-2}}\sigma$ for $l \geq 5$. Moreover, the cases of $l = 2, 3$ and 4 are easy to check for the endpoints of the intervals of σ, thereby completing our proof of $F(\sigma) < F_2(\sigma)$. $\square$

References

1. Austrin, P., Kaski, P., Koivisto, M., Määttä, J.: Space-time tradeoffs for subset sum: An improved worst case algorithm. CoRR, abs/1303.0609 (2013)
2. Barg, A.: Complexity issues in coding theory. Electron. Colloquium Computat. Complex (ECCC). **4**(46) (1997). http://eccc.hpi-web.de/eccc-reports/1997/TR97-046/index.html
3. Becker, A., Coron, J.-S., Joux, A.: Improved generic algorithms for hard knapsacks. In: Paterson, K.G. (ed.) EUROCRYPT 2011. LNCS, vol. 6632, pp. 364–385. Springer, Heidelberg (2011)
4. Becker, A., Joux, A., May, A., Meurer, A.: Decoding random binary linear codes in $2^{n/20}$: How $1 + 1 = 0$ improves information set decoding. In: Pointcheval, D., Johansson, T. (eds.) EUROCRYPT 2012. LNCS, vol. 7237, pp. 520–536. Springer, Heidelberg (2012)
5. Berlekamp, E., McEliece, R., Van Tilborg, H.: On the inherent intractability of certain coding problems (corresp.). IEEE Trans. Inf. Theory **24**(3), 384–386 (1978)
6. Bernstein, D.J., Lange, T., Peters, C.: Smaller decoding exponents: ball-collision decoding. In: Rogaway, P. (ed.) CRYPTO 2011. LNCS, vol. 6841, pp. 743–760. Springer, Heidelberg (2011)

7. Bernstein, D.J., Lange, T., Peters, C.: Attacking and defending the McEliece cryptosystem. In: Buchmann, J., Ding, J. (eds.) PQCrypto 2008. LNCS, vol. 5299, pp. 31–46. Springer, Heidelberg (2008)

8. Bisson, G., Sutherland, A.V.: A low-memory algorithm for finding short product representations in finite groups. Des. Codes Crypt. **63**(1), 1–13 (2012)

9. Bjorklund, A., Husfeldt, T., Kaski, P., Koivisto, M.: Computing the Tutte polynomial in vertex-exponential time. In: IEEE 49th Annual IEEE Symposium on Foundations of Computer Science, FOCS 2008, pp. 677–686. IEEE (2008)

10. Brakerski, Z., Vaikuntanathan, V.: Efficient fully homomorphic encryption from (standard) LWE. In: IEEE 52nd Annual Symposium on Foundations of Computer Science (FOCS), pp. 97–106. IEEE (2011)

11. Dinur, I., Dunkelman, O., Keller, N., Shamir, A.: Efficient dissection of composite problems, with applications to cryptanalysis, knapsacks, and combinatorial search problems. In: Safavi-Naini, R., Canetti, R. (eds.) CRYPTO 2012. LNCS, vol. 7417, pp. 719–740. Springer, Heidelberg (2012)

12. Faugère, J.-C., Otmani, A., Perret, L., Tillich, J.-P.: Algebraic cryptanalysis of mceliece variants with compact keys. In: Gilbert, H. (ed.) EUROCRYPT 2010. LNCS, vol. 6110, pp. 279–298. Springer, Heidelberg (2010)

13. Finiasz, M., Sendrier, N.: Security bounds for the design of code-based cryptosystems. In: Matsui, M. (ed.) ASIACRYPT 2009. LNCS, vol. 5912, pp. 88–105. Springer, Heidelberg (2009)

14. Hoffstein, J., Pipher, J., Silverman, J.H.: NTRU: a ring-based public key cryptosystem. In: Buhler, J.P. (ed.) ANTS 1998. LNCS, vol. 1423, pp. 267–288. Springer, Heidelberg (1998)

15. Howgrave-Graham, N., Joux, A.: New generic algorithms for hard knapsacks. In: Gilbert, H. (ed.) EUROCRYPT 2010. LNCS, vol. 6110, pp. 235–256. Springer, Heidelberg (2010)

16. Joux, A.: A tutorial on high performance computing applied to cryptanalysis. In: Pointcheval, D., Johansson, T. (eds.) EUROCRYPT 2012. LNCS, vol. 7237, pp. 1–7. Springer, Heidelberg (2012)

17. Koivisto, M., Parviainen, P.: A space-time tradeoff for permutation problems.In: Proceedings of the Twenty-First Annual ACM-SIAM Symposium on Discrete Algorithms, pp. 484–492. Society for Industrial and Applied Mathematics (2010)

18. Lee, P.J., Brickell, E.F.: An observation on the security of McEliece's public-key cryptosystem. In: Günther, C.G. (ed.) EUROCRYPT 1988. LNCS, vol. 330, pp. 275–280. Springer, Heidelberg (1988)

19. Lenstra, A.K., Lenstra, H.W., Manasse, M.S., Pollard, J.M.: The factorization of the ninth Fermat number. Math. Comput. **61**(203), 319–349 (1993)

20. Lyubashevsky, V., Micciancio, D., Peikert, C., Rosen, A.: SWIFFT: a modest proposal for FFT hashing. In: Nyberg, K. (ed.) FSE 2008. LNCS, vol. 5086, pp. 54–72. Springer, Heidelberg (2008)

21. May, A., Meurer, A., Thomae, E.: Decoding random linear codes in $\tilde{\mathcal{O}}(2^{0.054n})$. In: Lee, D.H., Wang, X. (eds.) ASIACRYPT 2011. LNCS, vol. 7073, pp. 107–124. Springer, Heidelberg (2011)

22. May, A., Ozerov, I.: On computing nearest neighbors with applications to decoding of binary linear codes. In: Oswald, E., Fischlin, M. (eds.) EUROCRYPT 2015. LNCS, vol. 9056, pp. 203–228. Springer, Heidelberg (2015)

23. McEliece, R.J.: A public-key cryptosystem based on algebraic coding theory. DSN Prog. Rep. **42**(44), 114–116 (1978)

24. Meurer, A.: A coding-theoretic approach to cryptanalysis. Ph.D. thesis (2013)

25. Regev, O.: On lattices, learning with errors, random linear codes, and cryptography. J. ACM (JACM) **56**(6), 34 (2009)
26. Stern, J.: A method for finding codewords of small weight. In: Cohen, G., Wolfmann, J. (eds.) Coding Theory and Applications. LNCS, vol. 388, pp. 106–113. Springer, Heidelberg (1989)
27. van Oorschot, P.C., Wiener, M.J.: Parallel collision search with cryptanalytic applications. J. Cryptology **12**, 1–28 (1999)
28. Wiener, M.J.: Efficient DES key search. School of Computer Science, Carleton Univ. (1994)
29. Woeginger, J.G.: Open problems around exact algorithms. Discrete Appl. Math. **156**(3), 397–405 (2008)

Truthful Strategy and Resource Integration for Multi-tenant Data Center Demand Response

Youshi Wang[1,4], Fa Zhang[2], and Zhiyong Liu[1,3(✉)]

[1] Beijing Key Laboratory of Mobile Computing and Pervasive Device,
Institute of Computing Technology, Chinese Academy of Sciences, Beijing, China
{wangyoushi,zyliu}@ict.ac.cn
[2] Key Laboratory of Intelligent Information Processing, ICT, CAS, Beijing, China
zhangfa@ict.ac.cn
[3] State Key Laboratory for Computer Architecture, ICT, CAS, Beijing, China
[4] University of Chinese Academy of Sciences, Beijing, China

Abstract. Data centers' demand response (DR) program has been paid more and more attention recently. As an important component of data centers, multi-tenant data centers (also called "colocation") play a significant role in the demand response, especially in the emergency demand response (EDR). In this paper, we focus on how the colocation can perform better in the EDR program. We formulate the "uncoordinated relationship" in the colocation which is the key problem affecting energy efficiency, and propose a reward system to motivate tenants to join the EDR program, and a truthful strategy is developed to ensure the authenticity of tenants' information. For achieving the overall coordination, we integrate tenants' resources to increase the colocation's resource utilization and optimize the whole colocation's energy efficiency, then devise two algorithms to solve the actual resource migration and integration problem. We analyze the complexity of allocation model and two algorithms. Experimental results show that our solution is practical and efficient.

Keywords: Colocation · Emergency demand response · Uncoordinated relationship · Truthful strategy design · Algorithm analysis

1 Introduction

Data center has become an integral part of our everyday lives. As the scale of the network is continuously extending, more and more data centers have been established, and consume huge amount of energy. The latest data suggests that only in 2013 the power consumption of data centers in the United States has increased to 91 billion kilowatt-hour and the prospective consumption in 2020 will reach 138 billion kilowatt-hour [1]. Because of the huge energy consumption, the demand response program becomes more important in date centers. Especially, the emergency demand response, in some extreme cases such as extreme weather or natural disaster, is for protecting the secure of State Grid by reducing

© Springer International Publishing Switzerland 2015
J. Wang and C. Yap (Eds.): FAW 2015, LNCS 9130, pp. 259–270, 2015.
DOI: 10.1007/978-3-319-19647-3_24

the energy consumption. As one of the major sources of energy consumption, the control of data centers' energy consumption plays an extremely important role in the emergency demand response.

Generally, the data center can be classified into three types: private data center, colocation data center and public cloud computing data center. By the data from National Resource Defense Council (NRDC), the private data center is still the major part, accounting for 53 % of the total, although, the colocation has developed rapidly in recent years and has occupied for about 40 % of the total. NRDC estimates that in many small office-based organizations with on-premise server rooms, as much as 30 % of their total electricity use may be directed toward powering and cooling servers running 24 h a day, even when performing little or no work. The energy wasted by colocation annually is equivalent to the output of seven medium-sized coal-fired power plants [2].

However, the colocation's business model is different from other data centers: the colocation operator is responsible for energy supply and cooling, and all servers are controlled by tenants. Due to losing the direct control of servers, colocation becomes very weak for the energy consumption management. We call the dispersion of management as "uncoordinated relationship" and it has become a serious problem for colocation's energy-saving. In the colocation, because tenants pay the bill based on their peak energy, they don't care the energy consumption and always keep servers working on high performance. Moreover, due to tenants' independence and the obstacles of communication between the operator and tenants, even if each tenant wants to optimize his own energy consumption independently, it is very difficult to get the global energy consumption optimization by each tenant's independent optimization.

Statistic results from large private data centers such as Google and Amazon [3], show that rack units spend about 80 % of their time using less than 65 % of their peak power, and for the whole cluster, it never runs above 72 % of its aggregate peak power. Data analysis from "Hotmail", "MSR" and "Wikipedia" shows that the average utilization of tenant's server is only 15 % [4]. So low resource utilization has become a critical problem need to be addressed in the colocation.

The energy-saving problem of the data center has been researched widely, such as Dynamic Voltage and Frequency Scaling (DVFS) [5], energy-efficient topologies ([6–8]), virtualization of computer resources ([9–12]), and traffic engineering ([13–17]). However, almost all of these techniques work on a hypothesis that data centers are collaborative, that is, under the uncoordinated situation like the colocation, they can't be used directly to solve the energy consumption problem.

Facing the problem of the "uncoordinated relationship" and low resource utilization in the colocation, we propose a framework by combining resource integration technology and economic incentives. Firstly, we design the reward system to encourage tenants to join the EDR program actively. We evaluate each tenant's resource migration cost and then return the benefits of energy-saving based on each tenant's cost. Secondly, for bids of tenants who join the EDR, we propose a truthful strategy to ensure the authenticity of tenants' information. Then we develop two algorithms to integrate tenants' resources, and reduce the whole colocation's energy consumption by increasing resource utilization. Using the integration of tenants' resources, we achieve the unified dispatch management in the colocation, and strengthen the cooperation of the whole colocation.

To summarize, our contributions in this paper are three-fold:

- We formulate the "uncoordinated relationship" problem in the colocation which is the main reason for the colocation's energy inefficiency problem.
- To solve the colocation's "uncoordinated relationship", we propose the reward system by combining the economic incentive and the resource migration method, then design a truthful strategy to ensure the authenticity of information from tenants.
- We prove the resource integration model is an NP-complete problem. For solve it, we develop Priority-based Resource Allocation Algorithm (P-RAA) and Dynamic Resource Allocation Algorithm (D-RAA), and give the time complexity of two algorithms.

The rest of paper is organized as follows. We first propose the reward system and the truthful strategy in Sect. 2. In Sect. 3, we analyze the complexity of the resource integration model, then propose P-RAA and D-RAA to solve it and analyze their time complexity. In Sect. 4, we show and analyze our simulation experimental results. Finally, we summarize and conclude in Sect. 5.

2 Reward System and Truthful Strategy Design

In this section, we first study how to encourage tenants to join the EDR program actively, and propose a reward system to solve the amotivational problem. Then, we design a truthful strategy to ensure the authenticity of tenants' information including the resource requests and energy-saving targets.

2.1 Rewards System

Firstly, the cost of tenants can be divided into two parts, the switch loss and the performance degradation. The main source of switch loss is turning servers into sleep/off mode and bringing them back to normal operation [18]. Because the switch loss is independent of other external factors, we can use a fixed value η to denote it in one server. Then, we will focus on how to measure the degradation of performance.

When allocating resources based on tenants' requests, a bad situation is that real-time resource demand is beyond the actual load. From the view of the computer system, when actual load is beyond capacity limitation, redundant traffic will queue and wait. We use λ_i and μ_i to denote the arrive rate and processing rate of tenant i's traffic and use λ_t and μ_t to denote the actual arrive rate and actual processing rate. The theory rate is equivalent to each tenant's resource request, meanwhile, every tenant would get resources based on his request regardless of the actual traffic arrive rate, so the processing rate can be fixed, expressed as $\mu_i = \mu_t$. Under the actual situation, λ_t is fluctuating so we use λ_i as standard to estimate λ_t. An M/M/1 queue model is used to calculate the theory delay: $d_i = \frac{1}{\mu_i - \lambda_i}$ for tenant i. Tenants' actual delay could be divided into two parts by the actual traffic arrive rate:

$$
\begin{cases}
d_1 = \frac{1}{\mu_i - \lambda_i} |_t\ t \in T(\lambda_t \le \lambda_i) \\
d_2 = \frac{1}{\mu_i - \lambda_t} |_t\ t \in T(\lambda_t > \lambda_i)
\end{cases}
\tag{1}
$$

We set d_1 as benchmark. Because μ_i has been fixed by each tenant's request, the main influence on delay d_1 is traffic arrive rate λ_i. So delay d_1 can be expressed as a function only about λ_i:

$$
d_1 = \frac{1}{\mu_i - \lambda_i} \Leftrightarrow \lambda_i = 1 \times d_{base}
\tag{2}
$$

where d_{base} denotes the unit delay benchmark. For d_2, we have:

$$
d_2 = (\lambda_t / \lambda_i) d_{base}
\tag{3}
$$

Delay d_1 is determined by tenants' resource requests, so it is set as a part of performance degeneration. The total performance degeneration cost can be expressed:

$$
d_i' = \int_{t_1} d_2 = \int_{t_1} \frac{\lambda_t}{\lambda_i} d_{base} \\
t_1 \in T(\lambda_t > \lambda_i)
\tag{4}
$$

For λ_t, we adopt the gaussian distribution (limiting the data within a certain range) to simulate its fluctuating. In the actual simulation, we restrict interval length within 2σ (σ denotes the standard deviation). This can ensure that the simulation data is closer to the truth.

Synthesizing the switch loss and the performance degradation, we can get a total delay expression:

$$
d_{i\_cost} = m_i \times \eta + \int_{t_1} \frac{\lambda_t}{\lambda_i} d_{base} \\
t_1 \in T(\lambda_t > \lambda_i)
\tag{5}
$$

where m_i is the number of tenant i's servers.

The reward function, which is used to motivate tenants to join the EDR program, is based on tenant cost and energy-saving revenue. We can express the reward function as:

$$
Rd_i = \frac{d_{i\_cost}}{\sum\limits_{i=1}^{M'} d_{i\_cost}} \times \sum_{i=1}^{M'} (\gamma \times pr_i)
\tag{6}
$$

where γ is determined by operators to measure the tenant's energy-saving revenue based on the actual energy consumption target pr_i and M' denotes the set of tenants selected.

2.2 Truthful Strategy

Since tenants lack of enough economic incentives to save energy, and they seldom or never cooperate with each other, operators need to exploit a pattern to enhance the synergy between tenants. Therefore, we encourage tenants to submit their resource demands in the form of bids and integrate tenants' requests by the operator. The bid includes two aspects: resource request $r_{i\_j,k}(j \in (1, R), k \in (1, m_i))$ (R denotes the kinds of resource) and energy-saving target pr_i. Here $r_{i\_j,k}$ means the jth resource demand of tenant i's kth server. We get an equivalent relation between the resource request $r_{i\_j,k}$ and the arrive rate λ_i and the processing rate μ_i:

$$\begin{cases} \mu_i = f_1(r_{i\_j,k}) \\ \lambda_i = f_2(r_{i\_j,k}) \end{cases} \tag{7}$$

f_1 and f_2 are both proportional functions.

Next, we talk about how to ensure the authenticity of tenants' bids. We assume that each tenant is "rational-economic man", which means that the tenant always maximizes his own benefit. At the same time, this assumption also ensures that tenants wouldn't lie purposely without any benefit. In order to maximize interests, tenants trend to get more resources and more earnings. So we need not only to avoid lying about the resource demand, but also to get a truthful energy-saving target as a constraint for resource integration. Migrating fewer resources to save more energy is an optimal solution for decreasing the overall energy consumption. We use a weight function to express this relationship:

$$\omega_i = \frac{pr_i}{\sum_{k=1}^{m_i} \sum_{j=1}^{R} r_{i\_j,k}} \tag{8}$$

From the view of operators, they would like to accept bids whose weight ω_i is higher, and from the view of tenants, they want to get higher priority by submitting higher energy conservation goals or requesting less resource. We use an optimal model, showed in next section, to calculate tenants' priorities, and apply the weight coefficient to adjust the objective function of the model. Less resource demand, more cost reduction. But the resource demand should have a lower bound to maintain the system stability and quality of service. So for every tenant, submitting his real demand is more conducive to get higher priority. However, if tenants set an excessive energy-saving goal than the actual energy-saving for getting higher priority, it may cause resource integration scheme inefficiency even losing efficacy. For avoiding this situation, we firstly define a penalty function to increase the cost of lying, and then we propose an offline algorithm P-RAA to allocate the resource based on tenants priorities, and an on-line algorithm to allocate the resource when some tenants lie about their demands.

When finishing tenant i's resource migration, the operator can get his actual energy-saving pr_{i_true}. By it, we can get the penalty function:

$$f(\beta) = |\beta(pr_i - (1 + p_{c1})pr_{i\_true}), 0|_{\max}$$
$$\beta \propto \frac{pr_i - (1 + p_{c1})pr_{i\_true}}{pr_{i\_true}} \tag{9}$$

where p_{c1} is the percentage of maximum permissible error which is used to judge whether the error is factitious between the actual and the expected energy-saving, and β is the punishment coefficient which is used to decide gradations of punishment when tenants submitted inveracious energy-saving targets.

3   Model Analysis and Algorithm Design

Using prr to denote the whole energy-saving demand, we can get the energy-saving constrain:

$$\sum_{i=1}^{M} x_i pr_i \geq prr \tag{10}$$
$$x_i \in (0,1)$$

where x_i denotes whether tenant i is selected.

Combined with the capacity constrain, we propose the selection model to determine whose bids can be accepted to join the EDR program:

$$(P_1) \min \sum_{i=1}^{M} x_i (m_i \times \eta + \int_{t_1} \frac{\lambda_t}{\lambda_i} d_{base}) \frac{1}{\omega_i}$$
$$s.t.$$
$$\sum_{i=1}^{M} x_i pr_i \geq prr$$
$$\sum_{i=1}^{M} (x_i \sum_{k=1}^{m_i} \sum_{j=1}^{R} r_{i\text{-}j,k}) \leq \bar{C} \tag{11}$$
$$x_i \in (0,1)$$
$$t_1 \in T(\lambda_t > \lambda_i)$$

where $\bar{C}$ is the operator's capacity, M is the number of tenants. This is a mixed integer linear problem. By (P_1), we can get a selection result vector $\bar{X} = (x_1, x_2, ..., x_M)$. For ensuring that (P_1) has a solution, we need to make a hypothesis: we can find the solution based on two constraints including the energy-saving constraint and the resource capacity constraint in (P_1). This hypothesis is critical to our objective of designing the resource integration algorithm. After getting $\bar{X}$, the next objective is to design algorithms for optimizing the resources allocation which will be discussed in Subsect. 3.2.

3.1   Complexity Analysis

We now analyze the computational complexity of the optimizing resource allocation problem. Here, this problem is taken analogous to 0–1 multi-knapsack problem. A recognized result is that this model is also NP-complete, and based on this result, we propose the following Lemma 1. 0–1 multi-knapsack problem is an extension of 0–1 knapsack problem that considers multi-knapsacks rather than one knapsack, which is known to be NP-complete.

Lemma 1. *Finding the optimal solution of allocation problem based on the selection model is an NP-complete problem.*

Proof. We assume that existing a polynomial time algorithm can solve the optimality of allocation problem. The optimizing model can be seen as a process of finding minimum n, and this process includes limited attempts to find the optimal result. For every attempt, we set a value for n and then judge n severs whether can hold all requests, and we call it a subproblem. In this subproblem, there are $N = (M \times m_i)$ resource requests, and for each request, the resource demand is $r_{i\_k} = \sum_{j \in (1,R)} r_{i\text{-}j,k}$, which is indivisible. The resource integration model is to find an optimal solution which can use minimum servers to hold all selected tenants' requests. It can be divided into multiple subproblems, and each of them is to judge whether all requests can be holden when the number of servers is fixed. We see every request as an item and the resource demand just is the item's weight. Meanwhile, operator's servers are seen as n knapsacks whose capacity are $(k_1, k_2, \ldots \ldots, k_n)$. When each request's value is 1, the resource integration model can be expressed as:

$$(P_2) \quad \max \sum_{i=1}^{n} \sum_{j=1}^{m} z_{ij} v_j$$

$$s.t.$$

$$\sum_{j=1}^{m} z_{ij} w_j \le k_i \; 1 \le i \le n$$

$$\sum_{i=1}^{n} z_{ij} \le 1 \; 1 \le j \le m$$

$$z_{ij} \in (0,1)$$

where z_{ij} denotes whether item j is put into knapsack i, and v_j is the value of item j. This is a 0–1 multi-knapsack problem. According to the assumption, the resource integration problem can be solved within polynomial time, the subproblem should also be solved within polynomial time. We get the contradictory result $P = NP$, so the original assumption is false. Therefore, Lemma 1 is true. □

3.2   The Off-line and On-line Algorithms

For solving the resource allocation problem, we adopt the greedy algorithm to develop an offline algorithm (Priority-based Resource Allocation Algorithm (P-RAA), see Algorithm 1 and an online algorithm (Dynamic Resource Allocation Algorithm (D-RAA), see Algorithm 2. When all tenants provide their truthful demands, P-RAA can solve the resource allocation problem offline. We divide this algorithm into three parts. Firstly, we initialize the tab of each resource request and its time complexity is $O(N)$. Then in (P_1), we calculate each tenant's value based on the objective function and sort all tenants' value, and the corresponding time complexity is $O(N + M\log_2 M)$. Finally, we use the greedy algorithm to allocate resources and its time complexity is $O(M'^2)$. So the total time complexity of P-RAA is within $O(M'^2 + 2N + M\log_2 M)$, and this shows that P-RAA is a polynomial time algorithm.

A specific situation must be considered when the actual energy-saving is below the energy-saving demand prr. Under this situation, we need to accept more bids from remaining tenants. For these tenants, we adopt the sequence

---

Algorithm 1. Priority-based Resource Allocation Algorithm (P-RAA)

1: **for all** $i \in M$ and $j \in m_i$ **do**
2:    Initialize parameter $flag_{ij} = 0$; as the tab of each resource request
3: **end for**
4: By (P_1), choose s set of tenants who can join in the resource migration process and get the vector $\overline{X} = (\overline{x_1}, \ldots, \overline{x_{M'}})$
5: Using the greedy algorithm with suitable greed factor
6: **for all** $i \in \overline{X}$ **do**
7:    Allocate tenant $i's$ resources to the servers based on the greedy algorithm
8:    $flag_{ij} = 1$; set tenant i's resource request j as unavailable
9: **end for**
10: **Output:** The solution of all resource requests allocation

---

from Algorithm 1 and then use the greedy algorithm to choose bids. Based on the rank, D-RAA is designed as an online algorithm for accepting bids dynamically. Because accepting more bids may need more operator's servers, so our greed aim is to use fewer servers to satisfy energy-saving constrain. Firstly, we consider to fill in the servers which has been used in the offline algorithm, and then we would use more severs to handle remaining resources. We assume that n' tenants are selected. We don't consider the time complexity of sorting, because it has been consisted in Algorithm 1. So the time complexity of D-RAA can be expressed as $O(n' \times n)$.

Algorithm 2. Dynamic Resource Allocation Algorithm (D-RAA)

1: **Input:** The resources allocation information from **P-RAA** and the servers used is $K = (k_1, k_2, \ldots, k_n)$
2: Sort tenant $i(\forall i \in M'')$ based on Algorithm 1; M'' represents a set of tenants whose bids aren't accepted
3: Check K and get each server's resource information
4: **while** Energy conservation constraints are not satisfied **do**
5:    **if** i's bid can be put into K **then**
6:       Using Greedy strategy to fill K
7:    **else**
8:       Put i's bid into a new server in order
9:    **end if**
10: **end while**
11: **Output:** $\overline{x_i}(\forall i \in M'')$

Using algorithms P-RAA and D-RAA, we achieve a robust model for resource integration. Combined with the truthful strategy, operators can use the information to achieve the overall coordination.

4   Experimental Results

We evaluate the performance of our strategy and algorithms using the simulation data. In this work, we generate the resource demand of each tenant following Gaussian distribution under three different resource utilizations (15 %, 20 % and 30 %), and implement two strategies with different greed factors. From the experiments, we want to obtain two objectives: (1) looking for the better greed factor (Line 5 in Algorithm 1) to solve the multi-knapsack problem, (2) evaluating the performance of our algorithms in energy-saving, and showing the percentage of energy-saving of our solution in different conditions.

Firstly, Based on three different resource utilizations, we use the Gaussian distribution to simulate the actual total resource demand of each tenant. The main reason why choose the Gaussian distribution is that we need the characteristic of fluctuating around the mean. Since the Gaussian distribution is unbounded, we limit the value interval for avoiding negative number. Then we adopt the uniform distribution to divide each tenant's total resource demand for getting the CPU and memory demands. Figure 1(a) shows the characters of the data distribution under the resource utilization of 15 %. Then we express the feasibility of our solution. Figure 1(b) shows the running time of our solution in cases of 30, 60, 200 tenants. As shown in Fig. 1(b), the running time will be increased with the growth of tenants, but even when tenants reach to 200, the running time of our solution is only 0.2 second. For the EDR program, the response time is requested to be smallest possible, so our solution is fast enough for responding the EDR program.

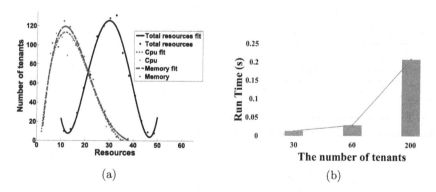

(a)   (b)

Fig. 1. (a) is the data distribution graph for the resource utilization of 15 %, (b) is the actual running time of the whole solution

Next, we verify the effect of different greed factors on performance of energy-saving. We use two strategies with different greed factors, one is to consider the size of resource requirements as the major factor, as shown with the strategy 1 in Fig. 2, the other tends to make resource requests to meet the resource supply structure of operators, as shown in strategy 2 in Fig. 2. Assuming that the resource is continuous and separable, the optimal solution can be got based on traditional linear programming with an absolute lower bound. Figure 2 shows that strategy 1 is better than strategy 2, and always under different resource utilizations, and very close to the optimal result.

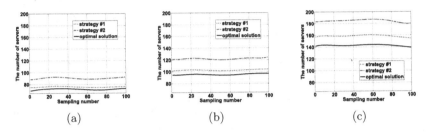

(a)   (b)   (c)

Fig. 2. Comparison of two kinds of greedy algorithms and the optimal solution based on three factors. (a), (b) and (c) signify the different resource utilization 15 %, 20 % and 30 %.

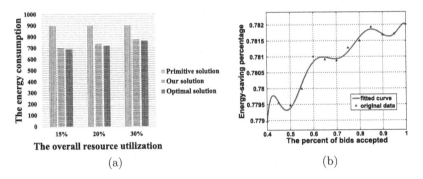

(a)   (b)

Fig. 3. Energy-saving figure: (a) is the comparison of energy-saving effect in three conditions; (b) shows the relationship between energy-saving and bids accepted

To further analyze the performance of our solution, we compare the energy consumption of our algorithm with the optimal solution, as shown in Fig. 3. In the Fig. 3(a), the actual energy consumption is very close to the optimal energy consumption, under different resource utilizations. However, how much the energy is saved depends on the accepted number of tenants' bids. Figure 3(b) describes the relationship between the accepted number of bids and the percent of energy-saving. We can find that, with the resource utilization increasing from 40 % to 100 %, our solution can get about 78 % average energy-saving. Also with the increasing of resources migrated, the energy-saving ratio takes on a growth trend but is not obvious, the reason is that more resources migrated need to take up more servers from operators. In this work, 78% average energy-saving is relatively stable, which suggests that we can obtain an efficient energy-saving solution by our algorithms.

5 Conclusion

Because of high energy consumption, the colocation plays an irreplaceable role in the EDR program. From its characteristics, we discover that the "uncoordinated relationship" and the low resource utilization problems are the keys for improving the colocation's energy efficiency. In this paper, we design a reward system to encourage tenants to submit their resource demands and energy-saving targets. The reward system firstly evaluate each tenant's migration cost and then return the benefits of energy-saving based on the cost. For ensure the authenticity of tenants' bids, we propose a truthful strategy including the design of the weight parameter and the build of the penalty function. Then we integrate all tenants' resource demands by two algorithms: an off-line algorithm P-RAA and an on-line algorithm D-RAA. By resource integration, we achieve the unified dispatch management in the colocation, and reduce the colocation's energy consumption by improving the overall resource utilization. Finally, we analyze the complexity of the resource integration model, and show the specific time complexity expression for P-RAA and D-RAA. Experimental results show that our solution is effective on energy-saving in the colocation and fast enough for responding the EDR program.

Acknowledgement. This research was supported in part by the National Natural Science Foundation of China (Grant No. 61221062 and 61202059).

References

1. Multi-tenant data centers need to play bigger energy efficiency role (2014). http://www.datacenterknowledge.com/archives/2014/08/26/data-center-energy-efficiency-role
2. Is cloud computing always greener? (2012). http://www.nrdc.org/energy/files/cloud-computing-efficiency-IB.pdf
3. Barroso, L.A., Hölzle, U.: The datacenter as a computer: an introduction to the design of warehouse-scale machines. Synth. Lect. Comput. Archit. **4**(1), 1–108 (2009)
4. Ren, S., Islam, M.: Colocation demand response: why do i turn off my servers. In: ICAC (2014)
5. Semeraro, G., Magklis, G., Balasubramonian, R., Albonesi, D.H., Dwarkadas, S., Scott, M.L.: Energy-efficient processor design using multiple clock domains with dynamic voltage and frequency scaling. In: Proceedings of the Eighth International Symposium on High-Performance Computer Architecture, pp. 29–40. IEEE (2002)
6. Hou, C., Zhang, F., Anta, A.F., Wang, L., Liu, Z.: A hop-by-hop energy efficient distributed routing scheme. ACM SIGMETRICS Perform. Eval. Rev. **41**(3), 101–106 (2014)

7. Abts, D., Marty, M.R., Wells, P.M., Klausler, P., Liu, H.: Energy proportional datacenter networks. ACM SIGARCH Comput. Architect. News **38**(3), 338–347 (2010). ACM

8. Huang, L., Jia, Q., Wang, X., Yang, S., Li, B.: Pcube: improving power efficiency in data center networks. In: 2011 IEEE International Conference on Cloud Computing (CLOUD), pp. 65–72. IEEE (2011)

9. Jin, X., Zhang, F., Hu, S., Liu, Z.: Risk management for virtual machines consolidation in data centers. In: 2013 IEEE Global Communications Conference (GLOBECOM), pp. 2872–2878. IEEE (2013)

10. Barham, P., Dragovic, B., Fraser, K., Hand, S., Harris, T., Ho, A., Neugebauer, R., Pratt, I., Warfield, A.: Xen and the art of virtualization. ACM SIGOPS Oper. Syst. Rev. **37**(5), 164–177 (2003)

11. Beloglazov, A., Buyya, R.; Energy efficient resource management in virtualized cloud data centers. In: Proceedings of the 2010 10th IEEE/ACM International Conference on Cluster, Cloud and Grid Computing, pp. 826–831. IEEE Computer Society (2010)

12. Mukherjee, T., Banerjee, A., Varsamopoulos, G., Gupta, S.K., Rungta, S.: Spatio-temporal thermal-aware job scheduling to minimize energy consumption in virtualized heterogeneous data centers. Comput. Netw. **53**(17), 2888–2904 (2009)

13. Wang, L., Zhang, F., Arjona Aroca, J., Vasilakos, A.V., Zheng, K., Hou, C., Li, D., Liu, Z.: Greendcn: a general framework for achieving energy efficiency in data center networks. IEEE J. Sel. Areas Commun. **32**(1), 4–15 (2014)

14. Wang, X., Yao, Y., Wang, X., Lu, K., Cao, Q.: Carpo: correlation-aware power optimization in data center networks. In: INFOCOM: 2012 Proceedings IEEE, pp. 1125–1133. IEEE (2012)

15. Heller, B., Seetharaman, S., Mahadevan, P., Yiakoumis, Y., Sharma, P., Banerjee, S., McKeown, N.: Elastictree: saving energy in data center networks. In: NSDI, vol. 10, pp. 249–264 (2010)

16. Shang, Y., Li, D., Xu, M.: Energy-aware routing in data center network. In: Proceedings of the First ACM SIGCOMM Workshop on Green Networking, pp. 1–8. ACM (2010)

17. Zhang, Y., Ansari, N.: Hero: hierarchical energy optimization for data center networks. In: 2012 IEEE International Conference on Communications (ICC), pp. 2924–2928. IEEE (2012)

18. Lin, M., Wierman, A., Andrew, L.L., Thereska, E.: Dynamic right-sizing for power-proportional data centers. IEEE/ACM Trans. Network. (TON) **21**(5), 1378–1391 (2013)

Local Search to Approximate Max NAE-k-Sat Tightly

Aiyong Xian[1], Kaiyuan Zhu[2], Daming Zhu[1]($^{\boxtimes}$),
and Lianrong Pu[1]

[1] School of Computer Science and Technology, Shandong University,
Jinan, China
dmzhu@sdu.edu.cn

[2] School of Informatics and Computing, Indiana University, Bloomington, USA

Abstract. A clause is not-all-equal satisfied if it has at least one literal assigned by T and one literal assigned by F. Max NAE-SAT is given by a set U of boolean variables and a set C of clauses, and asks to find an assignment of U, such that the not-all-equal satisfied clauses of C are maximized. Max NAE-SAT turns into Max NAE-k-SAT if each clause contains just k literals. Max NAE-k-SAT for $k = 2, 3$ and 4 can be approximated to 1.139 (1/0.878), 1.10047 (1/0.9087) and 8/7 respectively. When $k \geq 5$, little has been done in terms of algorithm design to approximate Max NAE-k-SAT. In this paper, we propose a local search algorithm which can approximate Max NAE-k-SAT to $\frac{2^{k-1}}{2^{k-1}-1}$ for $k \geq 2$. Then we show that Max NAE-k-SAT can not be approximated within $\frac{2^{k-1}}{2^{k-1}-1}$ in polynomial time, if P $\neq$ NP. Moreover, we extend the algorithm for Max NAE-k-SAT to approximate Max NAE-SAT where each clause contains at least k literals.

Keywords: Local Search · Algorithm · Complexity · Performance Ratio · Satisfiability

1 Introduction

The maximum satisfiability problem (abbr. Max-SAT) is the central problem in theoretical computer science. The maximum not-all-equal satisfiability problem (abbr. Max NAE-SAT) is regarded as a generalization of Max-SAT [1,11].

As Max-SAT is NP-Hard and in fact Max-*SNP*-hard [6,20], much attention has been devoted to approximating it. Johnson [14] first devised an approximation algorithm for Max-SAT with performance ratio at most 2. Chen, Friesen and Zheng [5] showed that the performance ratio of Johnson's algorithm is actually $\frac{3}{2}$. Many years passed before people began to approximate Max-SAT by linear and semi-definite programming relaxations [8,25]. The best performance ratio for approximating Max SAT is 1.25502 ($\frac{1}{0.7968}$) by now [1].

Max NAE-SAT is NP-Hard and Max-*SNP*-Hard as well [21]. Andersson and Engerbretsen [2] first investigated this problem and devised a semi-definite programming relaxation to approximate it to 1.3812 ($\frac{1}{0.7240}$) in 1998. Years later,

© Springer International Publishing Switzerland 2015
J. Wang and C. Yap (Eds.): FAW 2015, LNCS 9130, pp. 271–281, 2015.
DOI: 10.1007/978-3-319-19647-3_25

Han, Ye and Zhang improved the performance ratio for approximating it to 1.3335 $(\frac{1}{0.7499})$ [11].

The special versions where each clause contains bounded number of literals also play important roles in the algorithmic researches of Max-SAT and Max NAE-SAT. In this paper, Max-SAT is referred to as Max-k-SAT if each clause contains just k literals; Max-$[k]$-SAT if each clause contains at most k literals; and Max-(k)-SAT if each clause contains at least k literals. Likewise, Max NAE-SAT is referred to as Max NAE-k-SAT if each clause contains just k literals; Max NAE-$[k]$-SAT if each clause contains at most k literals; and Max NAE-(k)-SAT if each clause contains at least k literals.

As early as in 1974, Johnson [14] proposed an algorithm for Max-k-SAT with performance ratio $\frac{2^k}{2^k-1}$. The simple assignment of true or false to every variable with identical probability can also achieve an expected performance ratio $\frac{2^k}{2^k-1}$ for Max-k-SAT [17]. On the other hand, Hastad showed that if P $\neq$ NP, then Max-k-SAT cannot be approximated within $\frac{2^k}{2^k-1}$ for $k \geq 3$ [12]. This implies that either of Max-$[k]$-SAT and Max-(k)-SAT cannot be approximated within $\frac{2^k}{2^k-1}$, if P $\neq$ NP.

The semi-definite programming relaxation proves to be powerful to solve both Max-$[k]$-SAT [3,7–9] and Max NAE-$[k]$-SAT [1,2,11,15,26] when $k \leq 4$. However, as mentioned in [26], it seems very difficult to use semi-definite programming relaxation for solving Max NAE-k-SAT with $k \geq 5$.

Local search for Max-SAT, or SAT, the decision version of Max-SAT, is not uncommon. Many local search based SAT solvers can be found in the literature [4,10,13,18,19,22–24]. In some sense, those local search SAT solvers seem short at their unbounded performance ratios, if they are used for Max-SAT. Actually, Max-k-SAT can be approximated to $\frac{2^k}{2^k-1}$ [16] by *anonymous* local search. In terms of local search to approximate Max NAE-k-SAT, Zhu, Ma and Zhang proposed a local search algorithm with a performance ratio $\frac{k+1}{k}$. On the complexity of approximation, it is still open for how small a performance ratio Max NAE-k-SAT can be approximated to in polynomial time.

In this paper, we present an anonymous local search algorithm for Max NAE-k-SAT, which can achieve a performance ratio $\frac{2^{k-1}}{2^{k-1}-1}$. On the other hand, we show that, if P $\neq$ NP, then for any $k \geq 4$, Max NAE-k-SAT cannot be approximated within $\frac{2^{k-1}}{2^{k-1}-1}$. Our algorithm for Max NAE-k-SAT can be extended to solve Max NAE-(k)-SAT also with performance ratio $\frac{2^{k-1}}{2^{k-1}-1}$.

This paper is organized as follows. In Sect. 2, we present an anonymous local search algorithm for Max NAE-k-SAT with performance ratio $\frac{2^{k-1}}{2^{k-1}-1}$. In Sect. 3, we show that Max NAE-k-SAT cannot be approximated within $\frac{2^{k-1}}{2^{k-1}-1}$, if P $\neq$ NP. In Sect. 4, we extend the local search algorithm for Max NAE-k-SAT to approximate Max NAE-(k)-SAT to $\frac{2^{k-1}}{2^{k-1}-1}$. Section 5 is concluded by summarizing the whole paper and prefigure the future work for this problem.

2   Local Search for Max NAE-k-SAT

Let u be a boolean variable. Its positive and negative literals are represented as u and $\bar{u}$ respectively. If u is assigned a value T or F, either of its literals gets a value. In this paper, we do not distinguish the assignment of a boolean variable with that of its literals.

A *clause* is a set of boolean-variable literals. Let $C_t = \{x[1], x[2], \cdots, x[k]\}$ be a clause of k literals. If an assignment function, say $a(\bullet)$ for $x[1]$, ..., $x[k]$, makes $a(x[1]) \vee \cdots \vee a(x[k]) = T$, then C_t is *satisfied* under $a(\bullet)$; if $a(x[1]) \vee \cdots \vee a(x[k]) = T$ and $a(x[1]) \wedge \cdots \wedge a(x[k]) = T$, then C_t is *all-equal satisfied* under $a(\bullet)$; if $a(x[1]) \vee \cdots \vee a(x[k]) = T$ but $a(x[1]) \wedge \cdots \wedge a(x[k]) = F$, then C_t is *not-all-equal satisfied* under $a(\bullet)$. The maximum not-all-equal k satisfiability problem is given by a set of boolean variables and a set of clauses, where each clause contains k literals, and asks to find an assignment of all the variables, such that the not-all-equal satisfied clauses are maximized in number. It can be formalized as,

Instance. A set of boolean variables $U = \{u_1, \ldots, u_n\}$, a set of clauses $C = \{C_1, \ldots, C_m\}$, each of which has exactly k literals of the variables in U.

Objective. Find an assignment function a for the boolean variables in U: $U \rightarrow \{T, F\}$, such that the not-all-equal satisfied clauses in C are maximized.

Usually, this problem is abbreviated as Max NAE-k-SAT. It turns into Max-k-SAT, if the objective turns to ask for an assignment function of U, such that the satisfied clauses in C are maximized in number.

In this section, every clause is supposed to have k literals. Moreover, a clause cannot contain both the positive and the negative literal of a boolean variable, because if otherwise, it will be satisfied and not-all-equal satisfied by any assignment.

2.1   The Algorithm and Its Performance

Definition 1. *Given an assignment function of U, say $a(U)$, a clause is satisfied (resp. unsatisfied) by $a(u_j)$ or u_j, if it contains u_j with $a(u_j) = T$ (resp. F), or $\bar{u}_j$ with $a(u_j) = F$ (resp. T). A clause is i-satisfied under $a(U)$, if it is satisfied by exactly i literals under $a(U)$.*

Note that a clause is not satisfied, if it is 0-satisfied; a clause is all-equal satisfied, if it is k-satisfied. Under an assignment of U, let S_i be the set of i-satisfied clauses in C; $C[i, j]$ the set of i-satisfied clauses which are also satisfied by u_j; $N[i, j]$ the set of i-satisfied clauses which are unsatisfied by u_j. Moreover, to *flip* a boolean variable refers to that the variable is reassigned by the complement of its current value.

Because a clause does not contain both u_j and $\bar{u}_j$, flipping u_j must make an i-satisfied clause in $C[i, j]$ become $(i - 1)$-satisfied for $0 < i \leq k$, and an i-satisfied clause in $N[i, j]$ become $(i + 1)$-satisfied for $0 \leq i < k$.

Local search asks us to set an objective function which represents how many clauses are not-all-equal satisfied under an assignment of U. Then it will assign the boolean variables in U in the following way: (1)Assign T or F to each boolean variable randomly; (2)Select a variable and flip it such that the objective function can increase in quantitative value; (3)Repeat (2) until no boolean variable can be selected for flipping to increase the objective function in value. This generic method is referred to as *one-bit-flip* local search. Given an assignment, say $a(U)$ of U, then in C, $|S_1| + \cdots + |S_{k-1}|$ clauses are not-all-equal satisfied. Other than directly using $|S_1| + \cdots + |S_{k-1}|$ as the objective function of the one-bit-flip local search, our objective function for Max NAE-k-SAT is a weighted sum of $|S_1|, \cdots, |S_k|$, which can generally be specialized as,

$$F = \alpha_0|S_0| + \alpha_1|S_1| + \cdots + \alpha_{k-1}|S_{k-1}| + \alpha_k|S_k|, \tag{1}$$

where, α_i is the coefficient of $|S_i|$, $0 \leq i \leq k$. We aim to show that if F is used as the objective function, the one-bit-flip local search can make $|S_1| + \cdots + |S_{k-1}|$ arrive at a value not less than $(2^{k-1} - 1)(|S_0| + |S_k|)$.

Those clauses which are converted from i-satisfied to $(i - 1)$-satisfied by flipping u_j must lead to an increment $(\alpha_{i-1} - \alpha_i)|C[i, j]|$ $(1 \leq i \leq k)$ of F, while those clauses which are converted from i-satisfied to $(i + 1)$-satisfied by flipping u_j must lead to an increment $(\alpha_{i+1} - \alpha_i)|N[i, j]|$ $(0 \leq i \leq k-1)$ of F. Therefore, the increment of F due to flipping u_j can be summarized as,

$$\Delta[F, j] = \sum_{i=1}^{k}(\alpha_{i-1} - \alpha_i)|C[i, j]| + \sum_{i=0}^{k-1}(\alpha_{i+1} - \alpha_i)|N[i, j]|. \tag{2}$$

If flipping u_j leads to $\Delta[F, j] > 0$, the value of F could be improved. Thus the local search algorithm for Max NAE-k-SAT is given formally in Algorithm 1 and named as NAE-k-SAT(U, C). In the description of the algorithm, $\overline{a(u_j)}$ represents the complement of $a(u_j)$, and Rand($\bullet$) stands for the subroutine to generate T or F randomly.

Algorithm 1. NAE-k-SAT(U, C)

1: $a(u_j) \leftarrow$ Rand($\bullet$), $u_j \in U$;
2: While $(\exists u_j,$ s. t. $\Delta[F, j] > 0)$ do
3:      $a(u_j) \leftarrow \overline{a(u_j)}$;
4: End while
5: **return** $a(U)$

Those coefficients of F in (1) are in fact the key factors to affect the performance of NAE-k-SAT(U, C). Since the clauses in S_0, S_k are not not-all-equal satisfied, setting α_0 and α_k with zero seems reasonable. Here, we give the values of those coefficients α_i for $0 \leq i \leq k$ by a recurrent relation as follows.

$$\alpha_i = \begin{cases} 0 & i = 0 \\ \alpha_{i-1} + \dfrac{2^{k-1} - 1 - \sum_{j=1}^{i-1}\binom{k}{j}}{(k - i + 1)\binom{k}{i-1}} & 1 \leq i \leq k \end{cases} \tag{3}$$

Actually, the coefficients given in (3) are symmetrical in value, from which $\alpha_k = 0$ follows.

Property 1. *If $0 \leq i \leq k$, then $\alpha_i = \alpha_{k-i}$.*

Lemma 1. *Let U be assigned arbitrarily. Then $\sum_{j=1}^{n} |C[i,j]| = i|S_i|$, and $\sum_{j=1}^{n} |N[i,j]| = (k-i)|S_i|$.*

Proof. Let $a(U)$ be an arbitrary assignment function of U. Let $C_t = \{x[t,1], x[t,2], \ldots, x[t,k]\} \in S_i$ be i-satisfied under $a(U)$. Without loss of generality, let $a(x[t,1]) = a(x[t,2]) = \ldots = a(x[t,i]) = T$. Then C_t will become $(i-1)$-satisfied through flipping one of $x[t,1], \ldots, x[t,i]$. If $x[t,y] \in \{u_{j_y}, \overline{u}_{j_y}\}$ for $1 \leq y \leq i$, then $C_t \in C[i,j_y]$. If $j \notin \{j_y \mid 1 \leq y \leq i\}$, then $C_t \notin C[i,j]$. Namely, C_t occurs in just i sets of $C[i,j]$, $1 \leq j \leq k$. Finally, $\sum_{j=1}^{n} |C[i,j]| = i|S_i|$ follows from every i-satisfied clause occurs in just i sets of $C[i,j]$, $1 \leq j \leq k$.

Likewise, it can be shown that every i-satisfied clause must occur in just $k-i$ sets of $N[i,j]$, $1 \leq j \leq k$. Thus $\sum_{j=1}^{n} |N[i,j]| = (k-i)|S_i|$. □

Lemma 2. *Let $a(U)$ be the assignment function of U returned by NAE-k-SAT(U, C), while S_i the set of i-satisfied clauses under $a(U)$, then $|S_1| + \cdots + |S_{k-1}| \geq (2^{k-1} - 1)(|S_0| + |S_k|)$.*

Proof. Since $a(U)$ is returned by NAE-k-SAT(U, C), flipping any variable cannot increase the value of F for U assigned by $a(U)$. Thus $\Delta[F, j] \leq 0$ for $1 \leq j \leq n$. Namely,

$$\sum_{i=1}^{k} (\alpha_{i-1} - \alpha_i)|C[i,j]| + \sum_{i=0}^{k-1} (\alpha_{i+1} - \alpha_i)|N[i,j]| \leq 0, \tag{4}$$

$$j \in \{1, ..., n\}.$$

Adding these n inequalities, we have,

$$\sum_{j=1}^{n} \Delta[F, j] = \sum_{j=1}^{n} \sum_{i=1}^{k} (\alpha_{i-1} - \alpha_i)|C[i,j]| + \sum_{j=1}^{n} \sum_{i=0}^{k-1} (\alpha_{i+1} - \alpha_i)|N[i,j]|$$

$$= \sum_{i=1}^{k} (\alpha_{i-1} - \alpha_i) \sum_{j=1}^{n} |C[i,j]| + \sum_{i=0}^{k-1} (\alpha_{i+1} - \alpha_i) \sum_{j=1}^{n} |N[i,j]|$$

$$\leq 0. \tag{5}$$

By lemma 1, $i|S_i|$ and $(k-i)S_i$ can substitute $\sum_{j=1}^{n} |C[i,j]|$ and $\sum_{j=1}^{n} |N[i,j]|$ in (9) respectively. These substitutions lead to,

$$k\alpha_1|S_0| + \sum_{i=1}^{k-1} [(\alpha_{i-1} - \alpha_i)i + (\alpha_{i+1} - \alpha_i)(k-i)]|S_i|$$

$$+ k\alpha_{k-1}|S_k| \leq 0. \tag{6}$$

By (3) and property 1, we have, $\alpha_1 = \alpha_{k-1} = \frac{2^{k-1}-1}{k}$. Therefore, $k\alpha_1|S_0| + k\alpha_{k-1}|S_k| = k\alpha_1 (|S_0| + |S_k|) = (2^{k-1} - 1)(|S_0| + |S_k|)$. Hence inequality (6) becomes

$$-\sum_{i=1}^{k-1}[(\alpha_{i-1} - \alpha_i)i + (\alpha_{i+1} - \alpha_i)(k - i)]|S_i|$$

$$\geq (2^{k-1} - 1)(|S_0| + |S_k|). \tag{7}$$

The left side of this inequality is just $|S_1| + |S_2| + \ldots + |S_{k-1}|$, because $(\alpha_{i-1} - \alpha_i)i + (\alpha_{i+1} - \alpha_i)(k - i)$ always equals -1 for $1 \leq i \leq k - 1$, which can be verified by substituting $(\alpha_{i-1} - \alpha_i)$ and $(\alpha_{i+1} - \alpha_i)$ with the formulation in (3). □

Consequently, Lemma 2 implies that NAE-k-SAT(U, C) can approximate Max NAE-k-SAT to $\frac{2^{k-1}}{2^{k-1}-1}$ for $k \geq 2$.

2.2 The Time Complexity

In this subsection, the instances of Max NAE-k-SAT are also represented by U and C, where $|U| = n$, and $|C| = m$. Let $a(U)$ be an assignment function of U. It takes $O(k)$ time to calculate how many literals in a clause are assigned true by $a(U)$. Thus, it can take $O(km)$ time to construct the sets $C[i, j]$, $N[i, j]$ and S_i under $a(U)$. Finding a boolean variable u_j for $1 \leq j \leq n$ such that $\Delta[F, j] > 0$ takes $O(kn)$ time. Therefore, the time complexity of a while loop in NAE-k-SAT(U, C) is $O(k(m + n))$.

How many times the while loop in NAE-k-SAT(U, C) runs depends on the values of α_i, $1 \leq i \leq k - 1$.

Note that each coefficient of F is given by a fraction. Exactly, only when every coefficient of F acts as an integer, can the algorithm always output solutions with performance ratio no more than $\frac{2^{k-1}}{2^{k-1}-1}$. Thus for implementing the algorithm, we have to scale α_i up into an integer by multiplying it with a positive integer, say $\beta(k)$, for $1 \leq i \leq k - 1$. If we insist that $\beta(k)\alpha_1, \cdots, \beta(k)\alpha_{k-1}$ should all be positive integers, $\beta(k)\alpha_{\lfloor \frac{k}{2} \rfloor}$ must be the greatest one among them. Generally, $\beta(k)$ can be bounded by the following property.

Property 2. Let $\beta(k) = min\{\beta | \beta\alpha_i \text{ is positive integer}, 1 \leq i \leq k - 1\}$, where α_i is given by equation (3). Then $\beta(k) \leq \lfloor \frac{k}{2} \rfloor!$ if $k \geq 6$.

Theorem 1. For any Max NAE-k-SAT instance, NAE-k-SAT(U, C) can achieve a performance ratio no more than $\frac{2^{k-1}}{2^{k-1}-1}$ in $O(\beta(k)2^{k-1}k(m + n)m)$ time, where n is the number of boolean variables and m is the number of clauses. If $k \geq 6$, $\beta(k) \leq \lfloor \frac{k}{2} \rfloor!$.

3 The Complexity of Approximating Max NAE-k-SAT

The following theorem derives from Hastad in [12].

Theorem 2. *If $P \neq NP$, then for any $k \geq 3$, no polynomial time algorithm can approximate Max-k-SAT to $\frac{2^k}{2^k-1} - \varepsilon$, $\varepsilon > 0$.*

The focus of this section is on showing that Max NAE-k-SAT cannot be approximated within $\frac{2^{k-1}}{2^{k-1}-1}$ for $k \geq 4$. That is,

Theorem 3. *If $P \neq NP$, then for any $k \geq 4$, no polynomial time algorithm can approximate Max NAE-k-SAT to $\frac{2^{k-1}}{2^{k-1}-1} - \varepsilon$, $\varepsilon > 0$.*

Proof. The proof is a reduction from Max-l-SAT, $l \geq 3$. Let U, C as a whole be an instance of Max-l-SAT. We construct an instance, say U' and C' of Max NAE-$(l+1)$-SAT, as follows.

Firstly, we add a variable o to U to form U'. Formally, if $U = \{u_1, \ldots, u_n\}$, then $U' = \{u_1, \ldots, u_n, o\}$. Then, we add o into every clause in C to form the clauses of C'. Formally, let $C = \{C_1, \ldots, C_m\}$, where $C_t = (x[t,1], \ldots, x[t,l])$. Then corresponding to C_t for $1 \leq t \leq m$, we set a clause $D_t = (x[t,1], \ldots, x[t,l], o)$. Finally, $C' = \{D_1, \ldots, D_m\}$. We argue that for any assignment $a(U)$ of U, U' can be assigned to make as many clauses in C' as those clauses satisfied under $a(U)$ in C not-all-equal satisfied, and so works for the contrary side.

Let $a(u_1), \ldots, a(u_n)$ be an assignment of U, under which M clauses in C are satisfied. Assign the literal o by $a(o) = F$. Then exactly M clauses in C' are not-all-equal satisfied under $a(u_1), \ldots, a(u_n)$ and $a(o)$. This is because, if C_t is satisfied under $a(u_1), \ldots, a(u_n)$, then C_t must have literals with value T. Since $a(o) = F$, D_t is not-all-equal satisfied. If C_t is not satisfied under $a(u_1), \ldots, a(u_n)$, then since $a(o) = F$, D_t is not not-all-equal satisfied.

On the other hand, let $a(u_1), \ldots, a(u_n)$ and $a(o)$ be an assignment of U', under which M clauses in C' are not-all-equal satisfied. Then assigning $a(u_1) \oplus a(o), \ldots, a(u_n) \oplus a(o)$ to $u_1, \ldots, u_n$ must make M clauses in C satisfied exactly. This is because,

(1)If $a(o) = F$, then C' contains no all-equal satisfied clauses; and $a(u_i) \oplus a(o) = a(u_i)$ for $1 \leq i \leq n$. If D_t is not-all-equal satisfied under $a(u_1), \ldots, a(u_n)$ and $a(o)$, then D_t must have a literal, say $x[t, \bullet]$, with value T. Thus, C_t is satisfied by $a(x[t, \bullet])$. If D_t is not satisfied under $a(u_1), \ldots, a(u_n)$ and $a(o)$, C_t is not satisfied under $a(u_1), \ldots, a(u_n)$ trivially.

(2)If $a(o) = T$, then C' contains no not-satisfied clause, and $a(u_i) \oplus a(o) = a(u_i)$ for $1 \leq i \leq n$. If D_t is not-all-equal satisfied under $a(u_1), \ldots, a(u_n)$ and $a(o)$, then a literal, say $x[t, \bullet]$ in D_t is assigned by F. So C_t is satisfied by $a(x[t, \bullet])$. If D_t is all-equal satisfied under $a(u_1), \ldots, a(u_n)$ and $a(o)$, then C_t is not satisfied under $a(u_1), \ldots, a(u_n)$ trivially.

Let $OPT(U, C)$ be the maximum number of clauses in C which can be satisfied under some assignment of U, while $OPT(U', C')$ the maximum number of clauses in C' which can be not-all-equal satisfied under an assignment of U'. Then by the above mentioned argument, $OPT(U, C) = OPT(U', C')$.

If a polynomial time algorithm, which can approximate Max NAE-k-SAT to $\frac{2^{k-1}}{2^{k-1}-1}$ - ε, then for any Max NAE-k-SAT instance, say U' and C', this algorithm can always return such an assignment $a(U')$ of U', that at least $\frac{1}{2^{k-1}/(2^{k-1}-1)-\varepsilon}OPT(U',C')$ clauses are not-all-equal satisfied.

Then, at least $\frac{1}{2^{k-1}/(2^{k-1}-1)-\varepsilon}$ $OPT(U',C')$ clauses in C are satisfied under assigning $a(u_1)\oplus a(o), \ldots, a(u_n)\oplus a(o)$ to $u_1, \ldots, u_n$ respectively. This contradicts with Theorem 2. □

4   Local Search for Max NAE-(k)-SAT

In this section, a Max NAE-(k)-SAT instance is given by a set $U = \{u_1, \ldots, u_n\}$ of boolean variables, and a set $C = \{C_1, \ldots, C_m\}$ of clauses, each of which has k or more literals, and just like the objective of Max NAE-k-SAT, asks to find an assignment of U such that the not-all-equal satisfied clauses in C are maximized in number. Since the algorithm in Sect. 2 cannot be used for Max NAE-(k)-SAT instances to arrive at a substantial performance ratio, we aim to extend the algorithm NAE-k-SAT(U, C) to solve Max NAE-(k)-SAT in this section.

Under an assignment of U, let $S[b, i]$ be the set of i-satisfied clauses each of which contains b literals; $C[b, i, j]$ the set of i-satisfied clauses each of which contains b literals and is satisfied by u_j; $N[b, i, j]$ the set of i-satisfied clauses each of which contains b literals and is unsatisfied by u_j.

Suppose in C, a clause has at most k_{max} literals. In other words, a clause in C may contain $k, \ldots, k_{max}$ literals. To increase the number of not-all-equal satisfied clauses in C by anonymous local search, we pay attention to the following objective function:

$$G = \sum_{b=k}^{k_{max}} (\alpha_{b,1}|S[b,1]| + \ldots + \alpha_{b,b-1}|S[b, b-1]|). \tag{8}$$

One can notice that Eq. (8) arises from adding those objective functions of Max NAE-b-SAT for $k \leq b \leq k_{max}$, where the objective function of Max NAE-b-SAT has been given by Eq. (1). Following Eq. (3), the coefficients $\alpha_{b,i}$ ($0 \leq i \leq b$) in G can be specialized as

$$\alpha_{b,i} = \begin{cases} 0 & i = 0 \\ \alpha_{b,i-1} + \dfrac{2^{b-1} - 1 - \sum_{j=1}^{i-1}\binom{b}{j}}{(b-i+1)\binom{b}{i-1}}. & 1 \leq i \leq b \end{cases} \tag{9}$$

It follows from Property 1 that $\alpha_{b,i} = \alpha_{b,b-i}$, $0 \leq i \leq b$. Let U be assigned by $a(U)$. When flipping a variable, an i-satisfied clause in $S[b, i]$ can only move to $S[b, i+1]$ or $S[b, i-1]$. Similar to Eq. (2), flipping u_j will lead to the increment of G as,

$$\Delta[G, j] = \sum_{b=k}^{k_{max}} \sum_{i=1}^{b} (\alpha_{b,i-1} - \alpha_{b,i}) |C[b, i, j]|$$

$$+ \sum_{b=k}^{k_{max}} \sum_{i=0}^{b-1} (\alpha_{b,i+1} - \alpha_{b,i}) |N[b, i, j]|$$

$$= \sum_{b=k}^{k_{max}} [\sum_{i=1}^{b} (\alpha_{b,i-1} - \alpha_{b,i}) |C[b, i, j]|$$

$$+ \sum_{i=0}^{b-1} (\alpha_{b,i+1} - \alpha_{b,i}) |N[b, i, j]|]. \tag{10}$$

Therefore, the local search algorithm for Max NAE-(k)-SAT, except replacing $\Delta[F, j] > 0$ with $\Delta[G, j] > 0$ as the condition for flipping u_j, can be formulated in the same way as NAE-k-SAT(U, C), where $\Delta[G, j]$ is specialized by Eq. (10). Thus the algorithm must end with $\Delta[G, j] \leq 0$ for $1 \leq j \leq n$.

For convenience, we rename the algorithm NAE-k-SAT(U, C) in which $\Delta[G, j] > 0$ substitutes $\Delta[F, j] > 0$ as NAE-(k)-SAT(U, C).

Lemma 3. *Let U be assigned arbitrarily. Then $\sum_{j=1}^{n} |C[b, i, j]| = i|S[b, i]|$, and $\sum_{j=1}^{n} |N[b, i, j]| = (b - i)|S[b, i]|$.*

Lemma 4. *Let NAE-(k)-SAT(U, C) return an assignment $a(U)$ of U. Then $\sum_{b=k}^{k_{max}} \sum_{i=1}^{b-1} |S[b, i]| \geq \sum_{b=k}^{k_{max}} (2^{b-1} - 1)(|S[b, 0]| + |S[b, b]|)$.*

Since $C = \bigcup_{b=k}^{k_{max}} \bigcup_{i=0}^{b} S[b, i]$, Lemma 4 implies $\sum_{b=k}^{k_{max}} \sum_{i=1}^{b-1} |S[b, i]| \geq \frac{2^{k-1}}{2^{k-1}-1} |C|$. Namely, NAE-$(k)$-SAT$(U, C)$ can always return a solution with a performance ratio $\frac{2^{k-1}}{2^{k-1}-1}$. The running time of NAE-(k)-SAT(U, C) can be bounded following the analysis for NAE-k-SAT(U, C).

Theorem 4. *For every Max NAE-(k)-SAT instance of n variables and m clauses, the algorithm NAE-(k)-SAT(U, C) can always take $O(\sum_{b=k}^{k_{max}} \beta(b) 2^{b-1} b(m + n)m)$ time to return a solution with a performance ratio no more than $\frac{2^{k-1}}{2^{k-1}-1}$, $\beta(b) \leq \lfloor \frac{b}{2} \rfloor!$, if $b \geq 6$.*

5   Conclusion

In this paper, we have proposed a local search algorithm to solve Max NAE-k-SAT with performance ratio $\frac{2^{k-1}}{2^{k-1}-1}$, then shown that Max NAE-k-SAT cannot be approximated within $\frac{2^{k-1}}{2^{k-1}-1}$ for $k \geq 4$. Benefiting from the algorithm for Max NAE-k-SAT, we have proposed a local search algorithm for Max NAE-(k)-SAT with performance ratio $\frac{2^{k-1}}{2^{k-1}-1}$. The time complexity of the algorithms in this paper are still exponential functions of k. Improving the time complexity of local search for Max NAE-k-SAT to achieve a performance ratio $\frac{2^{k-1}}{2^{k-1}-1}$, remains interesting. Moreover, for scaling up the objective function to integer coefficients, the upper bound of the scaling parameter, $\beta(k)$ namely, cannot be estimated tightly in this paper, if $k > 8$. Obtaining a tighter bound of the scaling parameter may help analyze the time complexity of our algorithms more precisely.

Acknowledgements. We are grateful to the reviewers for giving us many suggestions to enhance the quality of the paper. This paper is supported by: National natural science foundation of China: 61472222, Natural science foundation of Shandong Province: ZR2012FZ002.

References

1. Avidor, A., Berkovitch, I., Zwick, U.: Improved approximation algorithms for MAX NAE-SAT and MAX SAT. In: Erlebach, T., Persinao, G. (eds.) WAOA 2005. LNCS, vol. 3879, pp. 27–40. Springer, Heidelberg (2006)
2. Andersson, G., Engebretsen, L.: Better approximation algorithms for set splitting and not-all-equal SAT. Inf. Process. Lett. **65**(6), 305–311 (1998)
3. Asano, T., Williamson, D.P.: Improved approximation algorithms for MAX-SAT. J. Algorithms **42**(1), 173–202 (2002)
4. Cai, S., Su, K.: Configuration checking with aspiration in local search for SAT. In: Proceedings of AAAI 2012, pp. 434–440 (2012)
5. Chen, J., Friesen, D., Zheng, H.: Tight bound on Johnson's algorithm for maximum satisfiability. J. comput. syst. sci. **58**(3), 622–640 (1999)
6. Cook S A, The complexity of theorem-proving procedures, In: Proceedings of the 3rd Annual ACM Symposium on Theorey of Computing, Shaker Heights, Ohio, USA, pp. 151–158. ACM, New York (1971)
7. Feige, U., Goemans, M.X.: Approximating the value of two power proof systems, with applications to MAX-2SAT and MAX-DICUT, In: Proceedings of the 3rd Israel Symposium on Theorey and Computing Systems, Tel Aviv, Israel, pp. 182–189. IEEE, Washington DC, USA (1995)
8. Goemans, M.X., Williamson, D.P.: New 3/4-approximation algorithms for the maximum satisfiability problem. Siam J. Discrete Math. **7**(4), 656–666 (1994)
9. Goemans, M.X., Williamson, D.P.: Improved approximation algorithms for maximum cut and satisfiability problems using semi-definite programming. J. ACM **42**(6), 1115–1145 (1995)
10. Gu, J.: Efficient Local search for very large-scale satisfiability problem. ACM SIGART Bull. **3**(1), 8–12 (1992)
11. Han, Q., Ye, Y., Zhang, J.: Improved approximation for max set splitting and max NAE SAT. Discrete Appl. Math. **142**(1–3), 133–149 (2004)
12. Hastad, J.: Some optimal inapproximability results. In: Proceedings of the 28th Annual ACM Symposium on Theorey of Computing, pp. 1–10. El Paso, Texas (1997)
13. Huang, W., Zhang, D., Wang, H.: An algorithm based on tabu search for satisfiability problem. J. Comput. Sci. Technol. **17**(3), 340–346 (2002)
14. Johnson, D.S.: Approximation algorithms for combinatorial problems. J. comput. syst. sci. **9**(3), 256–278 (1974)

15. Karloff, H., Zwick, U.: A 7/8-approximation algorithm for MAX 3SAT, In: Proceedings of the 38th IEEE Symposium on Foundations of Computer Science, Miami Beach, Florida. IEEE, Washington DC, USA (1997)
16. Khanna, S., Motwani, R., Madhu, S., Umesh, V.: On syntactic versus computational views of approximability. SIAM J. Comput. $28(1)$, 164–191 (1998)
17. Motwani, R., Raghavan, P.: Randomized Algorithms. Cambridge University Press, Cambridge UK (1995)
18. Mazure, B., Sais, L., Gregoire, E.: Tabu search for SAT, In: Proceedings of the 14th national conference on artificial intelligence (AAAI97), pp. 281–285. AAAI Press, Menlo Park, CA (1997)
19. McAllester, D., Selman, B., Kautz, H.: Evidence for invariants in local search, In: Proceedings Of AAAI 1997. pp. 321–326. AAAI Press, Menlo Park, CA (1997)
20. Papadimitriou, C.H., Yanakakis, M.: Optimization, approximation, and complexity classes. J. Comput. Syst. Sci. $43(3)$, 425–440 (1991)
21. Papadimitirou, C.H.: Computational Complexity. Addison Wesley, Boston (1994)
22. Selman, B., Levesque, H., Mitchell, D.: A new method for solving hard satisfiability problems, In: Proceedings of the 10th National Conference on Artificial Intelligence. pp. 440–446. Pasadena, Calefornia, USA (1992)
23. Selman, B., Kautz, H.A., Cohen, B.: Noise strategies for improving local search, In: Proceedings of the AAAI 1994, pp. 337–343. AAAI Press, Menlo Park, CA (1994)
24. Schurmans, D., Southey, F.: Local search characteristics of incomplete SAT procedures. Artif. Intell. $132(2)$, 121–150 (2001)
25. Yannakakis, M.: On the approximation of maximum satisfiability. J. Algorithms 17, 475–502 (1994)
26. Zwick, U.: Outward rotations: a tool for rounding solutions of semidefinite programming relaxations, with applications to MAX CUT and other problems, In: Proceedings of the 31th Annual ACM Symposium on Theory of Computing, pp. 679–687. Atlanta, Georgia (1999)
27. Zhu, D., Ma, S., Zhang, P.: Tight bounds on local search to approximate the maximum satisfiability problems. In: Fu, B., Du, D.-Z. (eds.) COCOON 2011. LNCS, vol. 6842, pp. 49–61. Springer, Heidelberg (2011)

Faster Computation of the Maximum Dissociation Set and Minimum 3-Path Vertex Cover in Graphs

Mingyu Xiao[✉] and Shaowei Kou

School of Computer Science and Engineering, University of Electronic Science
and Technology of China, Chengdu 611731, China
myxiao@gmail.com

Abstract. A dissociation set in a graph $G = (V, E)$ is a vertex subset D
such that the subgraph $G[D]$ induced on D has vertex degree at most 1.
A 3-path vertex cover in a graph is a vertex subset C such that every
path of three vertices contains at least one vertex from C. Clearly,
a vertex set D is a dissociation set if and only if $V \setminus D$ is a 3-path
vertex cover. There are many applications for dissociation sets and 3-
path vertex covers. However, it is NP-hard to compute a dissociation set
of maximum size or a 3-path vertex cover of minimum size in graphs.
Several exact algorithms have been proposed for these two problems
and they can be solved in $O^*(1.4658^n)$ time in n-vertex graphs. In this
paper, we reveal some interesting structural properties of the two problems, which allow us to solve them in $O^*(1.4656^n)$ time and polynomial
space or $O^*(1.3659^n)$ time and exponential space.

Keywords: Dissociation number · 3-Path Vertex Cover · Exact algorithms · Graph algorithms · Dynamic programming

1 Introduction

A subset of vertices in a graph is called a *dissociation set* if it induces a subgraph
with vertex degree at most 1. The maximum size of a dissociation set is called
the *dissociation number* of the graph. To compute a dissociation set of maximum
size or the dissociation number is NP-hard even in bipartite graphs or planar
graphs [23]. The complexity of this problem in more restricted graph classes has
been studied. It remains NP-hard even in C_4-free bipartite graphs with vertex
degree at most 3 [3]. But it is polynomially solvable in trees and some other graph
classes [1–3,5,6,9,12,14–16]. Computing the dissociation number can be helpful
in finding a lower bound for the 1-improper chromatic number of a graph; see [11].
In fact, dissociation set generalizes two other important concepts in graphs: independent set [20] and induced matching [22]. The MAXIMUM DISSOCIATION SET
problem, to find a maximum dissociation set is also a special case of the MAX-
IMUM BOUNDED-DEGREE-d problem [7], in which we are finding a maximum

Supported by NFSC of China under the Grant 61370071.

J. Wang and C. Yap (Eds.): FAW 2015, LNCS 9130, pp. 282–293, 2015.
DOI: 10.1007/978-3-319-19647-3_26

induced subgraph with degree bounded by d. The dual problem of MAXIMUM DISSOCIATION SET is known as the MINIMUM 3-PATH VERTEX COVER problem. A vertex subset C is called a 3-path vertex cover if every path of three vertices in a graph contains at least one vertex from C and MINIMUM 3-PATH VERTEX COVER is to find a 3-path vertex cover of minimum size. There are also some applications for MINIMUM 3-PATH VERTEX COVER [5,13]. It remains NP-hard to compute a special 3-path vertex cover C such that the degree of the induced graph $G[C]$ is bounded by any constant $d_0 \geq 0$ [21]. A more general problem, to find a minimum p-path vertex cover has been considered in the literature [4,5].

MAXIMUM DISSOCIATION SET and MINIMUM 3-PATH VERTEX COVER have been studied in approximation algorithms, parameterized algorithms and exact algorithms. For MINIMUM 3-PATH VERTEX COVER, there is a randomized approximation algorithm with an expected approximation ratio of $\frac{23}{11}$ [13] and an $O^*(2^k)$-time fixed-parameter tractable algorithm to compute a 3-path vertex cover of size k [19]. However, it is hard to compute a dissociation set of size at least k in approximation and parameterized algorithms. No approximation algorithms with constant ratio are known and it is W[1]-hard with parameter k [21]. In terms of exact algorithms, it does not make sense to distinguish these two problems. Kardoš et al. [13] gave an $O^*(1.5171^n)$-time algorithm to compute a maximum dissociation set in an n-vertex graph. Chang et al. [7] improved the result to $O^*(1.4658^n)$. Their algorithm was analyzed by the measure-and-conquer method. Although many fastest exact algorithms are obtained via the measure-and-conquer method, this paper will not use this technique and turn back to a normal measure. The reason is that if we use the measure-and-conquer method by setting different weights to vertices, we may not be able to use our dynamic programming algorithm to further improve the time complexity to $O^*(1.3659^n)$. It is also surprising that our polynomial-space algorithm using normal measure runs in $O^*(1.4656^n)$ time, even faster than the $O^*(1.4658^n)$-time algorithm analyzed by the measure-and-conquer method [7]. Our improvement relies on some new structural properties developed in this paper. Due to space limitation, some proofs and figures are omitted in this version, which can be found in the full version of this paper.

2   Preliminaries

We let $G = (V, E)$ denote a simple and undirected graph with $n = |V|$ vertices and $m = |E|$ edges. A singleton $\{v\}$ may be simply denoted by v. The vertex set and edge set of a graph G' are denoted by $V(G')$ and $E(G')$, respectively. For a subgraph (resp., a vertex subset) X, the subgraph induced by $V(X)$ (resp., X) is simply denoted by $G[X]$, and $G[V \setminus V(X)]$ (resp., $G[V \setminus X]$) is also written as $G \setminus X$. A vertex in a subgraph or a vertex subset X is also called a X-vertex. For a vertex subset X, let $N(X)$ denote the set of open neighbors of X, i.e., the vertices in $V \setminus X$ adjacent to some vertex in X, and $N[X]$ denote the set of closed neighbors of X, i.e., $N(X) \cup X$. Let $N_2(v)$ denote the set of vertices with distance exactly 2 from v. The degree of a vertex v in a graph G, denoted by

$d(v)$, is defined to be the number of neighbors of v in G. We also use $d_X(v)$ to denote the number of neighbors of v in a subgraph X. A vertex v is *dominated* by a neighbor u of it if v is adjacent to all neighbors of u. A vertex $s \in N_2(v)$ is called a *satellite* of v if there is a neighbor p_s of v such that $N[p_s] - N[v] = \{s\}$. The vertex p_s is also called a *parent* of the satellite s at v. If there is a neighbor u of v such that $|N[u] - N[v]| = 2$, then any vertex in $N[u] - N[v]$ is a *tw-satellite* of v, the two tw-satellites in $N[u] - N[v]$ are *twins*, and u is a *parent* of the tw-satellites at v. The set of all tw-satellites of a vertex v is denoted by T_v. A vertex subset V' is called a *dissociation set* of a graph if the induced graph $G[V']$ has maximum degree 1. In fact, in this paper, we will consider a general version of MAXIMUM DISSOCIATION SET PROBLEM, in which a specified vertex subset S is given and we are going to find a maximum dissociation set containing S. See the following definition.

GENERALIZED MAXIMUM DISSOCIATION SET PROBLEM (MDS)
Input: A undirected graph $G = (V, E)$ and a vertex subset $S \subset V$.
Output: A vertex set $D \supseteq S$ of maximum cardinality such that D is a dissociation set of G.

Our algorithm is a branch-and-search algorithm. In this kind of algorithms, we recursively branch on the current instance into several smaller instances to search for a solution. Assume we use w as the measure to evaluate the size of an instance, where w can be the number of vertices in a graph for graph problems. Let $C(w)$ denote the maximum number of leaves in the search tree generated by the algorithm for any instance with measure at most w. If a branch generates l branches and the measure in the i-th branch decreases by at least a_i, then the branch creates a recurrence

$$C(w) \le C(w - a_1) + C(w - a_2) + \cdots + C(w - a_l).$$

The largest root of the function $f(x) = 1 - \sum_{i=1}^{l} x^{-a_i}$ is called the *branching factor* of the recurrence. The running time of the algorithm can be bounded by $O^*(\gamma^n)$, where γ is the maximum branching factor among all branching factors in the algorithm. More details about the analysis can be found in the monograph [8].

The simplest branching rule in our algorithm is

(B1): *Branching on a vertex $v \in V \setminus S$ to generate two instances by either including v to S or deleting v from the graph directly.*

This rule is not often used, because for most cases it is not effective. Indeed, most of previous papers use the following branching rule

(B2): *Branching on a vertex $v \in V \setminus S$ to generate $d(v) + 2$ instances by either (i) deleting v from the graph (i.e., v is not in the solution set), or (ii) including v to S and deleting all of its neighbors from the graph, or (iii) including v and one of its neighbors to S.*

Branching rule (B2) catches more structural properties of the problem and it is possible to obtain an algorithm with running time better than $O^*(2^n)$.

We will give more branching rules based on special graph structures. With new branching rules, we can improve the running time bound to $O^*(1.4656^n)$. In our branch-and-search algorithm, we will guarantee that each sub-instance created by the algorithm is an induced subgraph of the original graph and the size of S is at most 4. This property allows us to further improve the running time bound by using dynamic programming.

3   Structural Properties

In this section, we always assume that $S = 0$. This means the properties in this section only hold for MAXIMUM DISSOCIATION SET, not for GENERALIZED MAXIMUM DISSOCIATION SET PROBLEM. In fact, the case that $S \neq \emptyset$ will be easy to deal with and we do not need to form some lemmas. Proofs of the lemmas in this section are omitted in this version due to space limitation.

Lemma 1. *If there is a vertex v in a graph G such that the induced graph $G[N(v)]$ is a clique, then there is a maximum dissociation set of G containing v.*

Lemma 2. *Let v be a vertex dominated by u. If there is a maximum dissociation set D' such that $v \in D'$, then there is a maximum dissociation set D containing both u and v.*

Based on Lemma 2, we design a branching rule to deal with dominated vertices.
    (B3): *Branching on a vertex v dominated by u to generate two instances by either deleting v from the graph or including both of v and u to the solution set.*

Lemma 3. *Let v be a degree-2 vertex in a graph G without dominated vertices. There is a maximum dissociation set D containing either v or $N(v)$.*

Lemma 3 allows us to design a branching rule to deal with degree-2 vertices.
    (B4): *Branching on a degree-2 vertex v to generate two instances by either including v to the solution set or including all vertices in $N(v)$ to the solution set and deleting v.*

Lemma 4. *Let G be a graph without dominated vertices. If a vertex v has a satellite s with its parent p_s, then there is a maximum dissociation set D such that at least one of the following holds (i) $v \notin D$, (ii) $v, p_s \in D$ and (iii) $v, u', s \in D$ for some neighbor u' of v not adjacent to s.*

Let s be a satellite of v and $N^s(v)$ be the set of neighbors of v not adjacent to s, where $N^s(v)$ is possibly empty. Based on Lemma 4, we design a branching rule to deal with vertices having satellites.
    (B5): *Branching on a vertex v with some satellite s to generate $2 + |N^s(v)|$ instances by either (i) deleting v from the graph, or (ii) including v and p_s to the solution set, or (iii) including v, u', s to the solution for each $u' \in N^s(v)$.*
    The following lemma will be used to deal with tw-satellites. Recall that we use T_v to denote the set of all tw-satellites of a vertex v.

Lemma 5. *Let G be a graph without dominated vertices and satellite vertices. If there is a vertex v with a neighbor u such that $N[u] - N[v] = \{s_1, s_2\}$, then there is a maximum dissociation set D such that at least one of the following holds: (i) $v \notin D$, (ii) $(\{v\} \cup T_v) \subseteq D$ and $D \cap N(v) = \emptyset$, (iii) $v, u \in D$, (iv) $v, u', s_{u'} \in D$ for some neighbor u' of v adjacent to only one of s_1 and s_2 and $s_{u'} \in \{s_1, s_2\}$ is not adjacent to u', (v) $v, u'' \in D$ for some neighbor u'' of v not adjacent to any of s_1 and s_2.*

Based on Lemma 5, we design a branching rule to deal with vertices having tw-satellites. Let v be a vertex having tw-satellites s_1 and s_2, where s_1 and s_2 are twins having a parent u. Let $W_1(v)$ denote the set of neighbors of v adjacent to only one of s_1 and s_2 and for any vertex $u' \in W_1(v)$, $s_{u'} \in \{s_1, s_2\}$ denote the vertex not adjacent to u'. Let $W_2(v)$ denote the set of neighbors of v not adjacent to any of s_1 and s_2.

(B6): *Branching on v to generate instances by either (i) deleting v from the graph, or (ii) including $\{v\} \cup T_v$ to the solution set and deleting $N(v)$, or (iii) including v, u to the solution, or (iv) including $v, u', s_{u'}$ to the solution set for each $u' \in W_1(v)$, or (v) including v, u'' to the solution for each $u'' \in W_2(v)$.*

Note that in (ii) of Rule (B6), we may include many vertices to S. In fact, we should avoid this to design a dynamic programming algorithm later. So we relax in this branch by including at most three vertices in T_v to the solution set. We revise (B6) to (B6') by replacing (ii) by

(ii') *including $\{v\} \cup T_v'$ to the solution set and deleting $N(v)$, where $T_v' = T_v$ if $|T_v| \leq 3$ and T_v' consists of arbitrary four vertices in T_v if $|T_v| \geq 4$.*

4   The Branch-and-Search Algorithm

We are going to design our algorithm based on the properties and branching rules in the above section. We use $\mathtt{mds}(G, S)$ to denote our algorithm for MDS, which takes a graph G and a vertex subset S as the input and computes a maximum dissociation set D containing S. Our algorithm is a recursive algorithm that consists of two parts. Part I is to deal with the case that $S \neq \emptyset$ and Part II is to deal with the case that $S = \emptyset$. Part I includes six reduction rules and one branching rule, and Part II includes two reduction rules and five branching rules, each of which will call the algorithm itself. Reduction rules simply reduce the instance by deleting some vertices of the graph or moving some vertices to S. Branching rules will generate recurrences with branching factor at most 1.4656. To avoid a confusing nesting of if-then-else statements, we assume that when we design a reduction rule or a branching rule, all previous rules are not applicable.

A vertex is *decided* if it is in S and *undecided* otherwise. We use U to denote the set of undecided vertices, i.e., $U = V \setminus S$. In this paper, we will use $n' = |U|$ as the measure to evaluate the size of the graph and then analyze our algorithm. Note that n' is not greater than the number n of vertices in the graph and this problem can be solved directly when $n' = 0$.

4.1   Part I of the Algorithm Where $S \neq \emptyset$

This part consists of seven steps of the algorithm.

Step 1 (Components). *If the graph G has $l \geq 2$ connected components $H_1, H_2, \ldots, H_l$, return $\cup_{i=1}^{l} \text{mds}(H_i, S_i)$, where $S_i = S \cap V(H_i)$.*

Step 2 (Simple Reduction 1). *If there is an S-vertex v with $d_S(v) \geq 2$, return $\perp$ to indicate no feasible solution.*

Step 3 (Simple Reduction 2). *If there is a U-vertex v adjacent to at least two S-vertices or adjacent to a S-vertex u with $d_S(u) = 1$, delete v from the graph by returning $\text{mds}(G \setminus v, S)$.*

Step 4 (Simple Reduction 3). *If there is a component H of G that contains only one or two S-vertices, return $\text{mds}(G \setminus H, S \setminus V(H)) \cup V(H)$.*

The correctness of these four steps is easy to observe and no new S-vertex is created in these steps. After Step 4, for each S-vertex v, there is no S-vertex in $N(v) \cup N_2(v)$. This property will be used in the following steps.

Step 5 (Special Degree-1 S-vertices). *If there is a degree-1 S-vertex v adjacent to a degree-2 vertex u, then we include u to the solution set directly by returning $\text{mds}(G \setminus N[u], S \setminus v) \cup \{v, u\}$.*

The reason for this step is based on the following observation: If the other neighbor v' of u ($v' \neq v$) is in the final dissociation set D', then we can replace v' with u in D' to get another solution; If $v' \notin D'$, then u should be included in D'. So we can include u to the solution set directly.

Step 6 (Dominated S-vertices). *If there is an S-vertex v dominated by a vertex u, then we include u to the solution set directly by returning $\text{mds}(G \setminus N[v], S \setminus v) \cup \{v, u\}$.*

The reason is similar to that for Step 5. Any solution D' contains at most one vertex $u^* \in N(v)$. If $u^* \neq u$, we can replace u^* with u in D' to get another solution. If $D' \cap N(v) = \emptyset$, then u could be added to D' directly to get a solution of larger size. So u can be included to the solution set.

Step 7 (Branching on Neighbors of S-vertices). *If $S \neq \emptyset$, branch on a neighbor u of an S-vertex v by deleting u from the graph or including u to the solution set, i.e., return one of the following with maximum size*

$$\text{mds}(G \setminus u, S) \quad and \quad \text{mds}(G \setminus N[\{v, u\}], S \setminus v) \cup \{v, u\}.$$

We analyze this branching operation. In the first branch, at least one U-vertex u is reduced. In the second branch, u is also reduced and all vertices in $N(\{v, u\})$ are reduced. We can see that $|N(\{v, u\})| \geq 2$ holds. It is impossible that $|N(\{v, u\})| = 0$ since Step 6 has been applied. If $|N(\{v, u\})| = 1$, then either v is a degree-1 S-vertex and u is degree-2 neighbor of v or v is a degree-2 S vertex dominated by u. For this case, either Step 5 or Step 6 would be applied. Therefore, in the second branch, at least $1 + |N(\{v, u\})| \geq 3$ U-vertices are reduced. We get a recurrence

$$C(n') \leq C(n' - 1) + C(n' - 3), \tag{1}$$

which has a branching factor of 1.4656.

Note that no new S-vertex is created in these steps. Next, we describe the second part of our algorithm.

4.2 Part II of the Algorithm Where $S = \emptyset$

After Step 7, it always holds that $S = \emptyset$. We still have 8 steps.

Step 8 (Trivial Cases). *If the graph has maximum degree at most 2, solve the problem directly in polynomial time and return a maximum dissociation set.*

Step 9 (Domination Reduction). *If there is a vertex v such that $N(v)$ induces a clique, then return $\mathtt{mds}(G, \{v\})$ by Lemma 1.*

Note that any degree-1 U-vertex satisfies the condition in this step. So after Step 9, the graph has no degree-1 vertex any more.

Step 10 (Domination Branching). *If there is a vertex v dominated by u, then branching on v with Rule (B3) by returning one of the following with maximum size*

$$\mathtt{mds}(G \setminus v, \emptyset) \quad and \quad \mathtt{mds}(G \setminus N[\{v, u\}], \emptyset) \cup \{v, u\}.$$

This step is based on Lemma 2. In the first branch, one U-vertex v is reduced. In the second branch, $|N[\{v, u\}]|$ U-vertices are reduced. Note that no of v and u can be a degree-1 vertex after Step 9. Then there are at least 3 vertices in $N[\{v, u\}]$. In this step, we can get (1) at least. After this step, the graph has no dominated vertex.

Step 11 (Degree-2 Vertices). *If there is a degree-2 vertex v, then branching on v with Rule (B4) by returning one of the following with maximum size*

$$\mathtt{mds}(G, \{v\}) \quad and \quad \mathtt{mds}(G \setminus \{v\}, N(v)).$$

The correctness of this rule is based on Lemma 3. Since $|N(v)| \geq 2$, there are at least 3 vertices reduced in the second branch. This step also gives (1).

After Step 11, the graph has no vertex of degree ≤ 2.

Step 12 (Vertices of Degree ≥ 4 having Satellites). *If there is a vertex v of degree ≥ 4 having a satellite s, then branch on v with Rule (B5) by returning one of the following with maximum size*

$$\mathtt{mds}(G \setminus \{v\}, \emptyset), \ \mathtt{mds}(G \setminus N[\{v, p_s\}], \emptyset) \cup \{v, p_s\}, \quad and$$
$$\arg \max_{u' \in N^s(v)} |\mathtt{mds}(G \setminus N[\{v, u'\}], \{s\}) \cup \{v, u'\}|. \tag{2}$$

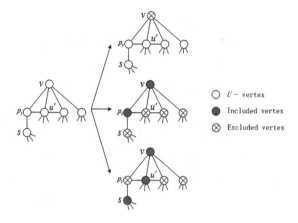

Fig. 1. Illustration for Step 12

Figure 1 gives an illustration for the operation in Step 12.

One U-vertex is reduced in the first branch, $d(v) + 2$ U-vertices are reduced in the second branch, and at least $d(v) + 3$ U-vertices are reduced in the other branches because each vertex $u' \in N^s(v)$ is adjacent to at least one vertex $t \in N_2(v) \setminus \{s\}$. Note that $|N^s(v)| \leq d(v) - 1$. Let $d = d(v)$. We get

$$C(n') \leq C(n' - 1) + C(n' - (d + 2)) + (d - 1) \cdot C(n' - (d + 3)), \qquad (3)$$

where $d \geq 4$. For the worst case that $d = 4$, the branching factor is 1.4602.

Step 13 (Vertices of Degree ≥ 4 having tw-satellites). *If there is a vertex v of degree ≥ 4 having two tw-satellites s_1 and s_2, where s_1 and s_2 are twins having a parent u, then branch on v with Rule (B6') by returning one of the following with maximum size*

$$\mathtt{mds}(G \setminus \{v\}, \emptyset), \quad \mathtt{mds}(G \setminus N[v], T'_v) \cup \{v\},$$
$$\mathtt{mds}(G \setminus N[\{v, u\}], \emptyset) \cup \{v, u\},$$
$$\arg \max_{u' \in W_1(v)} |\mathtt{mds}(G \setminus N[\{v, u'\}], \{s_{u'}\}) \cup \{v, u'\}|, \quad and$$
$$\arg \max_{u'' \in W_2(v)} |\mathtt{mds}(G \setminus N[\{v, u''\}], \emptyset) \cup \{v, u''\}|.$$

Figure 2 gives an illustration for the operation in Step 13.

Recall that W_1 is the set of neighbors of v adjacent to only one of s_1 and s_2, and W_2 is the set of neighbors of v not adjacent to any of s_1 and s_2. Let $d = d(v) \geq 4$, $x_1 = |W_1|$ and $x_2 = |W_2|$. We have that

$$x_1 + x_2 \leq d - 1.$$

In the first branch, one U-vertex is reduced. In the second branch, $|N[v]| + |T'_v| = d+1+|T'_v|$ U-vertices are reduced. In the third branch, $d+3$ U-vertices are reduced. In the fourth branch, there are at least $|N[\{v, u'\}]|+1 \geq d+4$ U-vertices

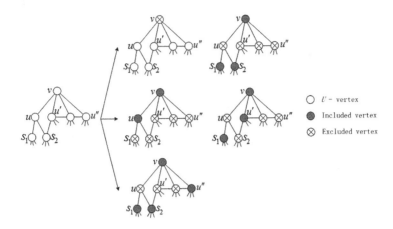

Fig. 2. Illustration for Step 13

reduced. In the last branch, the number of reduced U-vertices is $|N[\{v, u''\}]|$. We consider three cases according to the value of $|T'_v|$.

Case 1. $|T'_v| = 2$: then $|T_v| = 2$. For this case, we know that $x_1 = 0$, $x_2 \leq d - 1$ and $|N[\{v, u''\}]| \geq d + 4$ for each $u'' \in W_2(v)$. Then we can branch with a recurrence

$$C(n') \leq C(n' - 1) + C(n' - (d + 3)) + C(n' - (d + 3)) + (d - 1) \cdot C(n' - (d + 4)),$$

which has a maximum branching factor of 1.4460 when $d = 4$.

Case 2. $|T'_v| = 3$: then $|T_v| = 3$. For this case, we still have that $|N[\{v, u''\}]| \geq d + 4$ for each $u'' \in W_2(v)$. Then we can branch with a recurrence

$$\begin{aligned}
C(n') \leq\ & C(n' - 1) + C(n' - (d + 4)) + C(n' - (d + 3)) + x_1 \cdot C(n' - (d + 4)) + \\
& x_2 \cdot C(n' - (d + 4)) \\
\leq\ & C(n' - 1) + C(n' - (d + 3)) + d \cdot C(n' - (d + 4)),
\end{aligned}$$

which has a maximum branching factor of 1.4346 when $d = 4$.

Case 3. $|T'_v| = 4$: For this case, we may only have that $|N[\{v, u''\}]| \geq d + 3$ for each $u'' \in W_2(v)$. Then we get

$$\begin{aligned}
C(n') \leq\ & C(n' - 1) + C(n' - (d + 5)) + C(n' - (d + 3)) + x_1 \cdot C(n' - (d + 4)) + \\
& x_2 \cdot C(n' - (d + 3)) \\
\leq\ & C(n' - 1) + C(n' - (d + 5)) + d \cdot C(n' - (d + 3)),
\end{aligned}$$

which has a maximum branching factor of 1.4605 when $d = 4$.

After Step 13, no vertex of degree ≥ 4 has satellites or tw-satellites. Then each neighbor u of a vertex v of degree ≥ 4 is adjacent to at least three vertices in $N_2(v)$.

Step 14 (Vertices of Degree ≥ 5). *If there is a vertex v of $d(v) \geq 5$, then branch on v with Rule (B2) by returning one of the following with maximum size*

$$\text{mds}(G \setminus \{v\}, \emptyset), \text{mds}(G \setminus N[v], \emptyset) \cup \{v\},$$
$$\arg \max_{u \in N(v)} |\text{mds}(G \setminus N[\{u, v\}], \emptyset) \cup \{u, v\}|.$$

Note that for each neighbor u of v, there is $|N(u) - N(v)| \geq 3$. So in the branch that u and v are included to the solution set, at least $d + 4 = d(v) + 4$ U-vertices are reduced. This leads to the recurrence

$$C(n) \leq C(n - 1) + C(n - (d + 1)) + d \cdot C(n - (d + 4)). \tag{4}$$

For the worst case of $d = 5$, this recurrence has a branching factor of 1.4374.

Lemma 6. *After Step 14, $S = \emptyset$ and each connected component of the graph is a regular graph of degree 3 or 4.*

Step 15 (Regular Graphs of Degree 3 or 4). *Pick up an arbitrary vertex v and branch on v with Rule (B2) by returning one of the following with maximum size*

$$\text{mds}(G \setminus \{v\}, \emptyset), \text{mds}(G \setminus N[v], \emptyset) \cup \{v\},$$
$$\arg \max_{u \in N(v)} |\text{mds}(G \setminus N[\{u, v\}], \emptyset) \cup \{u, v\}|.$$

We do not analyze the branching factor for this step. In fact, we will show that this step will not exponentially increase the running time bound of the algorithm.

Theorem 1. *A maximum dissociation set in graph with n vertices can be computed in $O^*(1.4656^n)$ time and polynomial space.*

In fact, Algorithm $\text{mds}(G, \emptyset)$ satisfies this theorem. Note that each step of $\text{mds}(G, \emptyset)$, except Step 15, either reduces the graph directly or branches with a branching factor at most 1.4656. Note that any induced subgraph of a connected 4-regular (resp., 3-regular) graph is not a 4-regular (resp., 3-regular) graph. Then we can bound the number of times executing Step 15 and then prove this theorem. For more details of the proof, please refer to the full version of this paper.

It is also easy to observe that in each step of $\text{mds}(G, S)$, any created graph is an induced subgraph of the initial graph and the size of S is at most 4. This property allows us design a dynamic programming algorithm to improve the running time.

5   Reducing the Time Complexity via Dynamic Programming

In our recursive algorithm, the same subproblem I can appear many times. The smaller the size of the subproblem, the higher the probability of the subproblem appearing repeatedly. The idea is then to store the solutions to the subproblems of small size in a database in advance. The database can be implemented in such a way that the *query time* is logarithmic in the number of solutions stored. When our algorithm needs to solve a subproblem of small size, we do not solve it directly but we refer to our database, which ensures that a given subproblem will not be solved twice. This technique was firstly used by Robson [17] to reduce the running time bound for MAXIMUM INDEPENDENT SET and later it was used in many other problems [10, 18].

Let n_0 be the number of vertices in the initial graph and $\alpha = 0.1845$. We create a database *database* that contains solutions to all sub-instances $I = (G, S)$ such that the number of U-vertices in I is at most αn_0 and $|S| \leq 4$. We modify $\mathtt{mds}(G, S)$ to $\mathtt{dp}(G, S)$ by adding Step 0: If the number of U-vertices in the graph is at most αn_0, then return $D = database(G, S)$.

Lemma 7. *The database can be constructed in $O^*(1.3659^{n_0})$ time.*

A proof of this lemma can be found in the full version of this paper. We can interpret our improved algorithm as: first compute the solutions to the small instances in the database in a dynamic programming way, which takes $O^*(1.3659^n)$ time; second use a branch-and-search method to solve the instance with more than αn U-vertices, which has branching factors at most 1.4656 and then creates a search tree of size $O(1.4656^{(n-\alpha n)}) = O(1.3659^n)$. Furthermore, in each leaf of the search tree, the algorithm searches the databases in polynomial query time to find a solution directly. In total, the algorithm uses $O^*(1.3659^n)$ time and space.

Theorem 2. *A maximum dissociation set in an n-vertex graph can be found in $O^*(1.3659^n)$ time and space.*

6   Concluding Remarks

In this paper, we have presented improved polynomial-space and exponential-space algorithms for MAXIMUM DISSOCIATION SET. The polynomial-space algorithm is a branch-and-search algorithm, which takes $C(n') \leq C(n'-1) + C(n'-3)$ as the worst recurrences and then runs in $O^*(1.4656^{n'})$ time, where n' is at most the number of vertices in the initial graph. The improvement is mainly obtained by developing several new structural properties, such as Lemmas 4 and 5. With these properties we can adopt some branches in our search algorithm and then reduce the size of the search tree. Another technique to get the improvement is that we use the number n' of U-vertices (which can be regarded as 'undecided' vertices) as the measure instead of the number of total vertices in the graph. This idea helps us to greatly simplify our algorithm and its analysis. Furthermore, we will guarantee that the recursive algorithm only calls itself on induced subgraphs of the initial graph and the number of S-vertices (which are not counted in our measure n' but still in the graph) is always bounded by a constant. This allows us use classical dynamic programming to improve the running time bound with exponential space. Finally, we achieve the running time bound $O^*(1.3659^n)$.

References

1. Alekseev, V.E., Boliac, R., Korobitsyn, D.V., Lozin, V.V.: NP-hard graph problems and boundary classes of graphs. Theoret. Comput. Sci. **389**, 219–236 (2007)

2. Asdre, K., Nikolopoulos, S.D., Papadopoulos, C.: An optimal parallel solution for the path cover problem on P_4-sparse graphs. J. Parallel Distrib. Comput. **67**, 63–76 (2007)
3. Boliac, R., Cameron, K., Lozin, V.V.: On computing the dissociation number and the induced matching number of bipartite graphs. Ars Comb. **72**, 241–253 (2004)
4. Brešar, B., Jakovac, M., Katrenič, J., Semanišin, G., Taranenko, A.: On the vertex k-path cover. Discrete Appl. Math. **161**, 1943–1949 (2013)
5. Brešar, B., Kardoš, F., Katrenič, J., Semanišin, G.: Minimum k-path vertex cover. Discrete Appl. Math. **159**, 1189–1195 (2011)
6. Cameron, K., Hell, P.: Independent packings in structured graphs. Math. Program. **105**, 201–213 (2006)
7. Chang, M-S., Chen, L-H., Hung, L-J., Liu, Y-Z., Rossmanith, P., Sikdar, S.: An $O^*(1.4658^n)$-time exact algorithm for the maximum bounded-degree-1 set problem. In: Proceedings of the 31st Workshop on Combinatorial Mathematics and Computation Theory, pp. 9–18 (2014)
8. Fomin, F.V., Kratsch, D.: Exact Exponential Algorithms. Springer, Berlin (2010)
9. Göring, F., Harant, J., Rautenbach, D., Schiermeyer, I.: On F-independence in graphs. Discussiones Math. Graph Theor. **29**, 377–383 (2009)
10. Grandoni, F.: A note on the complexity of minimum dominating set. J. Discrete Algorithms **4**, 209–214 (2006)
11. Havet, F., Kang, R.J., Sereni, J.-S.: Improper colouring of unit disk graphs. Networks **54**, 150–164 (2009)
12. Hung, R.-W., Chang, M.-S.: Finding a minimum path cover of a distance-hereditary graph in polynomial time. Discrete Appl. Math. **155**, 2242–2256 (2007)
13. Kardoš, F., Katrenič, J.: On computing the minimum 3-path vertex cover and dissociation number of graphs. Theoret. Comput. Sci. **412**, 7009–7017 (2011)
14. Lozin, V.V., Rautenbach, D.: Some results on graphs without long induced paths. Inf. Process. Lett. **88**, 167–171 (2003)
15. Orlovich, Y., Dolgui, A., Finke, G., Gordon, V., Werner, F.: The complexity of dissociation set problems in graphs. Discrete Appl. Math. **159**, 1352–1366 (2011)
16. Papadimitriou, C.H., Yannakakis, M.: The complexity of restricted spanning tree problems. J. ACM **29**, 285–309 (1982)
17. Robson, J.M.: Algorithms for maximum independent sets. J. Algorithms **7**, 425–440 (1986)
18. Rooij, J.M.M., Bodlaender, H.L.: Exact algorithms for dominating set. Discrete Appl. Math. **159**, 2147–2164 (2011)
19. Tu, J.: A fixed-parameter algorithm for the vertex cover P3 problem. Inf. Process. Lett. **115**(2), 96–99 (2015)
20. Xiao, M., Nagamochi, H.: Exact algorithms for maximum independent set. In: Cai, L., Cheng, S.-W., Lam, T.-W. (eds.) Algorithms and Computation. LNCS, vol. 8283, pp. 328–338. Springer, Heidelberg (2013)
21. Xiao, M., Nagamochi, H.: Complexity and kernels for bipartition into degree-bounded induced graphs. In: Ahn, H.-K., Shin, C.-S. (eds.) ISAAC 2014. LNCS, vol. 8889, pp. 429–440. Springer, Heidelberg (2014)
22. Xiao, M., Tan, H.: An improved exact algorithm for maximum induced matching. In: Jain, R., Jain, S., Stephan, F. (eds.) TAMC 2015. LNCS, vol. 9076, pp. 272–283. Springer, Heidelberg (2015)
23. Yannakakis, M.: Node-deletion problems on bipartite graphs. SIAM J. Comput. **10**, 310–327 (1981)

Enumeration, Counting, and Random Generation of Ladder Lotteries

Katsuhisa Yamanaka[1]([⊠]) and Shin-ichi Nakano[2]

[1] Department of Electrical Engineering and Computer Science, Iwate University, Ueda 4-3-5, Morioka, Iwate 020-8551, Japan
yamanaka@cis.iwate-u.ac.jp
[2] Department of Computer Science, Gunma University, Tenjin-cho 1-5-1, Kiryu, Gunma 376-8515, Japan
nakano@cs.gunma-u.ac.jp

Abstract. A *ladder lottery*, known as "Amidakuji" in Japan, is one of the most popular lotteries. In this paper, we consider the problems of enumeration, counting, and random generation of the ladder lotteries. For given two positive integers n and b, we give algorithms of enumeration, counting, and random generation of ladder lotteries with n lines and b bars.

1 Introduction

A *ladder lottery*, known as "Amidakuji" in Japan, is one of the most popular lotteries for kids. It is often used to assign roles to members in a group. Imagine that a group of four members A, B, C, and D wish to determine their group leader using a ladder lottery. First, four vertical lines are drawn, then each member chooses a vertical line. See Fig. 1(a). Next, a check mark (which represents an assignment of the leader) and some horizontal lines are drawn, as shown in Fig. 1(b). The derived one is called a ladder lottery, and it represents an assignment. In this example the leader is assigned to D since the top-to-bottom route from D ends at the check mark. (We will explain about the route soon.) In Fig. 1(b), the route is drawn as a dotted line.

Formally, a *ladder lottery* is a network with $n \geq 2$ vertical lines (*lines* for short) and b horizontal lines (*bars* for short) each of which connects two consecutive vertical lines. We count the lines from left to right and call the i-th line from the left i-*th line*. See Fig. 2 for an example. The top ends of the lines correspond to a permutation $\pi = (p_1, p_2, \ldots, p_n)$ of $[n] = \{1, 2, \ldots, n\}$, and the bottom ends of the lines correspond to the identity permutation $\iota = (1, 2, \ldots, n)$ and they satisfy the following rule. Each p_i in π starts the top end of the i-th line, then goes down along the line; whenever p_i meets an end of a bar, p_i goes horizontally along the bar to the other end, and then goes down again. Finally, p_i must reach the bottom end of the p_i-th line. Each bar corresponds to a modification of the current permutation by swapping the two neighboring elements.

A ladder lottery appears in a variety of areas. First, a ladder lottery of the reverse permutation $(n, n-1, \ldots, 1)$ corresponds to a pseudoline arrangement

© Springer International Publishing Switzerland 2015
J. Wang and C. Yap (Eds.): FAW 2015, LNCS 9130, pp. 294–303, 2015.
DOI: 10.1007/978-3-319-19647-3_27

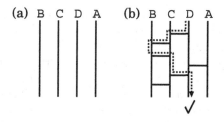

Fig. 1. An example of a ladder lottery.

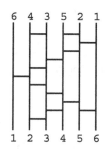

Fig. 2. A ladder lottery of (6,4,3,5,2,1).

in discrete geometry [7]. By replacing bars as intersections of pseudolines, ladder lotteries can be regarded as pseudoline arrangements, and it is observed that there is a one-to-one correspondence between pseudoline arrangements and "optimal" ladder lotteries of a reverse permutation [7]. A ladder lottery of a permutation is *optimal* if it has the minimum number of bars among ladder lotteries of the permutation. Second, it is strongly related to primitive sorting networks, which are deeply investigated by Knuth [3]. Third, in algebraic combinatorics, a reduced decomposition (by adjacent transpositions) of a permutation corresponds to a ladder lottery of the permutation with the minimum number of bars [4].

In this paper we consider the problems of enumeration, counting, and random generation of ladder lotteries. We propose three algorithms for these three problems. All the three algorithm are based on the code [1] of ladder lotteries.

2   Preliminaries

Code of Ladder Lotteries

In this subsection, we review a code of ladder lotteries in [1]. Using this code, we design three algorithms in this paper.

Let L be a ladder lottery with n lines and b bars. We first divide each bar of L into two horizontal line-segments, called *half-bars*. The left half of a bar is called an *l-bar* (left half-bar) and the right half of a bar is called an *r-bar* (right half-bar). We regard each original bar as a pair of an l-bar and an r-bar. Thus L

has $2b$ half-bars. The division results in n connected components, each of which consists of one line and some half-bars attached to the line.

We can encode how half-bars are attached to the i-th line, as follows. Let $\langle b_1, b_2, \ldots \rangle$ be the sequence of half-bars attached to the i-th line appearing from top to bottom. We replace b_i with 0 if b_i is an r-bar, and with 1 if b_i is an l-bar. Then appending a 0 to indicate the end-of-line. This results in the code of the i-th line, which is denoted by $C(i)$. Concatenating those codes $C(1), C(2), \ldots, C(n)$ results in the code $C(L)$ for L. For example, for the ladder lottery in Fig. 2 $C(1) = $ "10", $C(2) = $ "110110", $C(3) = $ "01001100", $C(4) = $ "1100100", $C(5) = $ "010010", $C(6) = $ "000", and

$$C(L) = \text{"10110110010011001100100010010000".}$$

Since the code contains two bits for each bar and one bit for each end-of-line, its length is $n + 2b$ bits.

Reconstruction from the Code

Now we explain how to reconstruct the original ladder lottery from the code.

In the code, a 0 represents either an r-bar or an end-of-line. Hence, we need to recognize the end-of-lines to partite $C(L)$ into $C(1), C(2), \ldots, C(n)$. After then, it is easy to reconstruct original bars by connecting the corresponding l-bars and r-bars, since the k-th l-bar of the i-th line and the k-th r-bar of the $(i+1)$-th line correspond to an original bar. Figure 3 shows an example of the reconstruction of the ladder lottery in Fig. 2 from its code.

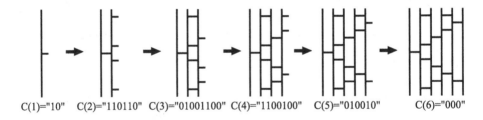

C(1)="10"   C(2)="110110" C(3)="01001100" C(4)="1100100" C(5)="010010"   C(6)="000"

Fig. 3. An example of the reconstruction from the code

We now explain how to recognize the end-of-lines. Since the first line has only l-bars, the first consecutive 1s correspond to the l-bars of the first line, so the first 0 is the end-of-line of the first line. Now we assume that the end-of-line for the $(i$-1$)$-th line is recognized and we are now going to recognize the end-of-line for the i-th line. We know the number, say k, of l-bars attached to the $(i$-1$)$-th line, and it equals to the number of r-bars attached to the i-th line. Then the end-of-line for the i-th line is the $(k+1)$-th line 0 after the end-of-line for the $(i$-1$)$-th line.

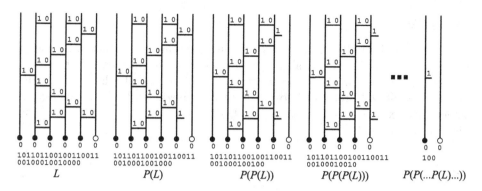

$$L \qquad P(L) \qquad P(P(L)) \qquad P(P(P(L))) \qquad P(P(...P(L)...))$$

Fig. 4. Pre-ladders derived from L.

Theorem 1 [[1]]. *Let L be a ladder lottery with n lines and b bars. One can encode L into a bitstring of length $n + 2b$. Both encoding and decoding can be done in $\mathcal{O}(n + b)$ time.*

Pre-ladder and its Code

Let L be a ladder lottery with n lines and b bars, and let $C(L)$ be the code of L. We define a substructure of L, as follows. Let $P(C(L))$ be the bitstring derived from $C(L)$ by removing the second last bit, and $P(L)$ be the substructure of L derived by "decoding" $P(C(L))$. Intuitively, $P(L)$ is the substructure of L only missing either a half-bar or an end-of-line, corresponding to the second last bit. Similarly $P(P(C(L)))$ is the bitstring derived from $P(C(L))$ by removing the second last bit, and $P(P(L))$ be the corresponding substructure of L. Similarly, we define $P(P(P(C(L)))), P(P(P(P(C(L))))), \dots$. We assume that a pre-ladder has at least two lines. See Fig. 4 for an example. In the figure, end-of-lines are depicted as black circles except for the end-of-line of the rightmost line, which is depicted as a white circle. We say each of those substructure (including L itself) a *pre-ladder* of L, and the sequence $L, P(L), P(P(L)), \dots$ the *removing sequence* of L. A pre-ladder possibly has unmatched l-bars only at the two rightmost lines.

3   Enumeration

Let $\mathcal{S}_{n,b}$ be the set of ladder lotteries with n lines and b bars. In this section, we consider the problem of enumerating all ladder lotteries in $\mathcal{S}_{n,b}$. We have presented an algorithm that enumerates all "optimal" ladder lotteries of a given permutation [7]. However, this algorithm cannot applied to the problem, since $\mathcal{S}_{n,b}$ includes (non-optimal) ladder lotteries. In this section, we propose a simple enumeration algorithm for $\mathcal{S}_{n,b}$.

Our enumeration algorithm is based on reverse search [2]. We first define a forest structure in which each leaf one-to-one corresponds to some ladder lottery in $\mathcal{S}_{n,b}$. Then, by traversing the forest, we can enumerate all leaves of the forest, and all corresponding ladder lotteries in $\mathcal{S}_{n,b}$. We designed several enumeration algorithms based on similar (but distinct) tree structures [5–7].

Family Forest

Let L be a ladder lottery in $\mathcal{S}_{n,b}$. By merging the removing sequence for every $L \in \mathcal{S}_{n,b}$, we have the forest, called *family forest* $F_{n,b}$, in which each leaf one-to-one corresponds to some ladder lottery in $\mathcal{S}_{n,b}$. We regard each edge corresponds to some parent-child relation between the two pre-ladders. Each root is a pre-ladder with exactly two lines and no half-bar attached to the second line. See Fig. 5 for an example.

Child Enumeration

We have the following lemma according to the parent-child relation.

Lemma 1. *Given any pre-ladder R in $F_{n,b}$, one can enumerate all child pre-ladders of R in $\mathcal{O}(1)$ time for each.*

By recursively enumerating all child pre-ladders of a derived pre-ladder in $F_{n,b}$, we have the following theorem.

Theorem 2. *One can enumerate all ladder lotteries with n lines and b bars in $\mathcal{O}(n+b)$ time for each. Our algorithm uses $\mathcal{O}(n+b)$ space.*

4   Counting

In this section we consider a counting problem. Given two positive integers $n \geq 2$ and $b \geq 0$, we wish to count the number of ladder lotteries with n lines and b bars. Using the enumeration algorithm in the previous section, we can count such ladder lotteries one by one, but very slowly. This method takes $\Omega(|\mathcal{S}_{n,b}|)$ time, which may be exponential on n and b. In this section, we propose an efficient counting algorithm. Our algorithm does not count ladder lotteries one by one, but counts each "type" of pre-ladders all together, and runs in polynomial time.[1]

We now define the type for each pre-ladder. A pre-ladder R is *type $t(\ell, h, p, q)$* if R satisfies the following conditions:

(a) R contains $\ell \geq 2$ lines;
(b) R contains $h \geq p + q$ half-bars (Each bar is counted as two half-bars);
(c) p unmatched l-bars are attached to the $(\ell\text{-}1)$-th line; and
(d) q unmatched l-bars are attached to the ℓ-th line.

For example, the pre-ladder $P(L)$ in Fig. 4 is type $t(6, 25, 1, 0)$. Note that any ladder lottery with n lines and b bars is type $t(n, 2b, 0, 0)$. We denote by $T(\ell, h, p, q)$ the set of pre-ladders of type $t(\ell, h, p, q)$. We give a useful recurrence for $|T(\ell, h, p, q)|$.

We have the following four cases.

---

[1] We assume that n and b are coded in unary codes.

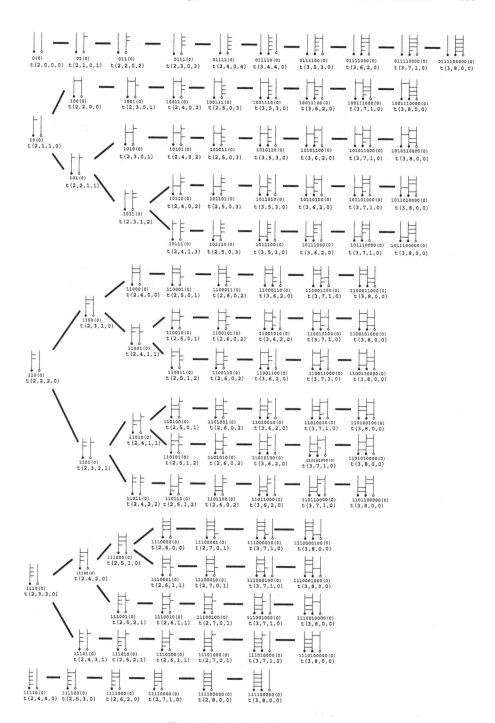

Fig. 5. The family forest $F_{3,4}$.

Case 1: $h < p + q$ or $\ell < 2$.
$|T(\ell, h, p, q)| = 0$ holds, since $h \geq p + q$ and $\ell \geq 2$ hold for any pre-ladder.

Case 2: $\ell = 2$, $q = 0$, and $h = p$
Clearly such pre-ladder is unique, so $|T(\ell, h, p, q)| = 1$ holds.

Case 3: $h \geq p + q$ and $q = 0$.
Let R be a pre-ladder of type $t(\ell, h, p, q)$. The second last bit of $C(R)$ is always 0. (Otherwise, ℓ-th line has an l-bar, a contradiction.) The second last bit 0 in $C(R)$ represents either an r-bar of ℓ-th line or the end-of-line of $(\ell\text{-}1)$-th line. For the former case $P(R)$ is type $t(\ell, h - 1, p + 1, 0)$. For the latter case $P(R)$ is type $t(\ell - 1, h, 0, p)$. For any distinct R_1 and R_2 of $t(\ell, h, p, q)$ with $h \geq p + q$ and $q = 0$, $P(R_1)$ and $P(R_2)$ are distinct. Thus $|T(\ell, h, p, 0)| = |T(\ell, h - 1, p + 1, 0)| + |T(\ell - 1, h, 0, p)|$ holds.

Case 4: $h \geq p + q$ and $q > 0$.
Let R be a pre-ladder of type $t(\ell, h, p, q)$. The second last bit in $C(R)$ is either 0 or 1. If the second last bit of $C(R)$ is 0, then it represents an r-bar attached to ℓ-th line. Thus, $P(R)$ is type $t(\ell, h-1, p+1, q)$. Otherwise, the second last bit of $C(R)$ is 1, then it represents an l-bar attached to ℓ-th line. Hence, $P(R)$ is type $t(\ell, h - 1, p, q - 1)$. Thus $|T(\ell, h, p, q)| = |T(\ell, h - 1, p + 1, q)| + |T(\ell, h - 1, p, q - 1)|$ holds.

For example, Fig. 6 shows the recurrence for $|T(3, 8, 0, 0)|$. By the recurrence, we have the following lemma.

Lemma 2. *For four non-negative integers ℓ, h, p, and q,*

$|T(\ell, h, p, q)|$

$$
= \begin{cases}
0 & if\, h < p + q\ \ or\, \ell < 2 \\
1 & if\, \ell = 2, q = 0,\ \ and\, h = p \\
|T(\ell, h - 1, p + 1, 0)| + |T(\ell - 1, h, 0, p)| & if\, h \geq p + q\ \ and\, q = 0 \\
|T(\ell, h - 1, p + 1, q)| + |T(\ell, h - 1, p, q - 1)| & if\, h \geq p + q\ \ and\, q > 0
\end{cases}
$$

Based on the recurrence above, **Algorithm 1** computes the number of ladder lotteries with n lines and b bars. **Algorithm 1** is a dynamic programming algorithm on the table of types. The number of entries is nb^3, and each entry is calculated in constant time, so the total running time is $\mathcal{O}(nb^3)$. As a byproduct the number of ladder lotteries with every $n' \leq n$ lines and every $b' \leq b$ bars are also computed.

Theorem 3. *The number of ladder lotteries with every $n' \leq n$ lines and every $b' \leq b$ bars can be calculated in $\mathcal{O}(nb^3)$ time in total.*

5   Random Generation

In this section we consider random generation of ladder lotteries. The recurrence in Lemma 2 generates a tree structure among the types (see an example

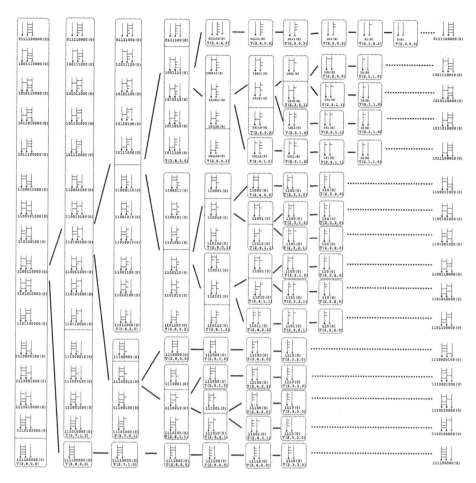

Fig. 6. The recurrence for $|T(3, 8, 0, 0)|$.

in Fig. 6), in which each path from the root to a leaf one-to-one corresponds to some ladder lottery of type $t(n, 2b, 0, 0)$. The choice of i-th generation type decides the meaning of the $(i+1)$-th last bit of the code. (Here the root belongs to the first generation.)

The table generated by **Algorithm 1** tells us the number of leaves in the subtree rooted at each type. We can choose a random path from the root to some leaf, by repeatedly choosing some child of the current type so that each leaf has an equal chance to be reached. Thus we can generate ladder lotteries, uniformly at random.

Our algorithm is shown in **Algorithm 2**. Suppose that we are now at a type $T(\ell, h, p, q)$ in the tree structure, and $T(\ell_1, h_1, p_1, q_1)$ and $T(\ell_2, h_2, p_2, q_2)$ are the two child types of $T(\ell, h, p, q)$. **Algorithm 2** computes a random value,

---

Algorithm 1. DP-COUNT(n, b)

---

1 **for** $\ell = 2$ **to** n **do**
2 $\quad$ **for** $h = 0$ **to** $2b$ **do**
3 $\quad\quad$ **for** $p = 0$ **to** h **do**
4 $\quad\quad\quad$ **for** $q = 0$ **to** h **do**
5 $\quad\quad\quad\quad$ **if** $h < p + q$ **then**
6 $\quad\quad\quad\quad\quad$ $|T(\ell, h, p, q)| = 0$
7 $\quad\quad\quad\quad$ **else if** $\ell = 2$, $q = 0$, *and* $h = p$ **then**
8 $\quad\quad\quad\quad\quad$ $|T(\ell, h, p, q)| = 1$
9 $\quad\quad\quad\quad$ **else if** $q = 0$ **then**
10 $\quad\quad\quad\quad\quad$ $|T(\ell, h, p, q)| = |T(\ell, h - 1, p + 1, 0)| + |T(\ell - 1, h, 0, p)|$
11 $\quad\quad\quad\quad$ **else if** $q > 0$ **then**
12 $\quad\quad\quad\quad\quad$ $|T(\ell, h, p, q)| = |T(\ell, h - 1, p + 1, q)| + |T(\ell, h - 1, p, q - 1)|$

Algorithm 2. RANDOM-GENERATION(ℓ, h, p, q)

---

1 **begin**
2 $\quad$ **if** $\ell = 2$, $h = p$, *and* $q = 0$ **then**
3 $\quad\quad$ **return** *the ladder lottery corresponding to the path from the root to the current leaf.*
4 $\quad$ **else**
5 $\quad\quad$ **if** $T(\ell, h, p, q)$ *has only one child, say* $T(\ell_1, h_1, p_1, q_1)$ **then**
6 $\quad\quad\quad$ RANDOM-GENERATION(ℓ_1, h_1, p_1, q_1)
7 $\quad\quad$ /* Let $T(\ell_1, h_1, p_1, q_1)$ and $T(\ell_2, h_2, p_2, q_2)$ be the two child types of $T(\ell, h, p, q)$. */
8 $\quad\quad$ Generate an integer x in $[1, |T(\ell, h, p, q)|]$ uniformly at random.
9 $\quad\quad$ **if** $x \leq |T(\ell_1, h_1, p_1, q_1)|$ **then**   /* Choose $T(\ell_1, h_1, p_1, q_1)$ */
10 $\quad\quad\quad$ RANDOM-GENERATION(ℓ_1, h_1, p_1, q_1)
11 $\quad\quad$ **else**   /* Choose $T(\ell_2, h_2, p_2, q_2)$. */
12 $\quad\quad\quad$ RANDOM-GENERATION(ℓ_2, h_2, p_2, q_2)

---

say x, in $[1, |T(\ell, h, p, q)|]$ uniformly at random, chooses $T(\ell_1, h_1, p_1, q_1)$ if $x \leq |T(\ell_1, h_1, p_1, q_1)|$ and $T(\ell_2, h_2, p_2, q_2)$ otherwise, then recursively call with the chosen type. Since **Algorithm 1** computes the table as the preprocessing, these numbers can be looked up in $\mathcal{O}(1)$ time. Thus we can generate ladder lotteries uniformly at random, as in the following theorem.

Theorem 4. *Given two integers $n \geq 2$ and $b \geq 0$, after computing the table of $|T(\ell, h, p, q)|$ by **Algorithm 1**, we can generate a ladder lottery with n lines and b bars in $\mathcal{O}(n + b)$ time for each, uniformly at random.*

6   Summary

We have designed three algorithms for enumeration, counting, and random generation of ladder lotteries with n lines and b bars. All the three algorithms are based on the code [1] of ladder lotteries.

Our enumeration algorithm enumerates all the ladder lotteries with n lines and b bars in $\mathcal{O}(n+b)$ time for each. Our counting algorithm counts the number of ladder lotteries with n lines and b bars in $\mathcal{O}(nb^3)$ time. Our random generation algorithm takes $\mathcal{O}(nb^3)$ time as a preprocessing, then generates ladder lotteries with n lines and b bars $\mathcal{O}(n+b)$ time for each, uniformly at random.

Acknowledgment. This work is partially supported by MEXT/JSPS KAKENHI, including the ELC project. (Grant Numbers 24106007 and 25330001).

References

1. Aiuchi, T., Yamanaka, K., Hirayama, T., Nishitani, Y.: Coding ladder lotteries. In: Proceedings of European Workshop on Computational Geometry 2013, Braunschweig, Germany, pp. 151–154, March 2013
2. Avis, D., Fukuda, K.: Reverse search for enumeration. Discrete Appl. Math. **65**(1–3), 21–46 (1996)
3. Knuth, D.E.: Axioms and Hulls. LNCS, vol. 606. Springer, Heidelberg (1992)
4. Manivel, L.: Symmetric Functions, Schubert Polynomials and Degeneracy Loci. American Mathematical Society, Providence (2001)
5. Nakano, S.: Efficient generation of triconnected plane triangulations. Comput. Geom. Theory Appl. **27**(2), 109–122 (2004)
6. Yamanaka, K., Nakano, S.: Listing all plane graphs. J. Graph Algorithms Appl. **13**(1), 5–18 (2009)
7. Yamanaka, K., Nakano, S., Matsui, Y., Uehara, R., Nakada, K.: Efficient enumeration of all ladder lotteries and its application. Theoret. Comput. Sci. **411**, 1714–1722 (2010)

Efficient Modular Reduction Algorithm Without Correction Phase

Haibo Yu[1]($\boxtimes$), Guoqiang Bai[2], and Huikang Hao[3]

[1] Department of Computer Science and Technology, Tsinghua University,
Beijing, China
yhb13@mails.tsinghua.edu.cn
[2] Institute of Microelectronics, Tsinghua University, Beijing, China
baigq@tsinghua.edu.cn
[3] School of Computer Science and Engineering,
Beihang University, Beijing, China
hhk@cse.buaa.edu.cn

Abstract. In this paper, we propose an efficient modular reduction algorithm without correction phase based on Barrett reduction algorithm. We eliminate the error between the estimated quotient and the real one, which leads that the proposed algorithm can perform modular reduction with less operation steps. On the other hand, it can also make for a decrease in extra subtraction circuits. According to the data flow diagram, it is crystal clear that the number of control steps of original Barrett algorithm is twice the proposed algorithm. Therefore, the throughput of our method is doubled compared with Barrett algorithm with same clock frequency. Additionally, the proposed method can be a good solution for high-speed applications.

Keywords: Public-key cryptography · Modular reduction · Barrett algorithm · Correction phase · High-speed hardware implementation

1 Introduction

Just shortly after Diffie and Hellman proposed the first public-key exchange protocol in 1976 [1], Rivest, Shamir and Adleman proposed the first practical public-key cryptosystem (PKC) in 1978 [2]. For now, elliptic curve is quite ubiquitous in mathematics and computing, and Elliptic Curve Cryptography (ECC) can offer equivalent security as RSA with much smaller key sizes [3,4], which has been proven on both theoretical and practical aspects. From a mathematic viewpoint, public-key encryption and decryption algorithms are strategically implemented in Residue Number System (RNS), where one of the cornerstones of PKC is modular arithmetic.

Because of modular addition being relatively simple, modular multiplication is still the core operation of many public-key cryptosystems, in which modular reduction is of importance. For high-speed hardware implementation, cryptography puts forward higher requirement of the throughput of modular multiplication [5,6]. Therefore, the high-throughput implementation of modular reduction is necessary.

J. Wang and C. Yap (Eds.): FAW 2015, LNCS 9130, pp. 304–313, 2015.
DOI: 10.1007/978-3-319-19647-3_28

The two most widely used algorithm for modular reduction are Montgomery reduction and Barrett reduction, which were both introduced in 1980's. [7] shows the detail of Barrett modular reduction algorithm. In 2000, J. F. Dhem and J.J. Quisquater optimized the original algorithm and named it as the improved Barrett method in [8,9] proposed a high-speed hardware implementation of RSA based on Barrett's modular reduction. After several years, [10] presented a modified Barrett algorithm with single folding in 2007. Miroslav Knežević et al. introduced modular reduction without precomputaional phase with some special modulus in 2009 [11], based on the original algorithm, and [12] proposed an efficient way for reduction of large integers by small modulus.

All contributions mentioned above have not changed the basic idea, replacing the exact quotient value with the estimated one, which could avoid complicated division operations. However, because of the error between the estimated value and the exact value, correction phase is essential to ensure reduction result being in the correct interval [8], which needs extra judgment and subtraction circuits for hardware implementation.

To satisfy the requirements for high-speed applications, the error between exact value and estimated value of quotient should be fixed or clear. Thus in this paper, we propose a high-speed modular reduction method without correction phase, which means the method can eliminate the estimated error to obtain reduction result. Additionally, because of only basic arithmetic operations involved, our method is suitable for hardware implementation especially.

The organization of this paper is given as follows: Euclidean division, the original Barrett reduction and the improved Barrett reduction algorithm are described in Sect. 2, which do a brief look back at the related work; in Sect. 3, the efficient modular reduction algorithm and the related hardware implementations are proposed, being the core part of this paper and Sect. 4 concludes the paper.

2   Preliminaries

To make the following explanations and descriptions easier, we show some notations used in this paper firstly. An n-bit integer X is represented in radix 2 representation as $X = (X_{n-1} \cdots X_0)_2$ and $X_{n-1} \neq 0$; X_i refers to the i-th bit of X, $0 \leq i \leq n - 1$ and $X_i \in \{0, 1\}$; $|X|$ refers to the bit length of X in radix 2 and $|X| = n$ based on description above. Note that all integers discussed in this paper are represented in radix 2.

2.1   Euclidean Division

Given two nonnegative integers U and V, with $U \neq 0$, the Euclidean division is the process of U and V producing unique integers q and r to satisfy the equation $U = qV + r$, with $0 \leq r < V$ [13].

The integer q, called *the quotient*, is denoted by the floor function as $q = \lfloor \frac{U}{V} \rfloor$, in which the floor function always rounds toward -∞. The integer r, called *the remainder*, is represented as $U \pmod V$. In addition, the floor function is suitable for hardware implementation if V is of type 2^w, which is only briefly touched in shifting operations.

2.2 Barrett Reduction

Barrett reduction algorithm, introduced by P. D. Barrett in 1986 [7], is a classical reduction method. This algorithm computes the remainder $r = N \pmod{M}$ for an input N and a fixed modulus M, with an estimated value of quotient.

With $|M| = m$ and $|N| = 2m$, we may write the quotient as

$$q = \left\lfloor \frac{N}{M} \right\rfloor = \left\lfloor \frac{\frac{N}{2^{m-1}} \cdot \frac{2^{2m}}{M}}{2^{m+1}} \right\rfloor$$

The estimation of q is recorded as

$$\hat{q} = \left\lfloor \frac{\left\lfloor \frac{N}{2^{m-1}} \right\rfloor \left\lfloor \frac{2^{2m}}{M} \right\rfloor}{2^{m+1}} \right\rfloor \tag{1}$$

where $\left\lfloor \frac{2^{2m}}{M} \right\rfloor$, being a constant for a fixed modulus M, can be precomputed.

Now we can get the reduction value of N modulo M, being written as $r = N - \hat{q}M$, that is

$$r = N - \left\lfloor \frac{\left\lfloor \frac{N}{2^{m-1}} \right\rfloor \left\lfloor \frac{2^{2m}}{M} \right\rfloor}{2^{m+1}} \right\rfloor M$$

Based on precomputational phase, Barrett reduction algorithm can take care of the estimated q-value without complicated division operation. Barrett reduction algorithm is shown in Algorithm 1.

Algorithm 1. Barrett reduction algorithm

Input: N, where $|N| = 2m$; M, being fixed and $|M| = m$; precomputing $\mu = \left\lfloor \frac{2^{2m}}{M} \right\rfloor$;
Output: $r = N \pmod{M}$;
1: $q_1 \Leftarrow \lfloor N/2^{m-1} \rfloor$;
2: $q_2 \Leftarrow q_1 \cdot \mu$;
3: $q_3 \Leftarrow \lfloor q_2/2^{m+1} \rfloor$;
4: $r_1 \Leftarrow q_3 \cdot M \pmod{2^{m+1}}$
5: $r \Leftarrow N \pmod{2^{m+1}} - r_1$
6: **if** $r < 0$ **then**
7:     $r \Leftarrow r + 2^{m+1}$;
8: **end if**;
9: **if** $r \geq M$ **then**
10:     $r \Leftarrow r - M$;
11: **end if**;
12: **return** r;

2.3   Improved Barrett Algorithm

Algorithm 1 only considers the case of $2m$-digit input data and m-digit modulus. However, when $|N| \neq 2|M|$, the error between exact value and estimated one may grow unpredictably. To improve the original algorithm, [8] introduced more parameters to minimize the error.

Rewrite the estimated value as

$$\hat{q} = \left\lfloor \frac{\left\lfloor \frac{N}{2^{m+\beta}} \right\rfloor \left\lfloor \frac{2^{m+\alpha}}{M} \right\rfloor}{2^{\alpha-\beta}} \right\rfloor \tag{2}$$

where α and β are parameters. Using the original Barrett reduction, m and -1 are in the roles of α and β respectively. For a fixed modulus M, $|M| = m$, assume arbitrary input data N with $|N| = m + \gamma$, such that $\gamma > 0$ and $N_{m+\gamma-1} \neq 0$. Finally

$$\left\lfloor \frac{N}{M} \right\rfloor - 2^{\gamma-\alpha} - 2^{\beta+1} + 2^{\beta-\alpha} - 1 < \hat{q} \leq \left\lfloor \frac{N}{M} \right\rfloor \tag{3}$$

The detail of derivation process can be found in [8], where the improved algorithm can guarantee the error limitation with appropriate parameters α, β and γ. To minimize the error, J. F. Dhem considered $\beta \leq -2$ and $\alpha > \gamma$, which result in $\hat{q} = q$ or $\hat{q} = q - 1$.

Both the original Barrett reduction and the improved Barrett reduction algorithms described above try their best to replace the real quotient with the estimated one and minimize the error. However, in order to perform modular reduction operation, they need a condition judgment to ensure reduction result being in the interval $[0, M)$ exactly, such as step 6 and step 9 in Algorithm 1, called *correction phase*.

It is evident that the reason of extra condition judgment is the error. From the implementation point of view, the appearance of error reduces the system throughput forcing one to add extra judgment circuit module and increase operation time, which has a serious impact on the circuit performance for high-speed system. Hence, both algorithms are not suitable for high speed applications. In next section we will show the proposed modular reduction algorithm without estimated error.

3   The Proposed Modular Reduction Algorithm

To meet the requirements of high-speed applications, the error between exact value and estimated one should be fixed or clear. For a fixed error, we can get the exact q-value with adding or subtracting the error from the estimated q-value. While elimination of error is the best way for high-speed modular reduction operation, which can overcome the shortcomings of two modular reduction method described above. In this section, we propose a high-speed modular reduction method without correction phase.

The mathematical algorithm of modular reduction is introduced firstly for which the estimated error in the original Barrett reduction or the improved Barrett reduction can be efficiently avoided. Besides, it is worth noting that the proposed algorithm is suitable for hardware implementation especially. Last but not least, we show the data flow diagram and the datapath of the proposed algorithm, which can indicate visually our advantages for high-speed applications.

3.1   Mathematical Algorithm of the Proposed Modular Reduction

Before describing the proposed algorithm, we first give some basic definitions to make the following explanations clearly. Let us, for now, assume that M and N are nonnegative integers whose lengths are m and $m + \gamma$ bits respectively, that is $|M| = m$ and $|N| = m + \gamma$, such that $\gamma > 0$, $2^{m-1} \leq M < 2^m$ and $2^{m+\gamma-1} \leq N < 2^{m+\gamma}$.

Based on Euclidean division, the quotient q is recorded as

$$q = \left\lfloor \frac{N}{M} \right\rfloor = \left\lfloor \frac{N \cdot \frac{2^{m+\alpha}}{M}}{2^{m+\alpha}} \right\rfloor \tag{4}$$

where α is a parameter and $\alpha > 0$. According to (4), we define $B = N \cdot \frac{2^{m+\alpha}}{M}$. It immediately follows that $q = \left\lfloor \frac{B}{2^{m+\alpha}} \right\rfloor$

Just the same as Barrett modular reduction, we take care of the estimated q-value instead of the real value. Based on (4), we define estimated quotient as

$$q' = \left\lfloor \frac{N \left\lfloor \frac{2^{m+\alpha}}{M} \right\rfloor}{2^{m+\alpha}} \right\rfloor \tag{5}$$

Let us define $B' = N \cdot \left\lfloor \frac{2^{m+\alpha}}{M} \right\rfloor$ and then we have $q' = \left\lfloor \frac{B'}{2^{m+\alpha}} \right\rfloor$.

Starting with the main idea of the proposed modular reduction algorithm, we present a lemma as follow, which shows the relation between the real quotient q and the estimated value q'.

Lemma 1. *Known that* $q = \left\lfloor \frac{B}{2^{m+\alpha}} \right\rfloor$ *and* $q' = \left\lfloor \frac{B'}{2^{m+\alpha}} \right\rfloor$, *the conclusion holds* $\left\lfloor \frac{B}{2^{m+\alpha}} - 2^{\gamma-\alpha} \right\rfloor \leq q' \leq q$.

Proof. Because we always have $\frac{U}{V} - 1 < \lfloor \frac{U}{V} \rfloor \leq \frac{U}{V}$ and $B' = N \left\lfloor \frac{2^{m+\alpha}}{M} \right\rfloor$, we may write

$$N \left(\frac{2^{m+\alpha}}{M} - 1 \right) < B' = N \left\lfloor \frac{2^{m+\alpha}}{M} \right\rfloor \leq N \frac{2^{m+\alpha}}{M}$$

$$N \frac{2^{m+\alpha}}{M} - N < B' \leq N \frac{2^{m+\alpha}}{M}$$

Since $B = N\frac{2^{m+\alpha}}{M}$, rewrite the inequality above

$$B - N < B' \le B \tag{6}$$

Dividing by $2^{m+\alpha}$ for the inequality (6), it is easy to obtain that

$$\frac{B}{2^{m+\alpha}} - \frac{N}{2^{m+\alpha}} < \frac{B'}{2^{m+\alpha}} \le \frac{B}{2^{m+\alpha}}$$

According to the definition of the floor function and input integer N, we have $N < 2^{m+\gamma}$, which leads to

$$\left\lfloor \frac{B}{2^{m+\alpha}} - 2^{\gamma-\alpha} \right\rfloor \le \left\lfloor \frac{B'}{2^{m+\alpha}} \right\rfloor \le \left\lfloor \frac{B}{2^{m+\alpha}} \right\rfloor \tag{7}$$

$$\left\lfloor \frac{B}{2^{m+\alpha}} - 2^{\gamma-\alpha} \right\rfloor \le q' \le q$$

It is the relation between q and q'.   ∎

Modular arithmetic is regarded as the cornerstone of public-key cryptography, in which modular multiplication is the core operation [9]. As the basic step of modular multiplication, the results of modular reduction are always not zero. Therefore, it is reasonable to assume $\frac{N}{M} = B/2^{m+\alpha}$ is not an integer, which means the value of $B/2^{m+\alpha}$ consists of two parts, integral part and fractional part.

Under the premise of the fractional part of $\frac{B}{2^{m+\alpha}}$ being greater than or equal to $2^{\gamma-\alpha}$, it is always true that $\left\lfloor \frac{B}{2^{m+\alpha}} - 2^{\gamma-\alpha} \right\rfloor = \left\lfloor \frac{B}{2^{m+\alpha}} \right\rfloor$ with the quality of floor function. In combination with (7), $\left\lfloor \frac{B}{2^{m+\alpha}} \right\rfloor$ is exactly equal to $\left\lfloor \frac{B'}{2^{m+\alpha}} \right\rfloor$ in this case. Recall the definition of q and q', we can get the conclusion:

$$q = q'$$

if the fractional part of $\frac{B}{2^{m+\alpha}}$ is not less than $2^{\gamma-\alpha}$.

Known that $\frac{N}{M} = B/2^{m+\alpha}$ and $|M| = m$, the minimum fractional part of $\frac{B}{2^{m+\alpha}}$ is 2^{-m}. Thus it requires

$$2^{-m} \ge 2^{\gamma-\alpha} \Rightarrow \alpha \ge m + \gamma \tag{8}$$

Hence, we draw a conclusion that q' is indeed a excellent estimate of q without estimated error and no correction step is needed to obtain $r = N \pmod{M}$.

The proposed modular reduction algorithm is presented in Algorithm 2. It is important to point out that we can have obviously less computational steps with the fixed modulus M. On the other hand, the proposed algorithm only involves in multiplication, shift and substraction operations, which is suitable for high-speed hardware and software implementations.

In contrast to the original Barrett modular reduction (shown in Algorithm 1) or the improved algorithm, our proposed algorithm does not include the correction steps. In the original Barrett reduction, the number of correction steps is at most 2 [7], and in the improved Barrett reduction, the number of correction steps is at most 1 [8]. While in our algorithm, this number can be zero, which reduces the number of redundant subtractions efficiently and brings a higher throughput. Thus it is suitable for high-speed applications.

Algorithm 2. The proposed modular reduction

Input: N, where $|N| = m + \gamma$; M, being fixed and $|M| = m$; precomputing $\mu = \left\lfloor \frac{2^{m+\alpha}}{M} \right\rfloor$;

Output: $r = N \pmod{M}$;

1: $q_1 \Leftarrow N \cdot \mu$;
2: $q_2 \Leftarrow \lfloor q_1/2^{m+\alpha} \rfloor$;
3: $r_1 \Leftarrow q_2 \cdot M$
4: $r \Leftarrow N - r_1$
5: **return** r;

3.2 Hardware Implementation

To verify our approach in practice, we show the data flow diagram and datapath of the proposed modular reduction in this subsection. The control and data flow is the basis of datapath synthesis [14], in which functional units and storage units are allocated to a special control step. Figure 1 shows the detail of the data flow diagrams of the proposed algorithm, on the left side, and the original Barrett algorithm, on the right side.

The control steps are numbered with t_i, where the number i is nonnegative and consecutive. For each control step, system can finish the related operations, such as arithmetic operations and logic operations. Time required for different operations can be very different from each other. Assigning the separate weighting factors to related control step or splitting a long control step into several ones can be a solution to balance control steps. For simplicity, every control step is deemed to be identical in terms of computation time in the following analysis.

In the proposed modular reduction, there are one multiplier named Mul1, one subtractor named Sub1 and three registers named R1, R2 and R3, respectively, where registers are represented as circles in the data flow diagram. As shown in Fig. 1, one right-shifter is required and one of registers can be used for the shifter, which can reduce the complexity of integrated circuits and simplifies circuit layout and wiring. In this case, R2 is the shift register. Note that the output results of multiplier always store in R2, which means the output terminal of Mul1 connects with the register R2 fixedly to save hardware cost of multiplexer.

The data flow diagram of original Barrett modular reduction is represented in the right part of Fig. 1. There are also one multiplier named Mul1', one subtractor/adder named Sub1' and four registers named R1', R2', R3' and R4'. An extra register is required to store the value of 2^{m+1} and the output of shift operation, compared with the proposed algorithm. Obviously, the correction phase in original Barrett modular reduction algorithm increases the number of control steps.

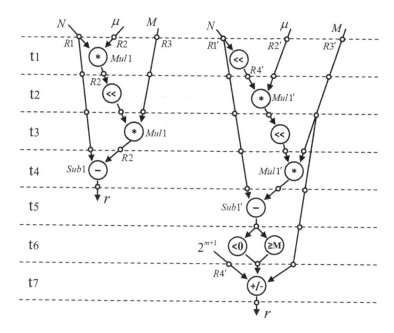

Fig. 1. Data flow diagrams of the proposed reduction algorithm and Barrett algorithm

Datapath of the proposed modular reduction algorithm, based on the data flow diagram, is shown in Fig. 2, which involves in three multiplexers. The more multiplexers are, the more complex circuit controlling signals becomes. A fixed connection between the output of Mul1 and the register R2, would bring not only small circuit scale but concise control codes in system. As mentioned above, the shifter in datapath can be replaced by a register with shifting operation, where the register R2 is the right choice.

According to the data flow diagram, it is crystal clear that the number of control step of original Barrett algorithm is twice the proposed algorithms. Based on many achievements of fast multipliers, multiplication on the critical path can be performed in one clock cycle with pipeline design, which means multiplication step can be identical with other control steps in terms of computation time. Accordingly, throughput of our method can be doubled with same clock frequency in the best case, compared with correction-needed Barrett reduction. In other words, for a given throughput, the clock frequency can be halved with respect to modular reduction using the proposed algorithm. Therefore, hardware implementation of the proposed algorithm could bring high throughput compared with the original method, which is suitable for high-speed applications.

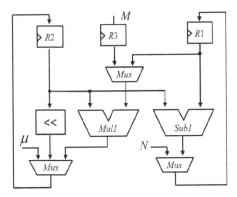

Fig. 2. Datapath of the proposed modular reduction algorithm

4   Conclusion

In this paper, we introduce an efficient modular reduction algorithm based on Barrett modular reduction. Because of high-precision multiplication operation, the value of remainder can be obtained in very simple control steps. Compared with the original Barrett reduction or the improved algorithm, the proposed algorithm does not include the correction steps. In the original Barrett reduction, the number of correction steps is at most 2, and in the improved Barrett reduction, the number of correction steps is at most 1. While in our algorithm, this number can be zero, which reduces the number of redundant subtractions efficiently and brings a higher throughput. Thus, the proposed algorithm is suitable for high-speed applications.

Acknowledgment. This work is supported by the National Natural Science foundation of China (No.61472208 and U1135004). The authors would like to thank the editor and reviewers for their comments.

References

1. Diffie, W., Hellman, M.E.: New directions in cryptography. IEEE Trans. Inf. Theory **22**(6), 644–654 (1976)
2. Rivest, R.L., Shamir, A., Adleman, L.: A method for obtaining digital signatures and public-key cryptosystems. Commun. ACM **21**(2), 120–126 (1978)
3. Miller, V.S.: Use of elliptic curves in cryptography. In: Williams, H.C. (ed.) CRYPTO 1985. LNCS, vol. 218, pp. 417–426. Springer, Heidelberg (1986)
4. Koblitz, N.: Elliptic curve cryptosystems. Math. Comput. **48**(177), 203–209 (1987)
5. Knežević, M., Vercauteren, F., Verbauwhede, I.: Speeding up bipartite modular multiplication. In: Hasan, M.A., Helleseth, T. (eds.) WAIFI 2010. LNCS, vol. 6087, pp. 166–179. Springer, Heidelberg (2010)
6. Knezevic, M., Vercauteren, F., Verbauwhede, I.: Faster interleaved modular multiplication based on barrett and montgomery reduction methods. IEEE Trans. Comput. **59**(12), 1715–1721 (2010)

7. Barrett, P.: Implementing the rivest shamir and adleman public key encryption algorithm on a standard digital signal processor. In: Odlyzko, A.M. (ed.) CRYPTO 1986. LNCS, vol. 263, pp. 311–323. Springer, Heidelberg (1987)

8. Dhem, J.-F., Quisquater, J.-J.: Recent results on modular multiplications for smart. In: Schneier, B., Quisquater, J.-J. (eds.) CARDIS 1998. LNCS, vol. 1820, pp. 336–352. Springer, Heidelberg (2000)

9. Großschädl, J.: High-speed RSA hardware based on Barret's modular reduction method. In: Paar, C., Koç, Ç.K. (eds.) CHES 2000. LNCS, vol. 1965, pp. 191–203. Springer, Heidelberg (2000)

10. Hasenplaugh, W., Gaubatz, G., Gopal, V.: Fast modular reduction. In: 18th IEEE Symposium on Computer Arithmetic, ARITH 2007, pp. 225–229. IEEE (2007)

11. Knezevic, M., Batina, L., Verbauwhede, I.: Modular reduction without precomputational phase. In: IEEE International Symposium on Circuits and Systems, ISCAS 2009, pp. 1389–1392. IEEE (2009)

12. Rutten, L., Van Eekelen, M.: Efficient and formally proven reduction of large integers by small moduli. ACM Trans. Math. Soft. (TOMS) **37**(2), 16 (2010)

13. PUBLIC-KEY CRYPTOGRAPHY. Primality testing and integer factorization in public-key cryptography (2009)

14. Paulin, P.G., Knight, J.P.: Force-directed scheduling for the behavioral synthesis of asics. IEEE Trans. Comput.-Aided Des. Integr. Circ. Syst. **8**(6), 661–679 (1989)

Super Solutions of Random Instances
of Satisfiability

Peng Zhang and Yong Gao[✉]

Department of Computer Science, Irving K. Barber School of Arts and Sciences,
University of British Columbia Okanagan, Kelowna V1V 1V7, Canada
yong.gao@ubc.ca

Abstract. We study the probabilistic behaviour of super solutions to random instances of the Boolean Satisfiability (SAT) and Constraint Satisfaction Problems (CSPs). Our analysis focuses on a special type of super solutions, the (1,0)-super solutions. For random k-SAT, we establish the exact threshold of the phase transition of the solution probability for the cases of $k = 2$ and 3, and upper and lower bounds on the threshold of the phase transition for the case of $k \geq 4$. For CSPs, by overcoming difficulties that do not exist in the probabilistic analysis of the standard solution concept, we manage to derive a non-trivial upper bound on the threshold for the probability of having a super solution.

1 Introduction

For many problems arising in uncertain, dynamic, or interactive environments, it is desirable to find solutions that can be modified at a low cost in response to changes of the environment. This requires that a solution has a certain degree of robustness or stability. For example, super solutions have been used to formalize the notion of a robust or stable solution to the Boolean Satisfiability problem and the constraint satisfaction problem [8,10]. An (a, b)-super solution to a CSP instance is a satisfying solution such that, if the values assigned to any set of a variables are no longer available, a new solution can be found by reassigning values to these a variables and at most b other variables.

In general, finding super solutions to SAT and CSPs is NP-complete. One of the fruitful approaches to such hard problems is to understand the typical-case complexity of a hard problem by studying the probabilistic behaviour of random instances [2,9]. By analyzing the threshold phenomena of the solution probability and the correlated easy-hard-easy pattern of the instance hardness of the standard solution concept for SAT and CSPs, much insight has been gained on the effectiveness of the many heuristics widely-used in practice to tackle these problems [6,7,9,11–13].

In this paper, we study the probabilistic behaviour of super solutions to random instances of SAT and CSPs. Our analysis focuses on a special (but highly non-trivial) type of super solutions, the (1,0)-super solutions. We denote the problems of finding $(1,0)$-super solution for k-SAT and CSPs by $(1,0)$-k-SAT and $(1,0)$-CSP respectively.

© Springer International Publishing Switzerland 2015
J. Wang and C. Yap (Eds.): FAW 2015, LNCS 9130, pp. 314–325, 2015.
DOI: 10.1007/978-3-319-19647-3_29

In Sect. 2, we establish the exact threshold for the probability of having a $(1,0)$-super solutions to random 3-SAT by making use of an observation on the equivalence between a $(1,0)$-k-SAT and a standard satisfying solution of a properly-constructed $(k-1)$-SAT instance. In Sect. 3, we establish upper and lower bounds on the threshold of random $(1,0)$-k-SAT for $k \geq 4$. In Sect. 4, we study possible upper bounds on the threshold of random $(1,0)$-CSPs.

2   Super Solutions for Boolean Satisfiability

Let $X = \{x_1, x_2, \cdots, x_n\}$ be a set of n Boolean variables. A literal is a variable or its negation. A k-*clause* is a disjunction of k different literals and a k-CNF formula is a conjunction of k-clauses. An assignment σ is a mapping $\sigma : X \to \{1,0\}^n$ and is said to satisfy a k-CNF formula F if each clause of F contains at least one literal that evaluates to true under σ. A satisfying assignment is also called a solution.

2.1   Equivalent Definitions of $(1,0)$-Super Solutions

As a special case of (a,b)-super solutions, a $(1,0)$-super solution for a k-SAT is a solution such that changing the value assigned to exactly one variable will not violate any clause. Equivalently, a $(1,0)$-super solution is an assignment such that every clause contains at least two literals that evaluate to true under the assignment. Another equivalent condition for a $(1,0)$-super solution is given below and plays a crucial role in our analysis.

Definition 1. *The projection of a clause $C = l_1 \vee \cdots \vee l_k$ is defined to be the conjunction of all $(k-1)$-clauses contained in C, i.e. $\pi(C) = \wedge_{i=1}^{k}(\vee_{j \neq i} l_j)$. We say that C projects onto $\pi(C)$ and call clauses in $\pi(C)$ siblings.*
  The projection of a CNF formula F is defined to be $\pi(F) = \wedge_{C_i \in F} \pi(C_i)$.

The following observation can be proved easily.

Lemma 1. *An assignment (1,0)-satisfies F if and only if it satisfies $\pi(F)$.*

The following theorem complements existing results on the worst-case complexity of super solutions [10].

Theorem 1. $(1,0)$-k-*SAT is in P for $k \leq 3$, and is NP-Complete otherwise.*

Proof. Any instance of (1,0)-3-SAT F can be solved by solving the 2-SAT instance of $\pi(F)$, which is in P. For $k \geq 4$, we first prove the NP-completeness of (1,0)-4-SAT via a reduction from 3-SAT. Note that, σ satisfies $(l_1 \vee l_2 \vee l_3)$ if and only if it $(1,0)$-satisfies $(l_1 \vee l_2 \vee l_3 \vee 1)$. For any 3-SAT F, we reduce it into a 4-SAT F' as following in three steps. First, create 4 additional variables, $Y = \{y_1, y_2, y_3, y_4\}$ and a 4-SAT F_y of all the possible $\binom{4}{2}$ clauses, where each clause has exactly two negations of variables. In order to $(1,0)$-satisfy F_y, we have $\sigma_y(y_i) = 1, 1 \leq i \leq 4$. Secondly, for each clause c_i in F, add $c_i' = (c_i \vee y_1)$ into F'. Finally, let F' be the conjunction of F' and F_y. Now, σ is a solution of F if and only if it is a $(1,0)$-solution of F'. Thus $(1,0)$-4-SAT is NP-complete. Similar methods can reduce any k-SAT instance to $(1,0)$-$(k+1)$-SAT instance.

2.2   Random Models of k-SAT

We denote by $F_k(n, m)$ the standard random model for k-CNF formulas on n variables, where the m clauses are selected uniformly at random without replacement from the set of all possible $2^k \binom{n}{k}$ k-clauses. We say that a sequence of events $\mathcal{E}_n$ occurs *with high probability* (w.h.p.) if $\lim_{n \to \infty} \mathbb{P}[\mathcal{E}_n] = 1$. As sometimes it is hard to directly analyze $F_k(n, m)$ due to the dependence created by selecting the clauses without replacement, we consider two related models. The first model selects from all $2^k \binom{n}{k}$ proper clauses with replacement. The second model selects each literal uniformly and independently with replacement. Both models may result in improper formula and the second model may have improper clauses. As long as k is fixed, the number of improper clauses and repeated clauses is $o(n)$. Therefore, with-high-probability properties of (1,0)-satisfiability hold in all these three models simultaneously. For notation convenience, we denote all three models by $F_k(n, m)$. When $k \le 3$, we use the first model. When $k \ge 4$, we use the second model. We also assume that k is fixed.

Due to Lemma 1, the probability for $F_k(n, m)$ to be (1,0)-satisfiable equals the probability for its projection $\pi(F)$ to be satisfiable. This, however, does not imply that the probability for a random $F_k(n, m)$ to be (1, 0)-satisfiable equals the probability for $F_{k-1}(n, km)$ to be satisfiable. The following result on the exact threshold of the solution probability of (1,0)-2-SAT is not hard to establish.

Theorem 2. $F_2(n, m)$ *is (1,0)-satisfiable w.h.p. when* $m = o(\sqrt{n})$ *and is (1,0)-unsatisfiable w.h.p. when* $m = \omega(\sqrt{n})$.

Proof. We say that two clauses are conflicting if some literal in one clause is the negation of some literal in the other clause. Note that a 2-CNF formula F is (1,0)-satisfiable if and only if no conflicting clauses exists. Let $F = C_1 \wedge \cdots \wedge C_m$ and $X_{i,j}$ be the indicator variable that C_i conflicts with C_j. Then, $\mathbb{E}[X_{i,j}] = \frac{2(2(n-1)-1)+1}{2^2 \binom{n}{2}} = \frac{4n-5}{2n(n-1)}$. Denote by $X = \sum_{(i,j)} X_{i,j}$ the number of conflicting pairs in F.

$$\mathbb{E}[X] = \binom{m}{2} \mathbb{E}[X_{i,j}] = \frac{m^2}{n}(1 - o(1))$$

When $m = o(\sqrt{n})$, $\lim_{n \to \infty} \mathbb{P}[X > 0] \le \lim_{n \to \infty} \mathbb{E}[X] = 0$.

Let $t = \binom{m}{2}$ and $p = \mathbb{E}[X_{i,j}]$, then $\mathbb{E}[X] = tp$. Note that, X^2 is composed of t^2 items of $X_{i,j} X_{i',j'}$. Group these items according to $h = |\{i, j, i', j'\}|$. We see that $\mathbb{E}[X_{i,j} X_{i',j'}]$ equals p when $h = 2$, and equals p^2 otherwise. Thus, $\mathbb{E}[X^2] = tp + (t^2 - t)p^2$. When $m = \omega(\sqrt{n})$,

$$\lim_{n \to \infty} \mathbb{P}[X > 0] \ge \lim_{n \to \infty} \frac{\mathbb{E}[X]^2}{\mathbb{E}[X^2]} = \lim_{n \to \infty} \frac{tp}{tp + 1 - p} = 1,$$

where the first inequality is due to the Cauchy-Schwarz inequality.

2.3   An Exact Threshold for the Solution Probability of (1,0)-3-SAT

We use the equivalence (Lemma 1) for a $(1,0)$-super solution to study the threshold for the solution probability of random $(1, 0)$-3-SAT. We upper bound (resp. lower bound) the probability for F to be (1,0)-unsatisfiable by the probability of some necessary (resp. sufficient) condition on the satisfiability of its projection $\pi(F)$ (a 2-CNF formula). The conditions were proposed in [4]. It is important to note that while $\pi(F)$ is a 2-CNF formula obtained from a random 3-CNF formula $F_3(n, m)$, $\pi(F)$ itself is not distributed as the random 2-CNF formula $F_2(n, m)$. This is the major obstacle we have to deal with in our analysis.

Theorem 3. $F_3(n, rn)$ is (1,0)-satisfiable w.h.p. if $r < 1/3$ and is (1,0)-unsatisfiable w.h.p. if $r > 1/3$.

The proof of the above result is presented in two lemmas. In the proof, we use F to denote a random formula $F_3(n, rn)$, $m = rn$, and write $N = 2^3 \binom{n}{3}$. A *bicycle* ([4]) of length $s \geq 2$, is a conjunction of $s + 1$ 2-clauses $C_0, \cdots, C_s$ defined on s variables $\{x_1, x_2, \cdots, x_s\}$ such that $C_i = \overline{l_i} \vee l_{i+1}, 0 < i < s$, $C_0 = u \vee l_1$, and $C_s = \overline{l_s} \vee v$, where

1. l_i is either x_i or $\overline{x_i}$, and
2. u and v are from $\{x_i, \overline{x_i} \mid 1 \leq i \leq s\}$.

It can be shown that if a 2-SAT is unsatisfiable, then it must contain a bicycle ([4]).

Lemma 2. $F_3(n, rn)$ is (1,0)-satisfiable w.h.p. when $r < 1/3$.

Proof. For any fixed bicycle $B = C_0 \wedge \cdots \wedge C_s$, we consider the number of 3-CNF formulae F such tht $B \subset \pi(F)$. Let $\mathcal{C} = \{C_1, C_2, \cdots, C_{s-1}\}$. Since clauses in $\mathcal{C}$ are defined on distinct literals, no two clauses in $\mathcal{C}$ can be siblings with respect to the projection of any 3-clause. Similarly, no three clauses from B can be siblings with respect to a 3-clause. The only possible siblings are (C_0, C_i) and (C_s, C_i) for some $0 \leq i \leq s$.

Denote by $g(s, l)$ the number of 3-CNF formulas F such that $B \subset \pi(F)$, where $l = 0, 1$, or 2 is the number of clause pairs that belong to the projection of a same 3-clause in F. We have

$$g(s, l) = \binom{N - (s + 1 - l)}{m - (s + 1 - l)} \cdot (2(n - 2))^{s + 1 - 2l}.$$

Let $p(s)$ denote the probability that a bicycle of length s over a given (ordered) set of s variables is part of $\pi(F)$. Then,

$$p(s) \leq \binom{N}{m}^{-1} (g(s, 0) + 2s \cdot g(s, 1) + g(s, 2))$$

$$\leq \binom{N}{m}^{-1} 2(s + 1) \binom{N - (s - 1)}{m - (s - 1)} \cdot (2(n - 2))^{s-3}$$

$$\leq \left(\frac{3r}{2(n - 1)}\right)^{s-1} \cdot \frac{s + 1}{2(n - 2)^2}$$

Let N_s denote the number of different bicycles of length of s and X be the number of bicycles in $\pi(F)$. As $N_s < n^s 2^s (2s)^2$, we have

$$\mathbb{E}[X] = \sum_{s-2}^{n} N_s p(s) \le \frac{4n}{(n-2)^2} \sum_{s=2}^{n} s^2(s+1)\left(\frac{3rn}{n-1}\right)^{s-1}$$

When $r < 1/3$, $\lim_{n \to \infty} \mathbb{P}[X > 0] \le \lim_{n \to \infty} \mathbb{E}[X] = 0.$ □

A *snake* of length $t \ge 1$ is the conjunction of $2t$ 2-clauses $C_0, C_1, \cdots, C_{2t-1}$ and has following structure.

1. $C_i = (\bar{l_i} \vee l_{i+1}), 0 \le i \le 2t - 1.$ $l_0 = l_{2t} = \bar{l_t}$
2. For any $0 < i, j < 2t - 1$, $l_i \ne l_j$ and $l_i \ne \bar{l_j}$.

If $\pi(F)$ contains a snake, then F is not $(1,0)$-satisfiable. We show that w.h.p. $\pi(F)$ contains a snake of length $\log_{3r} n$.

Lemma 3. $F_3(n, rn)$ *is $(1,0)$-unsatisfiable w.h.p. when $r > 1/3$.*

Proof. Let A be a snake of length t, X_A be the indicator variable that A occurs in F. Note that there only the two pairs, (C_0, C_{t-1}) and (C_t, C_{2t-1}), can potentially be siblings with respect to the projection of a 3-clause. Let $s = 2t - 1$ and let $p(s)$ be the probability that a snake of length t over a given set of variables is in $\pi(F)$. We have

$$p(s) = \binom{N}{m}^{-1} (g(s,0) + 2g(s,1) + g(s,2))$$

$$\approx \binom{N}{m}^{-1} 4g(s,2) \approx \left(\frac{3r}{2n}\right)^{s-1} \frac{1}{n^2}$$

Let X denote the number of snakes of length t in $\pi(F)$. $\mathbb{E}[X] = \binom{n}{s} s! 2^s p(s) \approx (3r)^s / n$. When $r > 1/3$ and $t = \omega(\log_{3r} n)$, $\lim_{n \to \infty} \mathbb{E}[X] = \infty$.

In order to apply the second moment method to X, we have to consider correlation between snakes. To satisfy a clause $(l_i \vee l_j)$, if $\bar{l_i}$ is false, then l_j must be true. This implication can be represented by two arcs $(\bar{l_i}, l_j)$, $(\bar{l_j}, l_i)$ in a digraph. The digraph for a snake of length t is a directed cycle $\bar{l_t}, l_1, l_2, \cdots, l_s, \bar{l_t}$. Two snakes are not independent if and only if there are some common arcs between the corresponding directed cycles. Let B be another snake of length t. Suppose B share i arcs with A and these arcs contain j vertices. Then, taking into consideration the fact that the dominating term is still the one where exactly two pairs in B are siblings in the projection of the formula, we have

$$\mathbb{P}[B|A] \le \frac{\binom{N-2t-(2t-i)}{m-2t-(2t-i)} \cdot (2(n-2))^{2t} \cdot (2(n-2))^{2t-i}}{\binom{N-2t}{m-2t} \cdot (2(n-2))^{2t}}$$

$$\le \left(\frac{m-2t}{N-2t} \cdot 2(n-2)\right)^{2t-i} \le \left(\frac{3r}{2n}\right)^{2t-i}$$

It is clear that those common i arcs comprise $(j - i)$ directed paths. Fixing A, there are L_1 number of choices for the shared j vertices to occur in B, and there are L_2 number of choices for the remaining $2t - j$ vertices to occur in B.

$$L_1 = \left(2 \cdot \binom{2t}{2(j-i)} \right)^2 \cdot (j-i)! \leq 4 \cdot (2t)^{4(j-i)}$$

$$L_2 \leq \binom{n - j + 1}{2t - j} (2t - j)! \cdot 2^{2t-j} \leq (2(n - j + 1))^{2t-j}$$

For a given A, let $\mathcal{A}(i, j)$ be the set of snakes sharing i arcs and j vertices with A, and write

$$p(i, j) = \sum_{B \in \mathcal{A}(i,j)} \mathbb{P}[B|A] = L_1 L_2 \mathbb{P}[B|A]$$

$$\leq \left(\frac{3r}{2n} \right)^{2t-i} 4(2t)^{4(j-i)} \, (2(n - j + 1))^{2t-j} \, .$$

If $i \leq t$, then $i + 1 \leq j \leq 2i$. If $t < i \leq 2t$, then $i + 1 \leq j \leq 2t$. Let $A \sim B$ denote the fact that A and B are dependent.

$$\sum_{A \sim B} \mathbb{P}[B|A] = \sum_{i=1}^{2t} \sum_{j=i+1}^{\min\{2i,2t\}} p(i, j) = \sum_{j=2}^{2t} \sum_{i=j/2}^{j-1} p(i, j)$$

$$\leq \sum_{j=2}^{2t} (2(n - j + 1))^{2t-j} 4 \sum_{i=j/2}^{j-1} \left(\frac{3r}{2n} \right)^{2t-i} (2t)^{4(j-i)}$$

$$\leq \sum_{j=2}^{2t} (2(n - j + 1))^{2t-j} 4 \cdot \frac{j}{2} \left(\frac{3r}{2n} \right)^{2t-j+1} (2t)^4$$

$$\leq \sum_{j=2}^{2t} 2j \left(\frac{3r}{2n} \right) (2t)^4$$

$$\leq \Theta(1) \cdot \frac{1}{n} t^6 = o(\frac{1}{n}(3r)^{2t}) = o(E[X]).$$

According to corollary 4.3.5 of [3], $\lim\limits_{n \to \infty} \mathbb{P}[X > 0] = 1$.

2.4   Thresholds for the Solution Probability of $(1, 0)$-k-SAT

Using Markov's inequality, the following upper bound on the threshold of the phase transition can be proved:

Theorem 4. *For all $k \geq 3$, $F_k(n, rn)$ is $(1, 0)$-unsatisfiable w.h.p. when $r > \frac{2^k}{k+1} \ln 2$.*

In the rest of this section, we establish a lower bound on the threshold for $k > 3$ and show that the ratio of the lower bound over the upper bound goes to 1 as k goes to infinity. Our analysis uses the techniques introduced in [2] for proving lower bounds on the threshold for the phase transition of standard satisfying solutions of random SAT, but the calculation we have to deal with is even more complicated. The idea is to use a weighting scheme on satisfying assignments when applying the second moment method to prove lower bounds on the threshold.

For a clause c, denote by $\mathcal{S}(c)$ the set of $(1,0)$-super solutions of c, $\mathcal{S}^0(c)$ (resp. $\mathcal{S}^1(c)$) the set of assignments that satisfies exactly 0 (resp. 1) literal of c. Define $H(\sigma, c)$ be the number of satisfied literals minus the number of unsatisfied literals. For an event A, let $\mathbf{1}_A$ be the indicator variable that A occurs. The *weight* of σ w.r.t. c is defined as $w(\sigma, c) = \gamma^{H(\sigma,c)} \mathbf{1}_{\sigma \in \mathcal{S}(c)}$, $0 < \gamma < 1$ and is determined by k. These definitions extend naturally to a formula F: $w(\sigma, F) = \gamma^{H(\sigma,F)} \mathbf{1}_{\sigma \in \mathcal{S}(F)} = \prod_{c_i} w(\sigma, c_i)$. Let $X = \sum_{\sigma} w(\sigma, F)$. F is $(1,0)$-satisfiable if and only if $X > 0$.

Note that by viewing an instance of $(1,0)$-k-SAT as a generalized Boolean satisfiability problem (Boolean CSP) and applying the conditions established in [5], random $(1,0)$-k-SAT has a sharp threshold. Therefore, to show $X > 0$ w.h.p., it is sufficient to prove that $\mathbb{P}[X > 0]$ is greater than some constant.

For a fixed σ and a random k-clause c, since σ (1-0)-satisfies c if at least two literals in c evaluate to true under σ, we have

$$\mathbb{E}[w(\sigma, c)] = \mathbb{E}\left[\gamma^{H(\sigma,c)}(1 - \mathbf{1}_{\sigma \in \mathcal{S}^0(c)} - \mathbf{1}_{\sigma \in \mathcal{S}^1(c)})\right]$$

$$= (\frac{\gamma + \gamma^{-1}}{2})^k - 2^{-k}\gamma^{-k} - k2^{-k}\gamma^{-k+2} = \phi(\gamma)$$

Thus, $\mathbb{E}[X] = \sum_{\sigma} \prod_{c_i} \mathbb{E}[w(\sigma, c)] = (2\phi(\gamma)^r)^n$.

We now consider $\mathbb{E}[X^2]$. Fix a pair of assignments σ, τ such that they overlap each other on $z = \alpha n$ variables. Consider a random k-clause c and write

$$f(\alpha) = \mathbb{E}[w(\sigma, c)w(\tau, c)] = \mathbb{E}\left[\gamma^{H(\sigma,c)+H(\tau,c)}\mathbf{1}_{\sigma, \tau \in \mathcal{S}(c)}\right].$$

We have the following equations for relevant events

$$\mathbf{1}_{\sigma, \tau \in \mathcal{S}(c)} = 1 - \mathbf{1}_{\sigma \notin \mathcal{S}(c)} - \mathbf{1}_{\tau \notin \mathcal{S}(c)} + \mathbf{1}_{\sigma, \tau \notin \mathcal{S}(c)},$$

$$\mathbf{1}_{\sigma \notin \mathcal{S}(c)} = \mathbf{1}_{\sigma \in \mathcal{S}^0(c)} + \mathbf{1}_{\sigma \in \mathcal{S}^1(c)},$$

$$\mathbf{1}_{\sigma, \tau \notin \mathcal{S}(c)} = \mathbf{1}_{\sigma \in \mathcal{S}^0(c), \tau \in \mathcal{S}^0(c)} + \mathbf{1}_{\sigma \in \mathcal{S}^0(c), \tau \in \mathcal{S}^1(c)}$$
$$+ \mathbf{1}_{\sigma \in \mathcal{S}^1(c), \tau \in \mathcal{S}^0(c)} + \mathbf{1}_{\sigma \in \mathcal{S}^1(c), \tau \in \mathcal{S}^1(c)},$$

and for mathematical expectations

$$\mathbb{E}\left[\gamma^{H(\sigma,c)+H(\tau,c)}\mathbf{1}\right] = (\alpha(\frac{\gamma^2+\gamma^{-2}}{2}) + 1 - \alpha)^k,$$

$$\mathbb{E}\left[\gamma^{H(\sigma,c)+H(\tau,c)}\mathbf{1}_{\sigma\notin S(c)}\right] = 2^{-k}((\alpha\gamma^{-2}+1-\alpha)^k$$
$$+ k(\alpha\gamma^{-2}+1-\alpha)^{k-1}(\alpha\gamma^2+1-\alpha)),$$

$$\mathbb{E}\left[\gamma^{H(\sigma,c)+H(\tau,c)}\mathbf{1}_{\sigma,\tau\notin S(c)}\right] = 2^{-k}(\alpha^k\gamma^{-2k}+2k\gamma^{-2k+2}\alpha^{k-1}(1-\alpha)$$
$$+ \gamma^{-2k+4}(k\alpha^k + k(k-1)\alpha^{k-2}(1-\alpha)^2)).$$

Therefore, the expectation of X^2 can be written as

$$\mathbb{E}\left[X^2\right] = \sum_{\sigma,\tau}\mathbb{E}[w(\sigma,F)w(\tau,F)]$$

$$= \sum_{\sigma,\tau}\prod_{c_i}\mathbb{E}[w(\sigma,c_i)w(\tau,c_i)] = 2^n\sum_{z=0}^{n}\binom{n}{z}f(z/n)^{rn}$$

The following lemma from [1] enables us to consider the dominant part of $\mathbb{E}\left[X^2\right]$.

Lemma 4. *Let h be a real analytic positive function on $[0,1]$ and define $g(\alpha) = h(\alpha)/(\alpha^\alpha(1-\alpha)^{1-\alpha})$, where $0^0 \equiv 1$. If g has exactly one maximum at $g(\beta)$, $\beta \in (0,1)$, and $g''(\beta) < 0$, then there exists constant $C > 0$ such that for all sufficient large n, $\sum_{z=0}^{n}\binom{n}{z}h(z/n)^n \le C \times g(\beta)^n$.*

Define $g_r(\alpha) = f(\alpha)^r/(\alpha^\alpha(1-\alpha)^{1-\alpha})$ and say $g_r(\alpha)$ satisfies the *dominant condition* if $g_r''(1/2) < 0$ and $g_r(1/2)$ is the unique global maximum. According to lemma 4 and $\phi(\gamma)^2 = f(1/2)$, if $g_r(\alpha)$ satisfies the dominant condition, then

$$\mathbb{P}[X > 0] > \frac{\mathbb{E}[X]^2}{\mathbb{E}[X^2]} = \frac{4^n f(1/2)^{rn}}{\mathbb{E}[X^2]}$$

$$> \frac{(2g_r(1/2))^n}{C \cdot (2g_r(1/2))^n} = \frac{1}{C},$$

where C is a constant when k is fixed.

If we can find suitable γ and r so that $g_r(\alpha)$ satisfies the dominant condition, then $X > 0$ w.h.p.. It is clear that the dominant condition implies $f'(1/2) = 0$. According to [2], a necessary condition for $f'(1/2) = 0$ is that the sum of vectors scaled by their corresponding weight is 0, i.e., $\sum_{\mathbf{v}\in\{0,1\}^k}w(\mathbf{v})\mathbf{v} = \mathbf{0}$. For $(1,0)$-k-SAT, this is $\sum_{i=1}^{k}\binom{k}{i}\gamma^{2i}(2i-k) = 0$. When $k = 4$, this equation requires $\gamma = 0$. Thus, the weighting scheme is not meaningful when $k = 4$. Therefore, we consider the case of $k > 4$ first and then the case of $k = 4$ in a different way.

It is too complicated to directly prove that $g_r(\alpha)$ satisfies the dominant condition, at least for small k. Therefore, we plot figures to show how $g_r(\alpha)$ changes when k is fixed. Figure 1 shows the case when $k = 5$. For each k, when r is smaller than some r_k^*, $g_r(\alpha)$ satisfies the dominant condition and $F_r(n, rn)$ is $(1,0)$-satisfiable w.h.p. Thus r_k^* is a lower bound for $F_k(n, rn)$. We do this analysis for k up to 11 and show the values in Table 1. We can see that the ratio of the lower bound over the upper bound of thresholds of $F_k(n, rn)$ goes to 1 as k becomes large. We still have to solve the case $k = 4$ separately. The weighting scheme, $w(\sigma, c) = \gamma^{H(\sigma,c)} \mathbf{1}_{\sigma \in \mathcal{S}(c)}$, does not work for any $\gamma > 0$. This is because $H(\sigma, F)$ is either 0 or positive. Thus, a compromise is to consider only those assignments which satisfy $H(\sigma, F) = 0$. Specifically, for each clause of F, exactly two literals are satisfied and exactly two literals are unsatisfied. And every satisfying assignment has the same weight, 1. By doing this, the likelihood for an assignment not to be in X is doubled. Therefore, the upper bound for such solutions becomes $\frac{2^{k-1}}{1+k} \ln 2$, half of the upper bound for $(1, 0)$-4-SAT. The remaining analysis is similar to the analysis of $k > 4$. The r_4^* we found is 0.602.

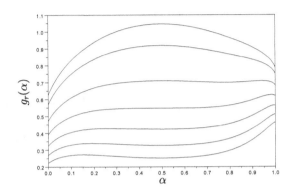

$g_r(\alpha)$

Fig. 1. $k = 5$, $r = 1, 1.2, 1.6, 2, 2.4, 2.8, 3.2$ (top down)

Table 1. Upper bound and lower bound for different k

k	4	5	6	7	8	9	10	11
Upper Bound	2.2	3.6	6.3	11.1	19.7	35.5	64.5	118.3
Lower Bound	0.6	1.6	3.7	7.8	15.8	30.9	59.3	113.4

3   Super Solutions for Random Binary CSPs

We consider random binary CSPs defined on a domain D of size $|D| = d$. A binary CSP $\mathcal{C}$ consists of a set of variables $X = \{x_1, \cdots, x_n\}$ and a set of binary constraints $(C_1, \cdots, C_m)$. Each constraint C_i is specified by its *constraint scope*, an unordered pair of two variables in X, and a *constraint relation* R_{C_i} that defines a set of incompatible value tuples in the binary relation $D \times D$ for the scope variables. An incompatible value tuple is also called a *restriction*. The constraint graph of a binary CSP is a graph whose vertices correspond to the set of variables and edges correspond to the set of constraint scopes. We use the following random CSP model $\mathcal{B}_{n,m}^{d,q}$ where the domain size is allowed to increase with the number of variables.

1. Its constraint graph is a random graph $G(n, m)$ where the m edges are selected uniformly at randomly from all the possible $\binom{n}{2}$ edges.
2. For each edge, its constraint relation is determined by choosing each value tuple in $D \times D$ as a restriction independently with probability q.

Proposed and studied in a series of papers by Xu, et al., this class of models for random CSPs is known as the Model RB [11–13]. In particular, the exact threshold of the phase transition of standard satisfiability has been established in [12] and the (resolution) complexity of random instances at the phase transition has been analyzed in [13].

Denote by $H(\sigma_1, \sigma_2)$ the set of variables being assigned different values by σ_1 and σ_2, i.e., $H(\sigma_1, \sigma_2) = \{x_i | \sigma_1(x_i) \neq \sigma_2(x_i), 1 \leq i \leq n\}$. Let σ be a fixed assignment and I be a random $\mathcal{B}_{n,m}^{d,q}$ instance. Define the following three events:

1. $S(\sigma)$: σ is a solution for I.
2. $S_i(\sigma)$: there exists another solution σ' for I such that $H(\sigma, \sigma') = \{x_i\}$.
3. $T(\sigma)$: σ is a $(1, 0)$-super solution for I.

It is clear that $\mathbb{P}[T(\sigma)] = \mathbb{P}[S(\sigma)]\mathbb{P}[\cap_{1 \leq i \leq n} S_i(\sigma)|S(\sigma)]$. Estimating the probability of a $(1, 0)$-super solution for a random CSP instance is, however, more complicated than estimating the probability of a satisfying assignment, largely due to the fact that the events $S_i(\sigma), 1 \leq i \leq n$, are not independent. This is the major hurdle we need to overcome. Note that in a random CSP instance, the selection of constraints and the selection of restrictions for each constraint are independent. Let $C \subset (X \times X)^m$ be the collection of all possible sets of m unordered pairs of variables. For a given set $e \in C$ of m unordered pairs, denote by $E(e)$ the event that e is selected as the set of constraints of the random instance I. Let m_i be the number of constraints x_i is involved with. Considering an assignment $\sigma', H(\sigma, \sigma') = \{x_i\}$, it is clear that $\mathbb{P}\left[\overline{S(\sigma')}|S(\sigma) \cap E(e)\right] = 1 - (1 - q)^{m_i}$. Let $D' = D \setminus \{\sigma(x_i)\}, \sigma'(x_i) = y, p = 1 - q$, then

$$\mathbb{P}[S_i(\sigma)|S(\sigma) \cap E(e)] = \mathbb{P}\left[\cup_{y \in D'} S(\sigma')|S(\sigma) \cap E(e))\right]$$
$$= \mathbb{P}\left[\cap_{y \in D'} \overline{S(\sigma')}|S(\sigma) \cap E(e))\right]$$
$$= 1 - \mathbb{P}\left[\cap_{y \in D'} \overline{S(\sigma')}|S(\sigma) \cap E(e))\right]$$
$$= 1 - (1 - (1 - q)^{m_i})^{d-1}.$$

This shows that, conditioned on $S(\sigma)$ and fixed constraint sets e, $S_i(\sigma)$ and $S_j(\sigma)$ are independent for any $i \neq j$.

$$\mathbb{P}[T(\sigma)] = \mathbb{P}[\cup_{e \in C} (E(e) \cap S(\sigma) \cap (\cap_{1 \leq i \leq n} S_i(\sigma)))]$$
$$= \sum_{e \in C} \mathbb{P}[E(e)]\mathbb{P}[S(\sigma)|E(e)]\mathbb{P}[\cap_{i \in [n]} S_i(\sigma)|S(\sigma) \cap E(e)]$$
$$= \binom{\binom{n}{2}}{m}^{-1} p^m \sum_{e \in C} \prod_{i=1}^{n} \left(1 - (1 - p^{m_i})^{d-1}\right). \tag{1}$$

Let Y_σ be an indicator variable of T_σ and $Y = \sum_\sigma Y_\sigma$ be the number of $(1, 0)$-super solutions. We have

$$\mathbb{E}[Y] = d^n \cdot \mathbb{E}[Y_\sigma] = d^n \cdot \mathbb{P}[T_\sigma]$$

We have the following lower and upper bounds on the threshold of solution probability.

Theorem 5. *Consider the random CSP $\mathcal{B}_{n,m}^{d,q}$ with $d = \sqrt{n}$, $p = 1 - q$, $m = c \cdot n \ln n$ where c is a positive constant.*

- *If $c > -\frac{1}{3 \ln p}$, $\lim_{n \to \infty} \mathbb{E}[Y] = 0$ and thus $\mathcal{B}_{n,m}^{d,q}$ is $(1,0)$-unsatisfiable w.h.p.*
- *If $c < -\frac{1}{10 \ln p}$ and $q = 1 - p < 0.43$, $\lim_{n \to \infty} \mathbb{E}[Y] \to \infty$.*

Proof. Due to space limit, we give a brief proof of the first part of the conclusion and omit the proof of the second part of the conclusion.

The right hand side of Eq. (1), subject to $\sum_{i=1}^{n} m_i = 2m$, achieves the global maximum when $m_i = \frac{2m}{n}$, $1 \leq i \leq n$. This can be proved by the method of Lagrange multipliers. Let $c = c' - \frac{1}{\ln p}$, then

$$\mathbb{E}[Y] \leq (d \cdot p^{c \ln n} \cdot (1 - (1 - p^{2c \ln n})^{d-1}))^n$$
$$= (d \cdot n^{c \ln p} \cdot (1 - (1 - n^{2c \ln p})^{d-1}))^n$$
$$\approx (n^{1/2-c'} \cdot (1 - (1 - n^{-2c'})^{n^{1/2}}))^n.$$

For any a, b satisfying $0 \leq a \leq 1$ and $ab < 1$, $(1-a)^b \geq 1 - ab$. If $c' > 1/3$, then

$$\mathbb{E}[Y] \leq (n^{1/2-c'} \cdot n^{-2c'} n^{1/2})^n = (n^{1-3c'})^n \to 0.$$

4   Conclusions

To the best of our knowledge, we have conducted (for the first time) a probabilistic analysis of super solutions of random instances of SAT and CSPs. While we have focused on the special (but already challenging) case of $(1,0)$-super solutions, some of our analysis extends to the case of $(a,0)$-super solutions for $a > 1$. For random instances of CSPs, new analytical methods and ideas are needed to obtain a more detailed characterization of the behavior of the super solutions, and we leave this as a future work. It is also highly interesting to conduct a systematic empirical analysis to fully understand the hardness of solving random instances of $(1,0)$-k-SAT as well as the hardness of solving the projected standard SAT instances, which may serve as suite of SAT benchmark with a unique structural properties. Finally, we wonder if our analysis can be extended to random instances of other problems such as graphical games where solution concepts similar to super solutions have been used.

References

1. Achlioptas, D., Moore, C.: The asymptotic order of the random k-sat threshold. In: Proceedings of the FOCS, pp. 779–788 (2002)
2. Achlioptas, D., Peres, Y.: The threshold for random k-sat is 2k (ln 2 - o(k)). In: STOC 2003, pp. 223–231. ACM, New York (2003)

3. Alon, N., Spencer, J.H.: The Probabilistic Method. Wiley, New York (2008)
4. Chvatal, V., Reed, B.: Mick gets some (the odds are on his side) (satisfiability). In: SFCS 1992, pp. 620–627. IEEE Computer Society, Washington (1992)
5. Creignou, N., Daud, H.: Combinatorial sharpness criterion and phase transition classification for random CSPS. Inf. Comput. **190**(2), 220–238 (2004)
6. Culberson, J., Gent, I.: Frozen development in graph coloring. Theoret. Comput. Sci. **265**(1–2), 227–264 (2001)
7. Gao, Y., Culberson, J.: Consistency and random constraint satisfaction models. J. Artif. Intell. Res. **28**, 517–557 (2007)
8. Ginsberg, M.L., Parkes, A.J., Roy, A.: Supermodels and robustness. In: AAAI 1998/IAAI 1998, Menlo Park, CA, USA, pp. 334–339 (1998)
9. Gomes, C., Walsh, T.: Randomness and structure. In: Rossi, F., van Beek, P., Walsh, T. (eds.) Handbook of Constraint Programming, pp. 639–664. Elsevier, Amsterdam (2006)
10. Hebrard, E., Hnich, B., Walsh, T.: Super solutions in constraint programming. In: Régin, J.-C., Rueher, M. (eds.) CPAIOR 2004. LNCS, vol. 3011, pp. 157–172. Springer, Heidelberg (2004)
11. Xu, K., Boussemart, F., Hemery, F., Lecoutre, C.: Random constraint satisfaction: easy generation of hard (satisfiable) instances. Artif. Intell. **171**, 514–534 (2007)
12. Xu, K., Li, W.: Exact phase transitions in random constraint satisfaction problems. J. Artif. Intell. Res. **12**, 93–103 (2000)
13. Xu, K., Li, W.: Many hard examples in exact phase transitions. Theoret. Comput. Sci. **355**, 291–302 (2006)

A Data Streams Analysis Strategy Based on Hadoop Scheduling Optimization for Smart Grid Application

Fengquan Zhou[1], Xin Song[2(✉)], Yinghua Han[2], and Jing Gao[2]

[1] XuJi Group Corporation, State Grid, Xuchang 461000, China
[2] Northeastern University at Qinhuangdao, Northeastern University,
Qinhuangdao 066004, China
bravesong@163.com

Abstract. The massive data streams analysis in the Smart Grids data processing system is very important, especially in the high-concurrent read and write environments where supporting the massive real-time streaming data storage and management. The computational and stored requirements for Smart Grids can be met by utilizing the Cloud computing. In order to support the robust, affordable and reliable power streaming data analysis and storage, in this paper, we propose a power data streams analysis strategy based on Hadoop scheduling optimization for smart grid monitoring application. The proposed strategy combined with the flexible resources and services shared in network, omnipresent access and parallel processing features of cloud computing. Finally, the simulation results show that proposed strategy can effectively improve the efficiency of computing resource utilization and achieve the massive information concurrent processing ability.

Keywords: Data streams analysis · Hadoop scheduling optimization · Smart grid application · Cloud computing

1 Introduction

With the wide application of new technology in the power grid system, the construction of the robust smart grid has become the inevitable trend of the power grid development. The monitoring and analysis platform will deal with the massive amounts of sensor streams data in smart grid system because of a wide variety of sensor devices on sensing layer. The heterogeneous sensor data objects include text, numbers, forms, and graphics etc. Unlike traditional data sets processing strategy, the power monitoring system put forward the high request for the real time and reliability in streams data storage and management method. However, the traditional computing system cannot solve the real-time calculation, comprehensive analysis and extensibility problem of power data streams. Smart grid data processing system need to store huge amounts of streams data from the endpoint sensor nodes. It needs have the higher concurrent read/write requests and the higher scalability. The emerging cloud computing service system provides the support for the complex smart grid applications [1]. It may be impossible to store an entire data stream or to scan through it multiple times due to the tremendous volume of the sensor data streams on the power monitoring application system. In front of the

© Springer International Publishing Switzerland 2015
J. Wang and C. Yap (Eds.): FAW 2015, LNCS 9130, pp. 326–333, 2015.
DOI: 10.1007/978-3-319-19647-3_30

amazing power data scale expanding, the power grid monitoring system arises some new technology challenges, such as massive information storage, real-time data processing, streams data retrieval, streams data mining and intelligent information processing etc. The smart grid system architecture is an urgent need to support the dynamic scalable characteristics of massive streams data processing for new storage calculation model, through constructing the massive streams data analysis processing platform based on Hadoop can effectively solve the above problems. MapReduce was developed as a distributed programming model on Hadoop by Google's Jeffrey Dean and Sanjay Ghemawat for the mass data processing; it is mainly used for parallel processing and realization of the big data sets. In this paper, we proposed a power data streams analysis strategy based on Hadoop distributed system scheduling optimization. The proposed strategy can realize the storage and fast parallel processing of the massive power stream data. Using Hadoop scheduling optimization strategy, the simulation results show that proposed strategy achieves more energy savings and also can ensure that the total amount of data to retain in memory or to be stored on disk is small.

The rest of this paper is organized as follows: in Sect. 2, we briefly review some closely related works. The proposed data streams multi-dimensional analysis strategy is derived and discussed in Sect. 3. The validity analysis and performance evaluation are presented in Sect. 4. Finally, the conclusions and future work directions are described in Sect. 5.

2 Related Works

There have been a few of studies on the storage and analysis of the smart grid data streams using cloud computing architecture. Reference [2] derived the near-optimal or suboptimal strategies of the two players in stackelberg game using convex optimization and simulated annealing techniques, it presented a stackelberg game-based optimization framework of the smart grid with distributed PV power generations and data centers, jointly accounting for the service request dispatch and routing problem in the cloud with the power data flow analysis in power grid. Reference [3] introduced the security and reliability issues of cloud computing architecture for the smart grid applications. Moreover, it presented cloud computing service based existing smart grid projects and open research issues. Chang, Sekchin et al. desired to use cognitive radio channels for communication among a wireless network of smart meters. They showed a framework for the utilization of a cloud computing smart grid [4]. It is important for cloud service providers to reduce electricity cost as much as possible. Reference [5] conceived a strategy for reducing the electricity cost by utilizing energy storage facilities which exist in data centers in smart grid environment. They proposed a dynamic energy storage control strategy based on the Q-Learning algorithm which did not assume any a priori information on the Markov process governing the energy management system of the data centers. For the data streams analysis of the power supply system, some researcher discussed how cloud computing model can be used for developing smart grid solutions. Flexible resources and services shared in network, parallel processing and omnipresent access are some features of Cloud Computing that are desirable for Smart Grid applications. Even though the Cloud Computing model is considered efficient for Smart Grids, it has some constraints such as security and reliability [6]. In the future smart grid application system, the implementation of a smart objects-oriented monitoring system is a complex challenge as distributed, autonomous, and heterogeneous sensor components at different levels of abstractions and granularity need to cooperate among themselves, with conventional networked infrastructures. Reference [7] proposed the integration of two complementary mainstream paradigms for large-scale distributed computing: Agents and Cloud. Authors introduced a cloud-assisted and agent-oriented IoT architecture that can be fit for the smart grid application for the massive data streams analysis processing. If the data centers for the power management were distributed multiple geographic area, the energy consumptions of data centers for smart grid were becoming unacceptably high, and placing a heavy burden. Various power management methodologies based on geographic load balancing have recently been proposed to effectively utilize several features of smart grid [8]. Smart grid is a power system with advanced communication technologies and information processing strategies integrated and leveraged. Reference [9] introduced an resource optimization framework of leveraging the cloud domain to reduce the cost of information management in the smart grid. A huge amount of row data was collected by smart meters and sensors from the end user and different part of the network to the computation system of smart grid. Subsequently, this considerably big amount of data must be processed, analyzed and stored in a cost effective ways. Researchers have been suggesting different solutions, Ref. [10] discussed the feasibility study of the handling of monitoring of renewable energy in smart grid on cloud computing framework retaining smart grid security, analysis of the availability of energy management software tools. Due to process the massive data stream for smart grid system, in this paper, we built a large-scale data stream processing system based on Hadoop scheduling optimization. The managing platform has both the distributed storage technology of massive sensor data streams and the data streams compression algorithm for smart grid application.

3   Implementation of Massive Data Streams Analysis Strategy for Smart Grid Monitoring System

The stream data are generated continuously in power enterprise, with huge volume, infinite flow, and fast-changing behavior. The ultimate goal of power data center construction is to support the massive data stream storage and management, such as data stream mining, data stream multi-dimensional analysis, and the decision support services etc. The default setting of Hadoop scheduler focused on homogeneous cluster computing resources. However, the heterogeneity of cluser node is inevitable due to the hardware upgrade or cloud computing resource virtualization. According to the characteristics of the smart grid data streams processing system, combined Hadoop parallel computing model, this section includes cloud architecture for smart grid, Hadoop scheduling optimization method, and power data streams analysis strategy. The strategy put forward on virtualization resources management platform using open source framework of Hadoop, and build a dynamic scalable distributed parallel computing scheme based on MapReduce computing model of Hadoop frame.

3.1   Cloud Computing Architecture for Smart Grid Application

The distributed MapReduce processing architecture of the power data stream in cloud computing environment is three levels, that is basic resource layer, data processing layer and business application layer, as shown in Fig. 1. The basic resource layer includes power data storage systems, servers and processors. The computer hardware resources were abstracted through virtualization technology to achieve the fine-grained management and allocation of resources for improving the efficiency of resources used and enhancing the scalability of resources. In addition, the power data were stored according to the different types of data for improving the efficiency of data reading and writing. The data processing layer includes algorithms, resource scheduling, real-time processing, batch implementation, etc. The core of the power data processing is MapReduce function that implements the different demand through the different functions. In order to realize the resource scheduling, the layer combined the basic resource scheduling algorithm, active queue management, neural network and genetic algorithm to realize the different scheduling function. The business application layer includes middleware, K/V storage and front page display function. Based on the business needs of the smart grid data, the business application layer will implement the calculation, analysis, simulation, optimization, planning, design and decision-making, etc.

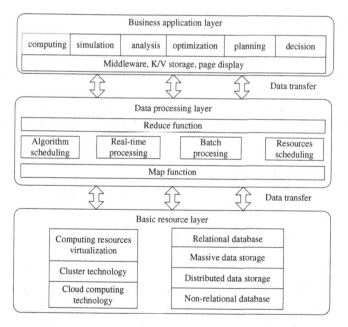

Fig. 1. The three layer architecture of smart grid data platform based on the MapReduce model

3.2   Power Data Streams Analysis Strategy Based on Hadoop Scheduling Optimization Method

Hadoop is a distributed system infrastructure, developed by the Apache foundation. The user can develop the distributed application program and make full use of the high speed computing and storage of the cluster in the case of not understand the distributed low-level details. Hadoop MapReduce is calculation model of the massive data stream processing. It has the characteristics of parallel processing, distributed computing, especially suitable for the calculation of big data. The strong open source implementation of the Hadoop can satisfy the need of most of the data computing and analysis.

Hadoop default scheduler is aimed at homogeneous cluster computing resources. However, even in a homogeneous environment, some tasks also may be backward in execution due to the resources competition and unfair distribution, thus affecting the progress of the whole job. In order to solve this problem, Hadoop adopted a Speculative Execution strategy, that is, the backward tasks were found in time and rescheduled by calculating the execution progress of each task in the process of execution. However, the Hadoop scheduling algorithm is inefficiency in heterogeneous cluster environment. Both the Map tasks and Reduce tasks, it will not to perform at a rate constant. In view of the heterogeneous cluster environment on smart grid application, this paper adopted rescheduling optimization algorithm based on history information. The main idea of the algorithm includes two aspects. (1) The inner subphase of the Map tasks and Reduce tasks is not fixed. The proportion of each stage was determined according to different application and the history information of job execution. In the process of execution, the proportion of each stage will be corrected according to the execution progress and the speed of execution will be recalculated. The backward task was computed the remaining completion time, then the task with the longest remaining time will be rescheduling. (2) The cluster node in the fast node pool is not necessarily meeting the demand of rescheduling, and ideal node needs to meet. The rescheduling task on the node performs faster than the original node and does not become a new backward task. The algorithm is described as follows:

Step 1. Before the application execution, the scaling factor of the Map tasks and Reduce tasks in each phase was made sure based on the history execution information.

Step 2. In the application execution process, the scaling factor in step 1 was corrected continuously according to the execution progress.

Step 3. According to the execution progress and speed of calculation tasks, the task which the execution progress is below *Slow_Task_Threshold* will be computed the remaining completion time. Then, the *speculative_task* with the longest remaining completion time need be rescheduled.

Step 4. The most appropriate computing nodes (*node_best*) were screened from the fast node pool. The cluster nodes in the fast node pool were screened according to the rescheduling cost to filter out the nodes that cannot shorten the execution time after rescheduling. Then, the most appropriate computing node was selected combined with the data locality principle and the optimal node performance principle.

Step 5. The *speculative_task* was rescheduled on *node_best*.

In addition, in order to make the strategy for computing resources have the universality, the execution speed standard deviation of task in different nodes was calculated for determining the heterogeneity size of computing resources. Then, in accordance with the size, three important parameters in algorithm were updated, that is, the slow node threshold *slow_node_threshold*, the slow task threshold *Slow_Task_Threshold*, maximum rescheduling task number *SpeculativeCap*.

The Map-Reduce task scheduling process and fast node selection strategy is shown in Fig. 2.

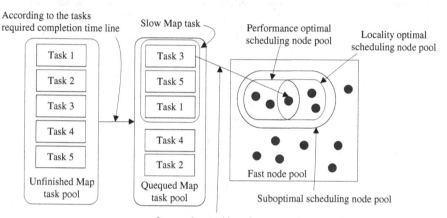

(a) Map task scheduling and node selection

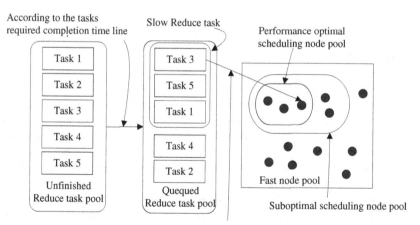

(b) Reduce task scheduling and node selection

Fig. 2. The MapReduce task scheduling process on heterogeneous grid environment

4   Performance Evaluation

In this section, the intermediate result local storage performance of proposed strategy was evaluated. The power data flow velocity is 1 MB/s (that is, the data is sent by 200 B each, and 5000 sequence /s), the scale of the intermediate result is 50 GB, Each test for 10 times, each time 10 min, the experimental results is the average value. The Fig. 3 shows the performance contrast for the LRU algorithm (Least Recently Used), the recent research RTMR algorithm [11], naive algorithm and the proposed power data streams analysis strategy based on Hadoop Scheduling Optimization (HSO). The memory read/write performance is improved 18 % by the proposed strategy. The external storage read/write and memory hit rate performance are improved 22.8 % and 12.1 % respectively. Entirety read/write performance is improved 21.8 %.

Performance indicators	Test method	Experimental results	Enhancing effect
Memory read/write	Read/write synchronization	75385.2t/s	18.0%
	synchronization elimination(RTMR)	84506.8t/s	
	synchronization elimination(HSO)	88956.3t/s	
The external storage read/ write	Big Table	4425.5t/s	22.8%
	RTMR	5111.4t/s	
	HSO	5432.6t/s	
Memory hit rate	LRU	66.7 %	12.1%
	RTMR	72.9%	
	HSO	74.8%	
Entirety read/write performance	Naive algorithm	73901.4t/s	21.8%
	RTMR	88608.5t/s	
	HSO	89975.6t/s	

Fig. 3. The performance optimization of the intermediate result storage

5   Conclusion

Cloud computing is applied to the smart grid for solving the unified management and scheduling problem of power information resources. In view of the request of concurrent processing massive data on the future smart grid system, in this paper, we have proposed and described a massive power data streams analysis strategy based on Hadoop scheduling optimization. The process model adopted the virtualization architecture to integrate and optimize large-scale heterogeneous information resources. Depending on the smart grid application, the proposed strategy can improve the heterogeneous computing resource utilization in smart grid and reduce the cost of grid monitoring system operation. It provided a dynamic scalable computing platform for the power data streams storage and processing based on cloud computing technology.

Acknowledgment. The research work was supported by the Fundamental Research Funds of the Central University under Grant no. N120323009, the Natural Science Foundation of Hebei Province under Grant No. F2014501055, the Program of Science and Technology Research of Hebei University No. ZD20132003, the Natural Science Foundation of Liaoning Province under

Grant No.201202073, and the National Natural Science Foundation of China under Grant No. 61403069, No.61473066 and No.61374097.

References

1. McDaniel P., Smith S. W.: Outlook: cloud computing with a chance of security challenges and improvements. In: Proceeding of the IEEE Computer and Reliability Societies, pp. 77–80 (2010)
2. Wang, Y.Z., Lin, X., Pedram, M.: A stackelberg game-based optimization framework of the smart grid with distributed pv power generations and data centers. IEEE Trans. Energy Convers. **29**(4), 978–987 (2014)
3. Yigit, M., Gungor, V.C., Baktir, S.: Cloud compting for smart grid applications. Comput. Netw. **70**, 312–329 (2014)
4. Chang, S., Nagothu, K., Kelley, B.: A beamforming approach to smart grid systems based on cloud cognitive radio. IEEE Syst. J. **8**(2), 461–470 (2014)
5. Zhang S.B., Ran Y.Y. and Wu X.M. et al.: Electricity cost optimization for data centers in smart grid environment. In: IEEE International Conference on Control and Automation, Taichung, TaiWan, pp. 290–295 (2014)
6. Markovicn, D.S., Zivkovic, D., Branovic, I., et al.: Smart power grid and cloud computing. Renew. Sustain. Energy Rev. **24**, 566–577 (2013)
7. Fortino G., Guerrieri A. and Russo W.: Integration of agent-based and cloud computing for the smart objects-oriented IoT. In: IEEE International Conference on Computer Supported Cooperative Work in Design, Hsinchu, TaiWan, pp. 493–498 (2014)
8. Rahman, A., Liu, X., Kong, F.X.: A survey on geographic load balancing based data center power management in the smart grid environment. IEEE Commun. Surv. Tutorials **16**(1), 214–233 (2014)
9. Fang, X., Yang, D.J., Xue, G.L.: Evolving smart grid information management cloudward: a cloud optimization perspective. IEEE Trans. Smart Grid **4**(1), 111–119 (2013)
10. Bitzer B., Gebretsadik E. S.: Cloud computing framework for smart grid applications. In: 48th International Universities Power Engineering Conference, Dublin, Ireland, (2013)
11. Qi, K.Y., Zhao, Z.F., Fang, J., Ma, Q.: Real-time processing for high speed data stream over large scale data. Chin. J. Comput. **35**(3), 477–490 (2012)

Author Index

Printed in the United States
By Bookmasters